Cinnamon and Cassia

The genus *Cinnamomum*

Medicinal and Aromatic Plants — Industrial Profiles

Individual volumes in this series provide both industry and academia with in-depth coverage of one major genus of industrial importance.

Edited by Dr Roland Hardman

Volume 1
Valerian, edited by Peter J. Houghton
Volume 2
Perilla, edited by He-ci Yu, Kenichi Kosuna and Megumi Haga
Volume 3
Poppy, edited by Jenö Bernáth
Volume 4
Cannabis, edited by David T. Brown
Volume 5
Neem, edited by H.S. Puri
Volume 6
Ergot, edited by Vladimír Křen and Ladislav Cvak
Volume 7
Caraway, edited by Éva Németh
Volume 8
Saffron, edited by Moshe Negbi
Volume 9
Tea Tree, edited by Ian Southwell and Robert Lowe
Volume 10
Basil, edited by Raimo Hiltunen and Yvonne Holm
Volume 11
Fenugreek, edited by Georgios Petropoulos
Volume 12
Gingko biloba, edited by Teris A. Van Beek
Volume 13
Black Pepper, edited by P.N. Ravindran
Volume 14
Sage, edited by Spiridon E. Kintzios
Volume 15
Ginseng, edited by W.E. Court
Volume 16
Mistletoe, edited by Arndt Büssing
Volume 17
Tea, edited by Yong-su Zhen
Volume 18
Artemisia, edited by Colin W. Wright
Volume 19
Stevia, edited by A. Douglas Kinghorn
Volume 20
Vetiveria, edited by Massimo Maffei

Volume 21
Narcissus and Daffodil, edited by Gordon R. Hanks
Volume 22
Eucalyptus, edited by John J.W. Coppen
Volume 23
Pueraria, edited by Wing Ming Keung
Volume 24
Thyme, edited by E. Stahl-Biskup and F. Sáez
Volume 25
Oregano, edited by Spiridon E. Kintzios
Volume 26
Citrus, edited by Giovanni Dugo and Angelo Di Giacomo
Volume 27
Geranium and Pelargonium, edited by Maria Lis-Balchin
Volume 28
Magnolia, edited by Satyajit D. Sarker and Yuji Maruyama
Volume 29
Lavender, edited by Maria Lis-Balchin
Volume 30
Cardamom, edited by P.N. Ravindran and K.J. Madhusoodanan
Volume 31
Hypericum, edited by Edzard Ernst
Volume 32
Taxus, edited by H. Itokawa and K.H. Lee
Volume 33
Capsicum, edited by Amit Krish De
Volume 34
Flax, edited by Alister Muir and Niel Westcott
Volume 35
Urtica, edited by Gulsel Kavalali
Volume 36
Cinnamon and Cassia, edited by P.N. Ravindran, K. Nirmal Babu and M. Shylaja
Volume 37
Kava, edited by Yadhu N. Singh
Volume 38
Aloes, edited by Tom Reynolds

Cinnamon and Cassia

The genus *Cinnamomum*

Edited by

P.N. Ravindran
*Centre for Medicinal Plants Research
Arya Vaidya Sala
Kottakkal, Kerala, India*

K. Nirmal Babu
*Indian Institute of Spices Research
Calicut, Kerala, India*

and

M. Shylaja
*Providence Women's College
Calicut, Kerala, India*

Medicinal and Aromatic Plants — Industrial Profiles

CRC Press is an imprint of the
Taylor & Francis Group, an **informa** business

CRC Press
Taylor & Francis Group
6000 Broken Sound Parkway NW, Suite 300
Boca Raton, FL 33487-2742

First issued in paperback 2019

© 2004 by Taylor & Francis Group, LLC
CRC Press is an imprint of Taylor & Francis Group, an Informa business

No claim to original U.S. Government works

ISBN-13: 978-0-415-31755-9 (hbk)
ISBN-13: 978-0-367-39467-7 (pbk)

This book contains information obtained from authentic and highly regarded sources. Reasonable efforts have been made to publish reliable data and information, but the author and publisher cannot assume responsibility for the validity of all materials or the consequences of their use. The authors and publishers have attempted to trace the copyright holders of all material reproduced in this publication and apologize to copyright holders if permission to publish in this form has not been obtained. If any copyright material has not been acknowledged please write and let us know so we may rectify in any future reprint.

Except as permitted under U.S. Copyright Law, no part of this book may be reprinted, reproduced, transmitted, or utilized in any form by any electronic, mechanical, or other means, now known or hereafter invented, including photocopying, microfilming, and recording, or in any information storage or retrieval system, without written permission from the publishers.

For permission to photocopy or use material electronically from this work, please access www.copyright.com (http://www.copyright.com/) or contact the Copyright Clearance Center, Inc. (CCC), 222 Rosewood Drive, Danvers, MA 01923, 978-750-8400. CCC is a not-for-profit organization that provides licenses and registration for a variety of users. For organizations that have been granted a photocopy license by the CCC, a separate system of payment has been arranged.

Trademark Notice: Product or corporate names may be trademarks or registered trademarks, and are used only for identification and explanation without intent to infringe.

Library of Congress Cataloging-in-Publication Data

Catalog record is available from the Library of Congress

Visit the Taylor & Francis Web site at
http://www.taylorandfrancis.com

and the CRC Press Web site at
http://www.crcpress.com

This volume is dedicated to Prof. (Dr.) K.V. Peter, Vice Chancellor, Kerala Agricultural University, Kerala, India, and Former Director, Indian Institute of Spices Research (IISR), Calicut, India, for his friendship, guidance and encouragement.

Contents

List of contributors		ix
Preface to the series		xi
Preface		xiii
Acknowledgements		xv

1 Introduction — 1
P.N. RAVINDRAN AND K. NIRMAL BABU

2 Botany and crop improvement of cinnamon and cassia — 14
P.N. RAVINDRAN, M. SHYLAJA, K. NIRMAL BABU AND
B. KRISHNAMOORTHY

3 Chemistry of cinnamon and cassia — 80
U.M. SENANAYAKE AND R.O.B. WIJESEKERA

4 Cultivation and management of cinnamon — 121
J. RANATUNGA, U.M. SENANAYAKE AND R.O.B. WIJESEKERA

**5 Harvesting, processing, and quality assessment
of cinnamon products** — 130
K.R. DAYANANDA, U.M. SENANAYAKE AND R.O.B. WIJESEKERA

6 Chinese cassia — 156
NGUYEN KIM DAO

7 Indonesian cassia (Indonesian cinnamon) — 185
M. HASANAH, Y. NURYANI, A. DJISBAR, E. MULYONO,
E. WIKARDI AND A. ASMAN

8 Indian cassia — 199
AKHIL BARUAH AND SUBHAN C. NATH

9 Camphor tree — 211
K. NIRMAL BABU, P.N. RAVINDRAN AND M. SHYLAJA

viii *Contents*

10 **Pests and diseases of cinnamon and cassia** 239
M. ANANDARAJ AND S. DEVASAHAYAM

11 **Pharmacology and toxicology of cinnamon and cassia** 259
K.K. VIJAYAN AND R.V. AJITHAN THAMPURAN

12 **Economics and marketing of cinnamon and cassia – a global view** 285
M.S. MADAN AND S. KANNAN

13 **End uses of cinnamon and cassia** 311
B. KRISHNAMOORTHY AND J. REMA

14 **Cinnamon and cassia – the future vision** 327
U.M. SENANAYAKE AND R.O.B. WIJESEKERA

15 **Other useful species of *Cinnamomum*** 330
M. SHYLAJA, P.N. RAVINDRAN AND K. NIRMAL BABU

Index 356

Contributors

Anandaraj, M.
Division of Crop Protection
Indian Institute of Spices Research
P.O. Box 1701, Calicut-673 012
Kerala, India

Asman, A.
Research Institute for Spices and
  Medicinal Plants
J.L. Tentara Pelajar, No. 3, Bogor
Indonesia

Akhil Baruah
Department of Botany
Darrang College (GU)
Tezpur-784 001
Assam, India

Dayananda, K.R.
Industrial Technology Institute
[Ceylon Institute of Scientific &
  Industrial Research]
P.O. Box 787
363, Baudhaloka Mawatha
Colombo 7, Sri Lanka

Devasahayam, S.
Division of Crop Protection
Indian Institute of Spices Research
P.O. Box 1701, Calicut-673 012
Kerala, India

Djisbar, A.
Research Institute for Spices and
  Medicinal Plants
J.L. Tentara Pelajar, No. 3, Bogor
Indonesia

Hasanah, M.
Research Institute for Spices and
  Medicinal Plants
J.L. Tentara Pelajar, No. 3, Bogor
Indonesia

Kannan, S.
Spices Board, Cochin-682 025, India

Nguyen Kim Dao
Department of Botany
Institute of Ecology & Biological
  Resources
National Centre for Natural Science &
  Technology (IEBR-NCST)
Hoang Quoc Viet Road
Cau Gray
Hanoi, Vietnam

Krishnamoorthy, B.
Division of Crop Improvement &
  Biotechnology
Indian Institute of Spices Research
P.O. Box 1701, Calicut-673 012
Kerala, India

Madan, M.S.
Division of Social Sciences
Indian Institute of Spices Research
P.O. Box 1701, Calicut-673 012
Kerala, India

Mulyono, E.
Research Institute for Spices and
  Medicinal Plants
J.L. Tentara Pelajar, No. 3, Bogor
  Indonesia

x *Contributors*

Subhan C. Nath
Regional Research Laboratory
Plant Sciences and Ecology Division
Jorhat-785 006
Assam, India

Nirmal Babu, K.
Division of Crop Improvement &
  Biotechnology
Indian Institute of Spices Research
P.O. Box 1701, Calicut-673 012
Kerala, India

Nuryani, Y.
Research Institute for Spices and
  Medicinal Plants
J.L. Tentara Pelajar, No. 3, Bogor
Indonesia

Ranatunga, J.
Industrial Technology Institute
[Ceylon Institute of Scientific &
  Industrial Research]
P.O. Box 787
363, Baudhaloka Mawatha
Colombo 7, Sri Lanka

Ravindran, P.N.
Indian Institute of Spices Research
P.O. Box 1701, Calicut-673 012
  Kerala, India
[P.a.: Centre for Medicinal Plants
  Research, Shatabdinagar,
  Kottakkal-676 503,
  Kerala, India]

Rema, J.
Division of Crop Improvement &
  Biotechnology
Indian Institute of Spices
  Research
P.O. Box 1701, Calicut-673012
Kerala, India

Senanayake, U.M.
Industrial Technology Institute
[Ceylon Institute of Scientific &
  Industrial Research]
P.O. Box 787
363, Baudhaloka Mawatha
Colombo 7, Sri Lanka
[P.a.: 38/3, S. De S Jayasinghe Mawatha
  Nugegoda, Sri Lanka]

Shylaja, M.
Department of Botany
Providence Women's College
Calicut-673 009, Kerala, India

Ajithan Thampuran, R.V.
Department of Pharmacology
Medical College, Calicut-673 008
Kerala, India

Vijayan, K.K.
Department of Chemistry
University of Calicut
Calicut University P.O. 673 635
Calicut, Kerala, India

Wijesekera, R.O.B.
Industrial Technology Institute
[Ceylon Institute of Scientific &
  Industrial Research]
P.O. Box 787
363, Baudhaloka Mawatha
Colombo 7, Sri Lanka.
[P.a.: National Science & Technology
  Commission
No. 2, Galpotta Road, Nawala
  Rajagiriya, Sri Lanka]

Wikardi, E.
Research Institute for Spices and
  Medicinal Plants
J.L. Tentara Pelajar, No. 3, Bogor
Indonesia

Preface to the series

There is increasing interest in industry, academia and the health sciences in medicinal and aromatic plants. In passing from plant production to the eventual product used by the public, many sciences are involved. This series brings together information which is currently scattered through an ever increasing number of journals. Each volume gives an in-depth look at one plant genus, about which an area specialist has assembled information ranging from the production of the plant to market trends and quality control.

Many industries are involved, such as forestry, agriculture, chemical, food, flavour, beverage, pharmaceutical, cosmetic and fragrance. The plant raw materials are roots, rhizomes, bulbs, leaves, stems, barks, wood, flowers, fruits and seeds. These yield gums, resins, essential (volatile) oils, fixed oils, waxes, juices, extracts and spices for medicinal and aromatic purposes. All these commodities are traded worldwide. A dealer's market report for an item may say "Drought in the country of origin has forced up prices".

Natural products do not mean safe products and account of this has to be taken by the above industries, which are subject to regulation. For example, a number of plants which are approved for use in medicine must not be used in cosmetic products.

The assessment of safe to use starts with the harvested plant material, which has to comply with an official monograph. This may require absence of, or prescribed limits of, radioactive material, heavy metals, aflatoxin, pesticide residue, as well as the required level of active principle. This analytical control is costly and tends to exclude small batches of plant material. Large scale contracted mechanised cultivation with designated seed or plantlets is now preferable.

Today, plant selection is not only for the yield of active principle, but for the plant's ability to overcome disease, climatic stress and the hazards caused by mankind. Such methods as *in vitro* fertilization, meristem cultures and somatic embryogenesis are used. The transfer of sections of DNA is giving rise to controversy in the case of some end-uses of the plant material.

Some suppliers of plant raw material are now able to certify that they are supplying organically farmed medicinal plants, herbs and spices. The Economic Union directive (CVO/EU No. 2092/91) details the specifications for the *obligatory* quality controls to be carried out at all stages of production and processing of organic products.

Fascinating plant folklore and ethnopharmacology leads to medicinal potential. Examples are the muscle relaxants based on the arrow poison, curare, from species of *Chondrodendron*, and the anti-malerials derived from species of *Cinchona* and *Artemisia*. The methods of detection of pharmacological activity have become increasingly reliable and specific, frequently involving enzymes in bioassays and avoiding the use of laboratory animals. By using bioassay linked fractionation of crude plant juices or extracts,

xii *Preface to the series*

compounds can be specifically targeted which, for example, inhibit blood platelet aggregation, or have anti-tumour, or anti-viral, or any other required activity. With the assistance of robotic devices, all the members of a genus may be readily screened. However, the plant material must be *fully* authenticated by a specialist.

The medicinal traditions of ancient civilisations such as those of China and India have a large armamentarium of plants in their pharmacopoeias which are used throughout South-East Asia. A similar situation exists in Africa and South America. Thus, a very high percentage of the World's population relies on medicinal and aromatic plants for their medicine. Western medicine is also responding. Already in Germany all medical practitioners have to pass an examination in phytotherapy before being allowed to practise. It is noticeable that throughout Europe and the USA, medical, pharmacy and health related schools are increasingly offering training in phytotherapy.

Multinational pharmaceutical companies have become less enamoured of the single compound magic bullet cure. The high costs of such ventures and the endless competition from "me too" compounds from rival companies often discourage the attempt. Independent phytomedicine companies have been very strong in Germany. However, by the end of 1995, eleven (almost all) had been acquired by the multinational pharmaceutical firms, acknowledging the lay public's growing demand for phytomedicines in the Western world.

The business of dietary supplements in the Western World has expanded from the health store to the pharmacy. Alternative medicine includes plant-based products. Appropriate measures to ensure the quality, safety and efficacy of these either already exist or are being answered by greater legislative control by such bodies as the Food and Drug Administration of the USA and the recently created European Agency for the Evaluation of Medicinal Products, based in London.

In the USA, the Dietary Supplement and Health Education Act of 1994 recognised the class of phytotherapeutic agents derived from medicinal and aromatic plants. Furthermore, under public pressure, the US Congress set up an Office of Alternative Medicine and this office in 1994 assisted the filing of several Investigational New Drug (IND) applications, required for clinical trials of some Chinese herbal preparations. The significance of these applications was that each Chinese preparation involved several plants and yet was handled as a *single* IND. A demonstration of the contribution to efficacy, of *each* ingredient of *each* plant, was not required. This was a major step forward towards more sensible regulations in regard to phytomedicines.

My thanks are due to the staff of Taylor & Francis who have made this series possible and especially to the volume editors and their chapter contributors for the authoritative information.

Roland Hardman, 1997

Preface

Cinnamon is one of the most popular spices used by humankind, as a glance through any cookbook will indicate. From breakfast rolls to spiced cookies, pudding and pies to quickbreads and chutneys, cinnamon finds its way into recipes for standard family fare as well as special treats. Cinnamon is the second most important spice (next to black pepper) sold in U.S. and European markets.

Cinnamon occupied a pre-eminent position in the ancient world and was much sought after. In the middle ages, the lure of spices tempted the Western powers to explore the unknown seas in search of the famed spice lands of the east. These explorations eventually led to the discovery of America and the sea route to India by Portuguese explorers. With those discoveries human history witnessed the transition from the medieval to the modern era. Imperialism and colonialism reigned the world scene in the next few centuries. It was the period when the world powers fought bitter wars for naval supremacy and for monopoly in the spice trade. In this struggle, cinnamon was the Holy Grail for foreign invaders, over which many a costly war was fought by Portugal, Holland, France and Britain.

Cinnamon and its close relative cassia are among the most popular spices. Cinnamon is often qualified as 'sensational cinnamon' and 'spice of life' because of the emotional attachment of Sri Lankan people with cinnamon; settlements, housing colonies and residential areas are often named after cinnamon.

The genus *Cinnamomum* has a centre of diversity in Western Ghats and the adjoining regions of south India. Two of the editors of this volume, P.N. Ravindran and M. Shylaja, carried out a detailed botanical study of the *Cinnamomum* species occurring in south India during the early 1980s. Subsequently, similar studies were also taken up by Baruah and his colleagues on the species occurring in north-east India, and Kostermans published a paper on the species occurring in south India. The third editor of this volume, K. Nirmal Babu, is involved in the collection, *in vitro* germplasm conservation, micropropagation and molecular characterisation of Cinnamon, Cassia and Camphor.

We took interest in this genus because of the fascinating history behind this spice and because no information was available on the species occurring in the region. Apart from our own work, we also had the occasion to be close to the research work being carried out at the Indian Institute of Spices Research, Calicut, where a good germplasm collection of cinnamon exists and that work is going on in the areas of crop management and improvement. The Senior editor also had the occasion to monitor and supervise the research work on cinnamon being carried out in various centres under the All India Coordinated Research Project on Spices. The research workers in all these centres have whole-heartedly collaborated with us during the production of this volume.

xiv *Preface*

When we approached the Ceylon Institute of Scientific and Industrial Research (Currently the Industrial Technology Institute (ITI)) its Director General Dr. WOB Wijesekera readily agreed to author the various chapters on Sri Lankan Cinnamon. We record here our deep appreciation to ITI and all the Scientists who wrote the chapters included in this volume on Ceylon cinnamon. We also received the collaboration of the Research Institute for Spices and Medicinal Plants of Indonesia and of the IEBR, Vietnam, in writing chapters on Indonesian and Chinese cassia respectively.

This volume contains fifteen chapters covering all aspects of cinnamon, camphor and on various cassia types and a chapter on other useful species on which information is available. Chapters on botany and crop improvement, economics and marketing, pharmacology and toxicology and end uses are common for both cinnamon and cassia. We have made sincere efforts to collect and collate as much information as we possibly could. This is the first monograph on cinnamon and cassia and we hope that this will remain as the main reference work on these sensational spices for many years to come. We hope that this volume will be useful to students and research workers in the areas of botany, economic botany, ethnobotany, horticulture, agriculture and allied fields, as well as to exporters, processors, planters and to all those who are interested in this spice of life.

Editors

Acknowledgements

The editors express their deep gratitude to all the contributors of this monograph who found time to collaborate in its production. We are especially thankful to Dr. WOB Wijesekera, former Director General, and to Dr. U.M. Senanayake, former Principal Scientist, Ceylon Institute of Scientific and Industrial Research (currently Industrial Technology Institute) for the interest they have taken in the preparation of the chapters on Ceylon cinnamon.

The editors are grateful to Dr. Roland Hardman, General Editor of the series on Medicinal and Aromatic Plants: Industrial profiles, for his constant encouragement and timely help during the production of this volume. I am thankful to him for accepting my proposal of editing this important monograph. He helped us by providing updated literature searches and photographs on cinnamon processing. His help and guidance were immensely helpful to the senior editor during the production of the earlier volumes in the series viz. Black pepper and Cardamom. Spices workers all over the world accepted these volumes as the most authentic and comprehensive publications on these crops.

We are extremely grateful to Professor K.V. Peter, former Director of Indian Institute of Spices Research and the present Vice Chancellor, Kerala Agricultural University, Kerala, India, for providing the necessary permission and facilities for the preparation of this volume and for allowing us to dedicate this volume to him. We are thankful to many of our friends and colleagues who helped us in the preparation of this volume. Our special thanks go to Mr. A. Sudhakaran for various drawings, Mr. K. Jayarajan and Mr. K.V. Tushar for computer work, Ms. P.V. Sali, Ms. Lovely and Ms. Nisha for typing work and Mr. P.A. Sheriff for library assistance. Our colleagues Ms. Geetha S. Pillai and Ms. Minoo Divakaran, provided us with invaluable support during various stages in the preparation of this volume. We are deeply grateful to them.

In the preparation of this volume, especially the chapter on Botany and Crop Improvement, we have made use of published information from many sources and by many authors. We acknowledge with gratitude all these authors, many are not with us now, but their contributions will continue to survive through this volume in the years to come. We salute all of them with reverence and gratitude.

We have sincere appreciation to Taylor & Francis Publishers for giving us the opportunity to edit this first monograph on Cinnamon and Cassia. We thank all our well-wishers and all those who helped us in the preparation of this volume.

P.N. Ravindran
K. Nirmal Babu
M. Shylaja

1 Introduction

P.N. Ravindran and K. Nirmal Babu

Cinnamon and cassia are among the earliest known spices used by humankind. Frequent references to these spices are available in both pre-biblical and post-biblical writings. The cinnamon of commerce is the dried inner bark of the tree *Cinnamomum verum* (= syn. *C. zeylanicum*), belonging to the family Lauraceae. It is native to Sri Lanka where it is grown on a large scale, and exported (known in trade as Ceylon cinnamon or Sri Lankan cinnamon). Cassia or cassia cinnamon comes from different sources, the important ones being the Chinese cassia (*C. cassia*, syn. *C. aromaticum*), and Indonesian cassia (*C. burmannii*). Chinese cassia is indigenous to the China–Vietnam region and is an important spice traded in the international market. Indonesian cassia is a native of the Sumatera–Java region of Indonesia and is exported on a large scale to the USA. The Indian cassia comes from *C. tamala* and a few other related species (*C. impressinervium*, *C. bejolghota*), which are indigenous to the north-eastern region of India.

The term *Cinnamomum* is derived from the Greek root *kinnamon* or *kinnamomon*, meaning sweet wood. This term possibly had a Semetic origin from the Hebrew *quinamom*. The Malayan and Indonesian name *kayu manis* also means sweet wood, and an ancient version of this term possibly might have contributed to Hebrew and Greek terminology. The Dutch (*kaneel*), French (*cannelle*), Italian (*cannella*) and Spanish (*canela*) names are derived from the Latin *canella* (meaning small tube or pipe, referring to the form of cinnamon quills). The Hindi name *dalchini*, meaning Chinese wood, refers originally to the Chinese cinnamon, which was popular in northern India before the Ceylon cinnamon became known. The name cassia seems to have derived from the Greek *kasia*, which probably has its roots in Hebrew *qeshiiah*. The name for cinnamon in different languages is given in Table 1.1.

There exists some confusion regarding the use of the terms cinnamon and cassia. In continental Europe and the UK, cinnamon applies only to *C. verum* (Ceylon cinnamon) while cassia to *C. cassia*. But in the USA cinnamon applies to the bark from both sources and also from *C. burmannii*. The US Food, Drug and Cosmetic Act officially permits the term cinnamon to be used for Ceylon cinnamon, Chinese cassia and Indonesian cassia. The delicately flavoured Ceylon cinnamon is rarely used now in the USA and cassia cinnamon has conquered the market almost wholly (Rosengarten, 1969).

Early History

The early history of cinnamon and cassia is fascinating. They are among the earliest spices used. They formed the ingredients of the embalming mixture in ancient Egypt, and were among the most expensive materials in ancient Greece and Rome; only royalty could afford

0-415-31755-X/04/$0.00 + $1.50
© 2004 by CRC Press LLC

Table 1.1 Cinnamon – terminology in different languages

Amharic	*K'erefa*
Arabic	*Qurfa, Darasini, Kerfa*
Assamese	*Dalchini*
Bengali	*Dalchini, Daruchini*
Burmese	*Thit-ja-bo-gank, Hminthin, Timboti kyobri*
Chinese	*Jou kuei, Yuk gwai*
Czech	*Skorice*
Danish	*Kanel*
Dutch	*Kaneel*
English	*Ceylon cinnamon; Sri Lanka cinnamon*
Estonian	*Tseiloni kaneelipuu*
Fante	*Anoater dua*
Farsi	*Dar chini*
Finnish	*Kaneli, Ceyloninkaneli*
French	*Canelle type ceylan, Cannelle*
German	*Zimt, Echter Zimt, Ceylon-Zimt, Zimtblute* (buds)
Greek	*Kanela, Kinnamon*
Gujarati	*Tuj*
Hebrew	*Kinamon, Quinamom*
Hindi	*Darchini/Dalchini*
Hungarian	*Fahej, Ceyloni fahej*
Icelandic	*Kanell*
Indonesian	*Kayu manis*
Italian	*Cannella*
Japanese	*Seiron nikkei, Nikkei*
Kannada	*Lavangapatta*
Malay	*Kayu manis, Kulit manis*
Malayalam	*Patta, Karuapatta, Ilavangam*
Mandarin	*You kwei*
Marathi	*Dalchini*
Nepalese	*Newari dalchini*
Norwegian	*Kanel*
Oriya	*Dalochini*
Pashto	*Dolchini*
Persian	*Darchini*
Portuguese	*Canela*
Romanian	*Scortisoara*
Russian	*Koritsa*
Sanskrit	*Tvak, Tvaka, Twak, Darusita*
Singhalese	*Kurundu*
Spanish	*Canela*
Swahili	*Madalasini*
Swedish	*Kanel*
Tamil	*Illavangam*
Telugu	*Lavangamu, Dalchini chekka*
Thai	*Op cheuy*
Turkish	*Tarchin*
Twi	*Anoatre dua*
Cassia (*C. cassia*)	
Arabic	*Darseen, Kerfee, Salikha*
Chinese	*Kuei, Rou gui pi*
Duch	*Kassie, Bastaard kaneel, Valse kaneel*
English	*Chinese cassia, Bastard cinnamon, Chinese cinnamon*
Estonian	*Hiina kaneelipuu*
Finnish	*Talouskaneli, Kassia*

French	*Casse, Canefice, Canelle de Chine*
German	*Chinesisches Zimt, Kassie*
Hungarian	*Kasszia, Fahejkasszia, Kinai fahej*
Icelandic	*Kassia*
Italian	*Cassia, Cannella della Cina*
Japanese	*Kashia keihi, Bokei*
Laotian	*Sa chouang*
Norwegian	*Kassia*
Russian	*Korichnoje derevo*
Spanish	*Casia, Canela de la China*
Swedish	*Kassia*
Thai	*Ob choey*
Urdu	*Taj*
Tejpat (*C. tamala*)	
Arabic	*Sazaj hindi*
Bengali	*Tejpat*
Burmese	*Thitchubo*
English	*Indian cassia, cassia lignea*
French	*Cannelle*
German	*Zimtbaum*
Hindi	*Tejpat, Tajpat, Taj-kalam*
Japanese	*Tamara Nikkei*
Nepalese	*Tejpat*
Persian	*Sazaj hind*
Sanskrit	*Tejapatra, Tamalapatra, Patra, Tamalaka*
Singhalese	*Tejpatra*
Tamil	*Perialavangapallai, Perialavangapattai, Talishappattiri*
Telugu	*Talispatri*
Urdu	*Tezpat*

Source: Compiled from various sources.

Notes
Cinnamon is used as an adjective in naming some plants. Such plants have no relationship with cinnamon or the genus *Cinnamomum*, e.g.:
Cinnamon fern: *Osmunda cinnamomea* L. (Osmundaceae).
Cinnamon rose: *Rosa majalis* Herrm. (Rosaceae).
Cinnamon vine: *Dioscoria batatas* Decne. (Dioscoriaceae).
Cinnamon wattle: *Acacia leprosa* Sieber ex DC (Mimosaceae).
Cinnamodendron Endl. Agenus in Canellaceae.
C. corticosum Miers (wl)-bark of this tree is used as a spice and a tonic.
Cinnamosma Baillon (Canellaceae). Indigenous to Madagascar. *C. fragrans* Baillon has highly scented wood, produces scented fumes on burning. Used in religious ceremonies.

them. History tells us that the Egyptian Queen Hatshepsut (around 1500 BC), sent out an expedition of five ships to bring spices and aromatics from the land of "Punt" (which was believed to have been the land on either side of the lower Red Sea and Gulf of Eden). These ships returned loaded with "fragrant woods of god's land, heaps of myrrh-resin of fresh myrrh trees, cinnamon wood, with incense, eye-cosmetic" (Parry, 1969). Rosengarten (1969) writes that the origin of cinnamon which Hatshepsut collected is uncertain as cinnamon trees are not indigenous to the land of "Punt". According to the historian Miller (1969), there are indications that as early as the second millennium BC, cassia and cinnamon from China and South-East Asia might have been brought from Indonesia to Madagascar in primitive canoes, along a "cinnamon route" which might have

4   *P.N. Ravindran and K. Nirmal Babu*

existed at that time. These aromatic barks were then transported northward along the East African coast to the Nile Valley and from there to the land of "Punt".

References to cinnamon and cassia exist in the Old Testament of the Bible. In Exodus, the Lord spoke to Moses on the top of Mount Sinai and gave instructions that the children of Israel should build a tabernacle, so that he might dwell among them; and further instructed for the preparation of an anointing oil for the tabernacle containing cinnamon and cassia with other things:

> The Lord spake unto Moses. Take thou also unto thee principal spices, of pure myrrah five hundred shekels, and of sweet cinnamon half so much . . . and of cassia five hundred shekels . . . . And thou shalt make of it an oil of holy anointment. And thou shalt anoint the tabernacle of the congregation therewith and the ark of testimony
>
> (Exodus 30:23–26).

This is the first biblical reference to cinnamon and cassia (Parry, 1969). The building of the tabernacle is believed to have taken place around 1490 BC, and the biblical reference indicates that these spices were well known and held in very high esteem at that time. Again later in Psalm 45 (Verse 8) cassia is mentioned as perfume: "All thy garments smell of myrrh, and aloes and cassia, out of the ivory palaces, whereby thy hand made thee glad". Cinnamon is mentioned in the beautiful passages of the Song of Solomon, where there are many references about spices:

> . . . . thy plants are an orchard of pomegranates, with pleasant fruits, camphire, with spikenard and saffron, calamus and cinnamon, with all trees of frankincense, myrrh and aloes, with all the chief spices
>
> (Chapter 4:13:14).

We also find in the Revelations that St. John the Divine foretells the fall of the great city of Babylon and the distress that ensues:

> . . . . and the merchants of the earth shall weep and mourn over her, ... and cinnamon, and odours, and odours and ointments . . . . thou shalt find them no more at all . . . . .
>
> (Revelation 18:13).

It is thus amply clear that cinnamon and cassia were held in high esteem in those ancient days. At one time it was more valuable than gold (Farrell, 1985). They were among the most valuable medicinal plants for ancient Greeks and Romans. Dioscorides records:

> Cinnamon provoked urine, it cleared the eyes and made the breath sweet. An extract of cinnamon would bring down the menses and would counteract the stings and bites of venomous beasts, reduce the inflammation of the intestines and kidneys, comfort the stomach, break wind, would aid in digestion and when mixed with honey would remove spots from the face that was anointed there with
>
> (Farrell, 1985).

Parry (1969) was of opinion that the founders of the spice trade between India and the rest of the world were either the Phoenicians or the Arabs. The Phoenicians were accomplished sailors, and they might have been responsible for transporting cinnamon from the east to the west. They were expert traders of all commodities, including spices. There are references of cassia and cinnamon in Ezekiel (Chapter 27), where there are high praises about the richness of Tyre, the capital of Phoenicians under the emperor Hiram, and also references about Arabian merchants carrying spices. The Phoenicians were probably the first to carry cinnamon and cassia to Greece, and along with the spice, its name "cinnamon" also passed on to the Greeks.

It is not always possible to correctly identify the plants mentioned in ancient writings. The reference about the aromatic wood collected by Hatsheput's expedition casts a shadow of doubt as to its identity. Most probably it was not the present cinnamon at all. The wood of cinnamon has not much fragrance, and on burning cinnamon wood does not give fragrant smoke, only its bark has the spicy taste and smell. Again there were references about Nero burning cinnamon in the funeral pyre of his wife, why? Was it not for the fragrance it was giving out? Then can it be some other wood, having fragrance that produces fragrant smoke on burning? If so, a possible case is the small tree species, *Cinnamosma fragrans* (Canellaceae) occurring in the eastern coastal African forests as well as in Madagascar. The wood of this tree is fragrant and produces fragrant smoke on burning. It was easier for the Egyptians or to the enterprising people of "Punt" to travel along the coast line of Africa to the eastern coastal forests and collect the fragrant wood of *Cinnamosma* rather than travelling to Ceylon or the Far East. The name probably came to be applied to cinnamon at a later time.

In ancient times, the south Arabian region was occupied by a nomadic race, the Sabians, whose vocation was sailing and sea trading. Their main merchandise was spices. It was possibly from these people the ancient Egyptians collected their requirement of spices. Probably it was the same people who carried spices from Gilead to Egypt, and it was possibly from Arabia that the Egyptian Queen Hatshepsut collected cinnamon and cassia over 3500 years ago (the land of "Punt"). The south Arabians held a virtual monopoly on the spice trade for a very long time. Historical evidence is quite insufficient to come to a conclusion on the relative roles played by the Arabians and Phoenicians in the ancient spice trade.

The central position occupied by Arabia also gave way to the belief that cassia and cinnamon were produced in that country. Herodotus and Theophrastus recorded the magical stories perpetuated by the Arabians about the source of cinnamon and cassia. Theophrastus describes south-west Arabia as the land of myrrh, frankincense and cinnamon. The aromatics are in such abundance there, says Strabo, that the people use cinnamon and cassia instead of sticks as firewood (Parry, 1969). The Sabean traders fabricated all sorts of stories about cinnamon and cassia, and they were successful for a long time in shrouding the source of cinnamon in mystery. One such story goes like this:

> Far away, in a distant land, said Arab traders to spice buyers from Europe, there is a great lake. It is surrounded by deep and fragrant woods and high cliffs. On those cliffs nests a great dragon-like bird with a scimitar-sharp beak and talons like Saladin's sword. The nest of this bird is made of only one material; the bark of a rare tree. And when it has made its nest, it does not lay eggs in it. Oh no. It does not need to. It flaps its wings so fast that the bark catches fire and then

the bird sits in the middle of its blazing pyre. And then, lo and behold, it emerges from the flames refreshed, renewed, rejuvenated. The name of the bird is Phoenix. And the name of the bark? Ah! It is cinnamon. And that is why, that is exactly why, cinnamon is so expensive.

Thus, the astute Arabs, not only enhanced the value of cinnamon but they also concealed its origin for many centuries. This is perhaps one of the best kept trade secrets of all time. It was only when the Europeans started sailing around the world, in search of pepper, that they discovered cinnamon and its cousin cassia and that they grew fairly easily in South, South-East and East Asia (Gantzer and Gantzer, 1995).

The Arab domination of the spice trade was broken by the rise of the Roman empire. Around AD 40, Mariner Hippalus discovered the trade wind systems in the Indian Ocean, hitherto known only to the Arabs. It is believed that he had travelled to India and back around AD 40, thereby opening up the direct trade route between Rome and the West Coast of India. As a result, by the end of the first century AD, the use of spices in Rome had grown miraculously. Cinnamon, cassia and cardamom occupied the pride of place among the spices. The extravagance is clear from references to the huge supplies of the aromatic spices that were strewn along the path behind the funeral urn bearing the ashes of Commander Germanicus. Emperor Nero is said to have burned a year's supply of Rome's cinnamon at his wife's funeral pyre (AD 66). It was also customary for men to be heavily perfumed and even "the legionaries reeked of the fragrances of the east" (Rosengarten, 1969). Even lamp oil was mixed with aromatics to keep harmful vapours away.

By the end of the third century AD, the Arabians had established trade relationships with China, mainly for trading in cassia. This aided them to trade not only in cassia, but also in spices that came from the far eastern countries (East Indies). In AD 330, the Roman emperor Constantine founded the city of Constantinople on the site of the ancient Byzantium, which became the capital of the Byzantine Empire. During this period cassia from China, nutmeg and cloves from Moluccas, cinnamon from Ceylon, pepper and cardamom from the Malabar coast of India reached the new city in large quantities. Ceylon and the Malabar Coast were the transshipment spots in this spice trade.

The course of history never runs smoothly for long. Alaric the Gothic invaded and captured Rome in AD 410. The hegemony and splendour of Rome and her supremacy over trade all came to an abrupt end. Once again the Arabs came in and soon became the masters of the spice trade. This supremacy continued till the fifteenth century, when the sea route to India was discovered and Vasco-da-Gama landed in the West Coast of India on 20 May 1498.

During the middle ages (between the fifth and fifteenth centuries) spices started reaching Western Europe and were among the choicest gifts to royalty and the privileged, especially to the monasteries and ecclesiastical establishments (Parry, 1969). The travelogues of Marco Polo, the most renowned traveller of the middle ages, give the most authentic information on the spice trade in the middle ages. He writes about the cassia cultivation in China, cloves of Nicobar, pepper, ginger, cardamom and cinnamon of the Malabar Coast and many seed spices, such as sesame.

The earliest recorded reference to cinnamon bark as a product of Ceylon dates back to the thirteenth century (Redgrove, 1933). Redgrove writes:

It seems quite probable that the Chinese, who traded with Ceylon were concerned in the discovery of the valuable qualities of the bark of Sinhalese tree, similar but superior, to the cassia of their own country . . . . At any rate when the Sinhalese product was imported into Europe, its superior character was soon recognised and the product fetched very high prices (1933).

By the thirteenth century, the East Indies became a busy trading centre in spices. Java was the main centre for trading in nutmeg, mace and cloves that came from the Moluccas Islands. From Java the Arabian ships carried these spices to the west. In fact, "the East Indies gradually eclipsed the Malabar Coast of India as the most important source of costly spices, and both places attracted the princes and merchants of Western Europe to bring in the fifteenth and sixteenth centuries, the greatest and brightest age of discovery the world has ever seen" (Parry, 1969).

China emerged as a major trader of spices during this period, trading in cassia and ginger and procuring large quantities of pepper and other spices from the Malabar Coast and the East Indies. At this time cassia and cassia buds became popular spices in Europe and England. Of course, spices were beyond the reach of common folk, as they were so costly, mainly because no direct trade links existed between Europe and the eastern spice-producing countries. The Arabs monopolised the trade in the east, and Venice controlled the trade in the Mediterranean.

Spices like pepper, cardamon, cinnamon and cassia contributed greatly to European cooking. Parry (1969) writes:

The coming of the highly aromatic and pungent spices of the orient was the greatest boon to the European food and cooking of all times. New methods of preserving food quickly came into existence; dishes took on a fullness of flavour previously unknown; beverages glowed with a redolent tang, and life experienced a new sense of warmth and satisfaction.

Spices were also used as medicines by the people of ancient and middle ages. Warren R Dawson made a collection of medical recipes of the fifteenth century, wherein some 26 spices were indicated for various ailments. Cinnamon and cassia were components of medicines recommended for coughs, chest pain, headache, digestion and gas problems. In Chinese traditional medicine, cassia bark (cortex cinnamomi – *Rou gui*) and dried twig (ramulus cinnamomi – *Guo shi*) are two separate drugs used differently.

Modern History

The modern chapter in the saga of spices begins with the discovery of the sea route to India and the landing of Vasco-da-Gama on the Malabar coast (near the present day Calicut in Kerala state) of India on 20 May 1498. This indeed was the beginning of the history of modern India too. The West European countries were compelled to establish a sea route to the eastern spice lands following the conquest of the Roman Empire and the closure of Constantinople for trade by the Ottoman Turks in 1453. The quest for spices opened up the era of great expeditions; Columbus discovered America and Vasco-da-Gama sailed eastwards round the Cape of Good Hope and arrived in India. In the decades that followed the Portuguese gradually established their hold on spice trading.

By the beginning of the sixteenth century, the Portuguese started direct trading in spices with the Malabar Coast, and with this the Arab supremacy over the spice trade came to an end. The Portuguese soon established a monopoly in the spice trade, not only with India but also with Ceylon and later with the spice islands of the Far East. It was during the time of Portuguese domination that the cinnamon trade began to attain considerable stature. In 1506, the Portuguese forced the Sinhala kings to undertake the supply of about 11,000 kg of cinnamon bark annually. In the process of fulfilling this undertaking the cinnamon forests came to be ruthlessly exploited. By the end of the century the Portuguese had secured for themselves a monopoly on the world's cinnamon trade (Wigesekera et al., 1975). This Portuguese monopoly led to a growing price rise of pepper and cinnamon, causing a wave of resentment in Western Europe. Soon the other West European countries wanted to break this monopoly and many expeditions were sent out to establish a route to the Malabar Coast and the spice islands of the East Indies. Cinnamon was the Holy Grail of foreign invaders to the island of Ceylon.

Soon the Dutch navigator Cornelius Van Hortman reached the East Indies (1596) and established trade relations with some of the islands in spite of the stiff resistance from the Portuguese. Many more expeditions followed and the Dutch gradually established their supremacy in the East Indies, virtually expelling the Portuguese from the scene. The Dutch gradually conquered all important spice-producing islands, and they became the masters of the spice trade. The Dutch captured the Portuguese establishments in Ceylon in 1658, and secured control of the island and its rich cinnamon trade. In 1663 the Dutch also conquered Cochin and Cannanore of the Malabar Coast. They commenced large scale monopolistic cultivation of cinnamon in Ceylon, until which their cinnamon had been growing in patches of forests in the south-western coast of the island. The bark was obtained by the government in the form of tributes. In 1767 the Dutch Governor Falk introduced a planned system of cinnamon cultivation by allocation and distribution of land and by the enactment of legislation to ensure cultivation. The entire stock produced and purchased by them was not, however, exported. Only a sufficient quality to meet the demand was permitted to be exported. Large quantities of cinnamon were frequently destroyed, the idea being to limit exports in order to maintain high prices (Wijesekera et al., 1975).

The seventeenth century saw the rise of British naval supremacy. Yet another chapter in the saga of cinnamon starts with the British occupation of Ceylon in 1796, and the monopoly of cinnamon trade changed hands once more. The prohibitive export duty, which had been prevailing was reduced and finally in 1843 it was abolished. This resulted in large shipments of cinnamon being moved out of the island to serve the European markets. The mean annual exports rose from 200,000 kg to 375,000 kg within the next few years (Wijesekera et al., 1975). Large scale plantations of cinnamon were established in Ceylon, and by 1850 about 40,000 acres of cinnamon were under cultivation (Rosengarten, 1969). Cinnamon was introduced into many islands in the tropics by British and Dutch colonists. Cinnamon from Ceylon was introduced into India in 1798 by Mr. Murdock Brown, who was then Superintendent of the East India Company in Anjarakandy (in the present Kannur District of Kerala). He established a cinnamon plantation in the Anjarakandy Estate, which still exists. The cultivation of cinnamon spread to other regions, especially in the islands of Seychelles, Madagascar and the West Indies, but Ceylon continued to be the major producer. In 1867 quills and chips were introduced as export products, and as a result the export from Ceylon rose to 450,000 kg. Soon afterward the cheaper substitute, cassia cinnamon, was introduced into the European markets (Wijesekara *et al.*, 1975).

The traditionally known cinnamon was the peeled cinnamon bark that was rolled into the form of 'quills' in order to facilitate storage and transportation. Cinnamon oil was not a commodity of commerce in those days, although oil distillation was known and cinnamon oil was used in cosmetics and pharmaceutical preparations. The first references to cinnamon oil are seen in the price ordinance in Berlin in 1574 and in Frankfurt in 1582 (Wijesekera *et al.*, 1975). It is believed, however, that the cinnamon oil distillation in Sri Lanka would have probably commenced during the Dutch regime.

The global production and trade in spices were affected drastically by the world wars. The Japanese occupation of the Dutch East Indies led to the devastation and decline of the spice production of the region and in the war-torn western countries imports declined sharply. The end of the Second World War witnessed the decline of colonialism and the spread of independence in the colonial countries. The old scenario changed rapidly. The spice producing eastern countries gained independence and spices became one of their major export earnings.

Cinnamon is associated with the lives of people of Sri Lanka, emotionally, socially and economically. For them it is the spice of life (Ratwatte, 1991). Sri Lanka's cinnamon groves are located in the western and south-western regions of the island. The tropical sunshine and abundant rain in these areas provide the ideal habitat for the growth of cinnamon. The sweetest, most prized variety grows in the 'Silver Sand' coastal belt of the Colombo District, just north of Colombo (Ratwatte, 1991).

Research and Development Efforts in Cinnamon and Cassia

Though cinnamon and cassia have played important roles in human cuisine from ancient times, efforts on research and development of these tree spices have not received the required attention. Some R&D efforts on cinnamon were initiated in Sri Lanka during the post-independence period by the Ceylon Institute of Scientific and Industrial Research and by the Department of Export Agriculture. These R&D efforts were mainly concentrated on chemistry, quality assessment, on developing agro-technology for cultivation and post-harvest processing. Certain elite lines (varieties) have been identified through selection. Refinement of the conventional practices were attempted and a package of practices has been brought out (Anon., 2000; Wijeskera *et al.*, 1975).

The only other country that has carried out some R&D efforts is India, where research on tree spices was initiated a couple of decades ago with the establishment of a Research Station at Calicut (Kozhikode) under the Central Plantation Crops Research Institute (Kasaragod, Kerala). This Research Station, established in 1975 (presently the Indian Institute of Spices Research-IISR), has initiated the collection and evaluation of germplasm. Cinnamon germplasm, both indigenous and introduced, were screened for quality and evaluated in the field. Subsequently two elite lines (*Navasree* and *Nithyasree*) were released (Krishnamoorthy *et al.*, 1996). Similar clonal selections were also carried out by Konkan Krishi Vidyapeeth in Dapoli (Maharashtra), Tamil Nadu Agricultural University (Horticultural Station, Yercaud), and at the Regional Research Laboratory, Bhubaneswar. So far five elite lines have been developed. The IISR has also established a germplasm collection of Chinese cassia and from this a few high quality lines have been identified. Little crop improvement work has gone into the Chinese and Indonesian cassias.

The Present Scenario

Cinnamon is grown mainly in Sri Lanka, whilst minor producing countries include Seychelles, Madagascar and India. It occurs naturally in Sri Lanka and southern India, and also in the Tenasserim Hills of Myanmar (De Guzman and Siemonsma, 1999). Sri Lanka produces the largest and the best quality of cinnamon bark, mainly as quills. The area under cultivation is estimated to be around 24,000 ha in Sri Lanka and 3400 ha in the Seychelles producing respectively around 12,000 and 600 t (Coppen, 1995). Cinnamon leaf oil is mostly produced in these countries, though the bark oil is distilled mostly in the importing countries. Sri Lankan export is to the tune of around 120 t of leaf oil and 4–5 t of bark oil.

As already mentioned cassia or cassia cinnamon is derived from different sources, such as:

Chinese cassia – *Cinnamomum cassia*
Indonesian cassia – *C. burmannii*
Indian cassia – *C. tamala*
Vietnam cassia – *C. cassia/C. loureirii*

In addition to cinnamon and cassia the genus also contains the camphor tree (*C. camphora*) which yields camphor and camphor oil. *C. cassia* or Chinese cassia occurs mainly in south China, Vietnam and also in Laos and Myanmar (Burma), and is grown commercially in China and Vietnam. In China the main production areas are in the Kwangsi and Kwangtung provinces in south China, the area being around 35,000 ha with a production of around 28,000 t of cassia bark annually. The UAE is the major buyer of cassia bark and cassia leaf oil. Cinnamaldehyde is the major component of bark and leaf oil. *C. burmannii* is the Indonesian cassia, sometimes called Padang cassia (Padang in west Sumatera) or Korintgi cassia (produced in the Kortingi mountain area of Indonesia). This is harvested from an area of about 60,000 ha, and the production is around 40,000 t. This is an important export product from Indonesia, mainly to the USA. The main component of the oil of bark and leaf is cinnamaldehyde.

Vietnam cassia or the Saigon cassia is also *C. cassia* and is cultivated in an area around 6,100 ha and the production is around 3,400 t which is exported mainly to the USA. The essential oil consists mainly of cinnamaldehyde. However, some doubts still exist regarding the correct botanical identity of Vietnam cassia. All the earlier literature identified Vietnam cassia with *C. loureirii* Nees. But this was reported to be a very rare species, and hence cannot be the source of Vietnam cassia. Recent reports (Dao, this volume) show that Vietnam cassia is nothing but *C. cassia*. The differences in the commercial samples of Vietnam and Chinese cassia are mainly due to the difference in harvesting and post-harvest treatment of the bark.

C. tamala, the Indian cassia, is distributed in the forests of north-eastern India and Myanmar (Burma). In the north-eastern region of India this is grown for the leaves that are extensively used in flavouring various dishes. The leaves are collected from forest grown trees as well as from cultivated ones. *C. tamala* is inferior to other cassias because of the lower oil content in the bark and leaves. The major component of oil is cinnamadehyde (in bark) and eugenol (in leaf). In addition there are other species of *Cinnamomum* in South and South-East Asian countries that are used as substitutes for cinnamon and cassia.

Sri Lanka has been the traditional producer and exporter of cinnamon and its value added products such as bark oil and leaf oil. Cinnamon bark oil is a very high value oil and Sri Lanka is the only supplier of this commodity with an annual production of only around 2.8 to 3 t. Western Europe is the major importer (especially France), followed by the USA in recent times.

The world demand for cinnamon leaf oil is around 150 t per annum, a demand met mainly by Sri Lanka. The USA and western Europe are the largest consumers of leaf oil. The eugenol-rich cheaper clove leaf oil poses severe competition for cinnamon leaf oil. The oil is useful for eventual conversion to iso-eugenol, another valuable flavouring agent.

Sri Lanka holds a virtual monopoly over the production of cinnamon bark and leaf oil. Small quantities are distilled in Madagascar, Seychelles and India to meet internal demands. Though a very costly oil, there is no international standard for cinnamon bark oil; the higher the cinnamaldehyde content, the higher the price. In USA the Essential Oil Association (EOA) standard specifies an aldehyde content of 55–78% (EOA, 1975). But in the case of leaf oil, international standards exist. In this case a phenol content of 75–85% has been specified for oil of Sri Lankan origin (ISO, 1977). In fact leaf oil from Seychelles has a higher eugenol content (about 90%). Cinnamaldehyde is another constituent of leaf essential oil contributing to the total flavour and the specification limits its content to 5%. In the USA the FMA (Fragrance Materials Association) specifies the eugenol content in cinnamon leaf oil in terms of its solubility in KOH (80–88%) (FMA, 1992).

Cassia oil is distilled from a mixture of leaves, twigs and fragments of bark and there is only one type of cassia oil, as mentioned earlier. Cassia oil is used mainly for flavouring soft drinks, confectionary and liquors and its use in perfumery is limited because of its skin sensitising properties (Coppen, 1995). The world trade in cassia oil is controlled by export from China and information on production is scarce. Imports into the USA are rising mainly due to the boom in the soft drinks industry. Cassia oil has specific ISO standards (ISO, 1974). The cassia oil imported into Japan is also re-exported to the USA. The total annual production of cassia oil is estimated to be more than 500 t (Coppen, 1995). The oils distilled in Indonesia (from *C. burmannii*) and Vietnam (*C. cassia*) are also being marketed as cassia oil, but they are less valued and are much less widely traded. Indonesian cassia has a good market in the USA.

Apart from cinnamon and cassia, there are other *Cinnamomum* species that could be exploited commercially. The potential of such species requires study because it is known that some species produce oil of differing compositions due to the existence of different chemotypes (Coppen, 1995). Such variations are well known in the case of camphor tree, where trees yielding camphor, linalool, safrole or cineole as the major constituent are available (Wang-Yang *et al.*, 1989). The genus holds a lot of promise in the future search for aroma chemicals. *C. tamala* leaf, for example, is widely used all over north India for flavouring a variety of vegetable, meat and fish preparations. There are populations of *C. tamala* producing predominantly cinnamaldehyde or eugenol. Recently two other species (*C. pauciflorum* and *C. impressinervium*) have been reported to possess high oil content and quality (Nath and Baruah, 1994; Nath *et al.*, 1996). From Malaysia, *C. molissimum* has been reported to contain safrole and benzyl benzoate as major constituents (Jantan and Goh, 1990). *C. rigidissimum* and *C. petrophyllum* are the other species having high safrole (Lu *et al.*, 1986). *C. petrophyllum* (syn. *C. pauciflorum*), growing in China in the Sichuan province, is now being commercially exploited for safrole production.

12 P.N. Ravindran and K. Nirmal Babu

Until recently, attention has not been given to quality improvement through breeding, although some elite lines have been identified in Sri Lanka and India, and some varieties have been evolved. In addition to screening for quality, crossing involving distinct populations and interspecific hybridization involving *C. verum* and *C. cassia* have been contemplated by IISR, Calicut, India.

Conclusion

Spices like pepper and cinnamon tempted explorers such as Vasco-da-Gama to sail round the storm tossed Cape of Good Hope in search of a new spice route to the spice lands of the east when the Ottoman Turks closed Constantinople to trade in 1453. It was indeed the same quest that made Columbus sail westwards to the unknown seas, when he discovered America. Cinnamon was the "rich bride Helen" for whom the Netherlands and Portuguese had for so many years contended (Ratwatte, 1991). It was precious enough to these colonial powers to wage bitter and frequent wars and to sacrifice many human lives.

Cinnamon/cassia continue to enchant us thorough its varied uses in a variety of food; it is the classic flavour for apple pie, Madeira cake, doughnuts, and for a host of cookies and pastries. It is the mainstay of sweet pudding spice and cinnamon cakes, which are essential items in Christmas celebrations. Cinnamon is a savoury spice too, and in Sri Lanka it is always added to vegetables, fish and meat curries, fancy rice dishes and sweet meats. In Mexico, cinnamon tea is a popular beverage. Cinnamon leaf and bark oils are essential ingredients in a variety of soft drinks. The tallow from cinnamon fruits makes the sweet scented candles used in Greek Orthodox churches.

In the Vietnamese language of flowers, cinnamon is translated as meaning "my fortune is yours". In Austria, lovers would exchange a posy containing cinnamon as a symbol of affection and love (Morris and Mackley, 1999).

Spices like cinnamon, cassia and pepper have influenced the course of human history. Parry (1969) writes:

> It is difficult for us who buy our supplies of pepper, cassia, cinnamon, cloves, ginger, mace and nutmeg, so casually and so cheaply, to believe that there was ever a time when these spices were so eagerly sought after and represented so much wealth and power that destiny itself was indivisible from them.

References

Anonymous (2000) Cassia (*Cinnamomum cassia* Blume). Accessed from Internet in July 2000.

Coppen, J.J.W. (1995) *Flavours and Fragrances of Plant Origin*, FAO, Rome.

De Guzman, C.C. and Siemonsma, J.S. (1999) (eds) *Plant Resources of South East Asia* Vol. 13 *Spices*. Backhys Pub., Lieden.

EOA (1975) Oil of Cinnamon bark. Ceylon EOA No.87. Essential Oil Association of USA, Washington.

Farrell, K.T. (1985) *Spices, Condiments and Seasonings*. The AVI Pub. Co., USA.

FMA (1992) Cinnamon leaf oil. FMA Monographs Vol. 1. Fragrance Materials Association of the US, Washington.

Gantzer, H. and Gantzer, C. (1995) Cinnamon: The Phoenix Spice. *Spice India*, 8(5), 4–5.

ISO (1974) Oil of cassia. ISO 3216–1974 (E.) International Organization for Standardization, Paris.

ISO (1977) Oil of cinnamon leaf ISO 3524–1977 (E.) International Organization for Standardization, Paris.

Jantan, I. and Goh, S.H. (1990) The essential oils of *Cinnamomum mollissimum* as natural sources of safrole and benzyl benzoate. *J. Tropical Forest Sci.*, 2, 252–259.

Krishnamoorthy, B., Rema, J., Zachariah, T.J., Abraham, J. and Gopalam, A. (1996) Navashree and Nithyashree – two high yielding and high quality cinnamon (*Cinnamom verum* Bercht & Presl.) selections. *J. Spices & Aromatic Crops*, 5, 28–33.

Lu, B., Li, Y., Mai, L., Sun, B. and Shu, L. (1986) Chemical constituents of essential oil from *Cinnamomum rigidissium*, a new natural source of safrole. *Chemistry and Industry of Forest Products*, 6(4), 39–44.

Miller, J.I. (1969) *The Spice Trade of the Roman Empire*. Clarendon Press, Oxford.

Morris, S. and Mackley, L. (1999) *Choosing and Using Spices*. Sebastian Kelly, Oxford.

Nath, S.C. and Baruah, K.S. (1994) Eugenol as the major component of the leaf oils of *Cinnamomum impressinervium* Meissn. *J. Essent. Oil. Res.*, 6, 211–212.

Nath, S.C., Hazarika, A.K. and Baruah, A. (1996) Cinnamaldehyde, the major component of leaf, stem bark and root bark oil of *Cinnamomum pauciflorum* Nees. *J. Essent. Oil. Res.*, 8, 421–422.

Parry, J.W. (1969) *Spices*, Vol. 1 Chemical Pub. Co., New York.

Ratwatte, F. (1991) *The Spice of Life – Cinnamon and Ceylon*. Accessed from Internet in August 2001.

Redgrove, H.S. (1993) *Spices and Condiments*. Sir Issac Pitman & Sons Ltd., UK.

Rosengarten, Jr, F. (1969) *The Book of Spices*. Livingston Pub. Co., Philadelphia.

Wang-Yang, S., Wei, H., Guang-gu, W., Dexuam, G., Guang Yuan, L and Yin-gou, L. (1989) Study on the chemical constituents of essential oil and classification of types from *C. camphora*. *Acta Bot. Sinica*, 31, 209–214.

Wijesekera, R.O.B., Punnuchamy, S. and Jayewardene, A.I. (1975) *Cinnamon*. Ceylon Institute of Scientific and Industrial Research, Columbo, Sri Lanka, p. 48.

2 Botany and Crop Improvement of Cinnamon and Cassia

P.N. Ravindran, M. Shylaja, K. Nirmal Babu and B. Krishnamoorthy

Genus *Cinnamomum*

The genus *Cinnamomum* Schaeffer, comprises evergreen trees and shrubs, found from the Asiatic mainland to Formosa, the Pacific Islands, Australia and in tropical America. The American species have only recently been recognised (formerly they were included mostly in *Phoebe*, an entirely Asiatic genus). There are 341 reported binomials in the genus, which according to Kostermans (1957) could be reduced considerably in a revision. According to Willis (1973) the genus comprises of 250 species. In a later publication Kostermans (1964) lists 452 binomials including synonyms under the genus.

Cinnamomum was earlier thought to be a purely Asiatic genus occurring only in the eastern hemisphere, specifically in the Asia-Pacific region. Later taxonomists, especially Kostermans (1957, 1961), transferred species from neotropical genera such as *Phoebe* to *Cinnamomum*. Flowers of *Phoebe* do not show any differential characters and are very similar to those of *Cinnamomum*. In fact even the earlier taxonomists (Nees, *Syst. Laur.* 1836) recognised the closeness between new world *Phoebe* and Asiatic *Cinnamomum*. Meissner (in DC Prodr. 15(1) 1864) went a step further and coined the subgenus *Persoideae* for the Asiatic and subgenus *Cinnamoideae* for the American species of *Phoebe*. Kostermans (1961) felt that the closely knit genera of Lauraceae could only be separated on minor characters, which can result in good natural groups. He was of the opinion that the nature of the perianth tube as found in the fruit represented a very natural and useful character for classification. A swollen, shallow fruit cup combined with a swollen pedicel with remnants of tepals (partly or entire) is a common feature in Asiatic *Cinnamomum*. Thus from a purely Asiatic genus (as thought earlier), *Cinnamomum* now occupies a pantropical status, occurring in both hemispheres, consisting of both Asiatic and new world species.

In *Cinnamomum*, the basal part of the tepals with the tube persists under the fruit. Here the abscission line goes half way through the tepals and results in a cup crowned with six truncate lobes. In a later publication Kostermans (1980) has given the main identifying characters of *Cinnamomum* species as: (i) the length of the two basal or sub-basal ascendant veins; (ii) the indumentum (covering of hairs); and (iii) the thalamus cup under the fruit. The number of anther cells of the third whorl of stamens was also considered to be a useful character.

The true cinnamon or Sri Lankan cinnamon (C. verum *Berchthold & Presl.*)

Berchthold & Presl., Priroz. Rostlin 2:36 at 37–44, t. 7, 1825; Sweet, Hort. Britt. 344, 1827; Loujon, Hort. Britt. 160, 1830; Arbor. & Fructic. Britt. 3: 1305, 1838;

Heynold, Nom. Bot. Hort. 197, 1840; Steudel, Nom., ed. 2, 1:366, 1840; Buchanin, Trans. Hort. Soc. Ser. 2.2: 168, 1842; Baillon, Hist. Pl. 2:462, 1872; Hager's Handb. Pharm. Praxix 1: 1019, 1849; Eichler, Blutendiagar. 2: 131, 1854; Kostermans, Bibl. Lau. 360, 1964.

Syn: *C. zeylanicum*, Breyne; Nat. Curdec. 1, ann. 4: 139, 1666: *C. zeylanicum* (Garc.) Bl. Blume, Bijdrm Fl. Ned. Ind., 568, 1826; Th. Nees, Pl.of.t.128, 1828; Wallich, Cat. No.2573, 1830; Th. Nees & Ebermayer, Med. Pharm. Bot. 2:420 et 427, 1831 (*zeylonicum*); C.G. Nees in Wallich, Pl. As. rar. 2: 74, 1831; 3:32, 1832; in Flora 15(2): 580, 1831; Progr. Grat. Laur. Expos. 9, 1833; Syst. Laur 45, 95, 664, 1836; in Linnaea 21: 487, 1848 (*Ceylanicum*); Miquel, Leerb. Artsenijgew. 228–230, 1839; Fl. Ind. bat. 1(1):898, 900, 1858; Graham, Cat. Pl. Bombay, 173, 1839; Wight, Icon 1: t. 123, 129, 134, text No.VII (*ceylanicum*) 1839; Presl., Wseobecny Rostalinopsis z: 1301, 1846; Bentham in Hooker, Niger Fl. 498, 1849; Thwaites, Enum. Pl. Zeyl.252, 1861; Meissner, in DC., Prodr. 15(1): 13, 20, 31, 1864; Birdwood, Cat. Veg. Prod. Bombay, ed. 2: 72, 1865; Balfour, Timber trees of India and S.E. Asia, ed. 3:75, 1870; Beddome, Fl. Sylv. Ind. T. 262, 1872; Stewart and Brandis, For. Fl. N.W. India, 375, 1874; Kurz, For. Fl. Br Burma, 2:287, 1877; Hooker, Fl. Br. India, 5:131, 1886; Talbot, Cat. Trees & Shrubs, Bombary, 167, 1894; Dalgado, Fl. Goa, 161, 1898; Gamble, Man. Indian timb. 305, 1881; Fl. Madras Pr. 1224, 1925; Ridley, Flora Malay. Pen 3: 97, 1924; Rao, Flow. Pl. Travancore, 341, 1914; Fischer in Rec. bot. Sur.Ind. 9: 153, 1921; 12(2): 128, 1938; Kostermans in Meded. Bot. Mus. Uterecht. 25:50, 59, 1936; in Notul. Syst. 8:120, 1939; in Humbert, Fl. Mad. Fam. 81:86, 1950; Commun. For. Res. Inst., Bogor, 57:8, 21, 24, 41, 58, 1957; in Reinwardtia 4:200, 213, 216, 233, 250, 1957; Bib. Lau., 364, 1964.

Nomenclatural notes

In 1825 Berchthold & Presl (*Priroz. Rostlin* 2:36–44, 1825) coined the binomial *C. verum*. Blume, almost immediately after this, described the same plant using the binomial *C. zeylanicum* (Blume, in *Rumphia* 1:1826). *C. zeylanicum* Blume was accepted by many later workers (Wallich, Cat.No.2573, 1830; Nees and Ebermeyer, *Med. pharm. Bot.* 2: 1831; Nees, *Systema Laurinarum*, 1836 etc.). Linnaeus (Sp.pl.I, 369, 1753) earlier described the material as *Laurus cinnamomum*, which according to Kostermans (1980) might have derived from *Laurus malabatrum*, given for 'Katou karua' of Rheede (Burman, *Thesaurus Zeylanicus*, 64, 1739). Kostermans was also of the opinion that what Burman described under *C. perpetuoflorens* was nothing but *C. verum*, and the one described by Nees von Esenbeck (*Syst. Laurinarum*, 1836) as *L. malabatrum* was also *C. verum*.

Even earlier to this, Breyne (*Nat. Curdee. 1*, 1666) described the plant and *C. zeylanicum*. Breyne was accepted by many earlier workers (Miquel, Fl. Ind.Bat.1(1): 898, 900, 1858; Meissner, in DC. Prodr. 15(1): 13, 20, 31, 1864; Hooker, 1885; Bourdillon, 1908; Rao, 1914 etc.). But Breyne's name was not valid as it was prior to the publication of *Genera Plantarum*. The first author to use the binomial *C. zeylanicum* after 1753 was Blume. But Berchthold & Presl's publication of the binomial *C. verum* precedes and hence the valid name for ceylon cinnamon or true cinnamon is *Cinnamomum verum* Bercht. & Presl. *C. cassia, C. camphora* and *C. malabatrum* are the three other species which Berchthold & Presl. described in their publication (1825).

16   *P.N. Ravindran* et al.

Taxonomical description

C. verum is a native of Sri Lanka and south India. It is a moderately sized, bushy, evergreen tree. The dried inner bark of semi-hardwood shoot is the true cinnamon of commerce. Under cultivation, the shoots are coppiced regularly almost at ground level, which results in the formation of dense bushes. The following taxonomical description is adapted from Kostermans (1983, 1995):

> Moderately sized tree, up to 16 m tall, up to 60 cm diameter at breast height. Buttresses up to 60 cm high, out 70 cm, thin. Bark smooth, light pinkish brown, thin; live bark brown, up to 10 mm thick with a strong pleasant cinnamon smell and a spicy, burning taste. Branchlets slender, often compressed, glabrous; end bud partly finely silky; flush almost glabrous, bright or light red. Leaves opposite or sub opposite, glabrous, thinly to stiffly coriaceous, oval or elliptic to lanceolate-oval or narrowly elliptic, $3 \times 7-8 \times 25$ cm (1.5×5 cm in the inflorescence), shortly or broadly acuminate, base acutish or cuneate, triplinerved (or with very slender, additional basal nerves: 5-nerved); upper surface dark green, shining, smooth; lower surface paler, dull, the 3 main nerves prominant on both surfaces; the basal or sub-basal ones running out near the base of the acumen; secondary nerves faint, bent in the middle, more or less parallel. Petiole rather stout, 10–20 mm, slightly concave above. Inflorescence is paniculate cymose, the initial branching of the inflorescence paniculate, with alternate or opposite branches, while the flowers are arranged in cymes (Type two of Van der Werff and Richter 1996). Panicles axillary, up to 20 cm long (or longer or shorter), consisting of a long main peduncle and a few, stiff short branches. Flowers pale yellowish green; perianth ca. 8 mm, silky; tube short, campanulate, tepals oblong-lanceolate, acutish or obtuse, up to 3 mm long, persistent. Fruit ellipsoid to oblong-ovoid, dark purple, up to 12.5 mm long; cup hemispheric, ribbed, topped by the indurate, enlarged tepals (Fig. 2.1 and Fig. 2.2).

The flowering time varies from October to February, and the fruit ripens in May to June. Trees start putting forth flushes in the monsoon period, around July to September. They look very attractive during the flushing period, dressed in purple flushes. In a population, the flush colour varies from green to deep purple. The flowers when open have a pleasant smell, and are visited by a number of insects, especially bees. Flowers exhibit protogynous dichogamy, the male and female phases are separated by almost a day.

The Chinese cassia (C. cassia (L.) Berchthold & Presl.)

Berchthold & Presl., Priroz. Rostlin 2: 36 et 44–45, 5, 6, 1825; Blume, Bijdr. Fl. Ned. Ind., 11 stuk 570, 1826; Sweet, Hort. Brit. 344, 1827; Th. Nees, Pl. Off. 1: 5. 129, 1829; London, Hort. Brit. 160, 1830; Nees & Ebermayer, Handb. Med. pharm. Bot. 2:424, 427, 1831; CG Nees in Wallich, Pl. As. Rar. 2: 73 et 74, 1831; Syst. Laur. 42 et 52, 1836; Lindley, Fl. Med. 330, 1838; Miquel, Fl. Ind. Bat. 1(1): 896, 1858: Meissner, in Dc., Prodr. 15(1): 12, 18, 466, 1864; Balfour, Timb. Trees & For. India & SE Asia ed. 3. 74, 1870; Kurx. For Fl. Brit. Burma 2: 288, 1877; Gamble, Man. Ind. Timb. 306, 1881 ed 2.560, 1902; Hooker, Fl. Brit. India 5: 130, 1886; Staub, Ge Schichte Genus *Cinnamomum* 11, 19, 27, 31, 32, 39, 40, 1905; Dunn & Tutcher, in Kew Bull., Add. Ser. 10:223, 1912; Kostermans, in J. Sci. Res. Indon, I: 84, 85, 1952;

Figure 2.1 (a) A cinnamon tree – about five years old with fruits and flowers. (b) A close view of a flowering branch. (c) A twig with inflorescence.

Commun. For. Res. Inst. Bogor, 57: 24, 1957; in Reinwardtia r: 216, 1956; Chopra *et al.*, Gloss. Ind. Med. Pl. 65, 1956; Wood, in J. Arnold Arb. 39: 335, 1958; Kostermans, Bib. Lau., 276, 1964.

Syn. *C. aromaticum* Nees. Nees, in Wallich, Pl. As. Rar. 2: 74, 1831; in Flora 15 (2): 585, 1831; Syst. Laur, 52, 1836; Nees & Ebermayer, Handb. Med. Pharm. Bot., 526, 1832; Lindley, Fl. Med. 330, 1838; Wight, Icon. 136, 1839. *Laurus cassia* Wight in Madras J.Lit. & Sci. 9: 130–135, 1839. Presl., Wseobecny, Rostalinopsis, 2:1302, 1846; Miquel, Fl. Ind. Bat. 1(1): 896, 1858; Meissner, in Dc: Prodr. 15(1): 12, 1864; Balfour, timber Trees & For. India & E Asia, ed. 3: 74, 1870; Lukmanoff, Nomencl. Icon. Canell. Comphr. 11, 1889, Kostermans, Bib. Lau, 253, 1964.

C. cassia is a native of China and Vietnam, being cultivated in those regions as well as in the Malayan Archipelago and the north-eastern Himalayan region. It is a small tree, the bark of which forms the cassia bark or Chinese cassia or Chinese cinnamon. Leaves are opposite, glabrous above, minutely hairy below, hairs microscopic, oblong-lanceolate, three-ribbed from about 5 mm above the base, side veins ascending to the apex, exstipulate. The length/breadth value is 4.26. Inflorescence is axillary panicle (panicled cyme), exceeding the leaves, many flowered, peduncle long, flowers with long pedicel and minutely hairy. Floral characteristics are similar to those of *C. verum*. Fruit is an ovoid one-seeded berry, seated in an enlarged perianth cup with truncate perianth lobes. Flowering from October to December (see also Chapter 6).

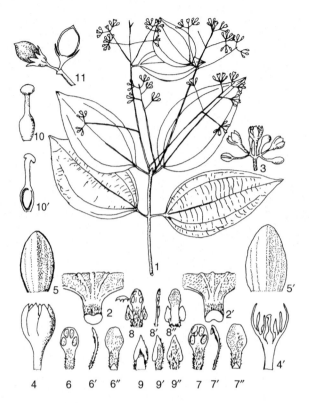

Figure 2.2 General characteristics of *C. verum*. 1 – Flowering branch (right hand leaf lower surface, left hand one upper surface). 2 – Base of the upper leaf surface, upper side of the petiole glabrous, grooved; 2' – Base of lower leaf surface. 3 – Flowers in cymes. 4 – Flower and 4' – flower in sectional view. 4'' – outside pilosity of flower. 5 – Interior tepal, inside hairy and punctulate; 5' – outer tepal. 6, 6', 6'' – Stamens of whorl 1. 7, 7', 7'' – Stamens of whorl 2. 8, 8', 8'' – Stamens of whorl 3. 9, 9', 9'' – Staminodes of whorl 4. 10 – Pistel; 10' – Pistel in L.S. 11 – Fruit (Adapted from Kostermans, 1983).

Nomenclatural notes

Berchthold & Presl were the first to describe *C. cassia* (*Priroz. Rostlin* 2:36 ct 44–45, 1825) followed by Blume (*Pl. Ned. Ind.*, 11, 570, 1826). Nees described the same species under *C. aromaticum* Nees, (in Wallich, *Pl. As. rar.* 2: 74, 1831., in Flora 15(2): 585, 1831) and this name became popular with the publication of *Systema Lauracearum* in 1836. But *C. cassia* precedes *C. aromaticum*. Hence the correct name for Chinese cassia should be *C. cassia* (Linn.) Bercht. & Presl. The original description by Linneaus was under *Laurus cassia* L. (Sp.pl.528). Later it was included under *Cinnamomum*, retaining the species epithet that Linneaus used. For a complete listing of nomenclature and synonyms see Kostermans (1964).

This is known as *Kwei* or *Kui* in China and as *cassia lignea* on the European market. All of the plant parts are aromatic, though only its leaves and bark are exploited commercially. The essential constituent of bark and leaf oil is cinnamaldehyde. Under cultivation, trees are coppiced periodically keeping the height to 3–4 m. Uncut trees of 15–20 years have a central bole of 15–20 m. Because of the acute angle at which the branches are produced, the tree generally has a conical shape.

The bark is widely used as spice and is an important Chinese medicine. For medicinal use bark from mature trees grown in the wild is preferred. For spice the inner bark from young shoots as well as bark from older trees are used, which is peeled or scraped and dried. The inner bark contains 1.5–4% essential oil, having up to 98% cinnamaldehyde. The bark produced from higher altitudes (*Kwangsi* bark) is of better quality, having higher oil content than that produced from plants grown in lower elevations (*Kwantung* bark). The immature fruits are often dried and sold in market as cassia buds (see Chapter 6 for details).

Saigon cassia or Vietnam cassia – C. cassia or C. loureirii – A case of mistaken identity

This is a much-disputed species. Allen (1939) has commented on this species as follows:

> 'The material, which seems to answer the description given by Loureiro, comes from near the type locality in Indochina. It has a sweet sandalwood odour. Loureiro mentions the fact that it is fairly rare, which is certainly true, if one can judge from the scarcity of herbarium specimens. The *C. loureirii* from Japan cited by Nees, bears no resemblance to the Indochinese specimens that conform to Loureiro's original description. The Japanese specimens have a sweet spicy odour …At this point it seems pertinent to discuss the so-called Saigon cinnamon used commercially….As Merrill (1920) commented in his discussion on the subject, the commercial cinnamon must necessarily be a widespread species, that it cannot have escaped the notice of collectors and taxonomists for all these years. Chevalier (Allen, 1939) believes that it is either purchased from Chinese or Annamese merchants and thus brought into the port of Saigon, or else it is furnished by *C. loureirii*. This latter belief would indicate that *C. loureirii* is a widespread species, a fact belied by the scarcity of herbarium specimens…'

Dao (this volume, Chapter 6), who spent more than twenty years studying the *Cinnamomum* species of Vietnam, came to the conclusion that the Vietnam cassia is nothing but *C. cassia. C. loureirii* was previously wrongly identified as Vietnam cassia (see in this volume Chapter 6 on Chinese and Vietnam cassia).

Indonesian (Java) cassia (C. burmannii C.G. Th. Nees)

C. burmannii (C.G. Th. Nees) Bl. Blume, Bijdr. Fl. Ned. Ind., II st UK: 569, 1826; Nees & Ebermayer, Handb. Med. Pharm. Bot. 31; 525, 1832; Nees, in Wallich, Pl. As. Rar. 2:75, 1831; in Flora 15(2): 587 et 600, 1831; Syst. Laur. 67, 1836; Lindley, Fl. Med. 330, 1838; Presl, Wseobecny, Rostalinopsis, 2:1303, 1846; Miquel, Fl. Ind. Bat. 1(1):910, 1858; Ann. Mus. Bot. Lugd. Bat. 1:266 et 270, 1864; Meissner, in DC. Prodr. 15(1):16, 1864; Hance, Suppl. Fl. Hongkong, 31, 1872; in J. Linn. Soc. 13: 119, 1872; Franchet & Savatier, Enum. Pl.Japan. 1:410, 1875; Bokorny, in Flora N.R.40: 359, 1882; Hooker f., Fl. Brit. India 5: 136, 1886; Pax in Engler & Pr., Nat. Pfl. Fam. 3(2): 114, 1889; Hemsley, in J. Linn. Soc. 26: 371, 1891; Holmes, Catal. Hanbury, Herb. 99, 1892; Cat. Med. Pl. 129, 1896; Pharmac. J. 12 May 1894; Exkurs, Fl. Java,

20   *P.N. Ravindran* et al.

21: 263, 1912; Allas Baumarten, Java 2: 266, 1914; Deane, in Pro. Linn. Soc. N.S. Wales, 25: 6. 37 (1), 1901; Merrill, Review spec. Blanco 73, 1905; Enum. Phillip. Flow. Pl. 2:187, 1923; in Lingnan Sci. J. 5:79, 1927; Staub, Gesch. Gen. Cinnamomum 20, 29, 31, 39, 40,. T. 1,2, 1905; Matsumura, Index Pl. Java, 2(2): 135, 1912; Chevalier, in Sudania 2:47 et 53, 1914; Expl. Bot. Afr. Occ. Fr. 1:543, 1920; Chung, in Mem. Sci. Soc. China 1(1): 58, 1924; Heyne, Nutt. Pl. Ned. Ind. ed. 2, 1: 649, 1927; ed. 3, 1: 649, 1950; Santos, in Phillip. J.Sci. 43(2):348–53, t. 11, 18, 19, 1930; Gimlette & Burkill, in Garden's Bull. S.S. 6: 340 et 446, 1930; D. Bois, Pl. aliment. 3662, 1934; Burkill, Dict. Econ. Prod. Malay Pen. 1:546, 1935; Kostermans, in Natul. Syst. Paris 8:120, 1939; Comm. For. Res.Inst., Bogor, 57, 11:21, 1957; in Reinwardtia 4, 203, 213, 1957; Backer, Fl. Java, Fam. 27: 8, 1941; Chow & Wang, Cat. Pl. Kwangsi 22, 1955; Kostermans, Bib. Lau., 256, 1964. (See Kostermans 1964 for full nomenclature citation).

Indonesian cassia is commonly imported to the USA and the bark has high oil content. All plant parts are aromatic, but only the bark is commercially exploited. This plant (known as *kaju manis* in Indonesia) is grown on a large scale in certain areas and based on quality. Two types have been recognised: one is known as Korintji cassia and the other Padang (or Batavia) cassia. The former is grown in higher altitudes and is much superior in quality having a higher oil content (about 4% in the bark), the main constituent of which is cinnamaldehyde. The leaf on steam distillation yields about 0.5% oil, the main constituent of which is also cinnamaldehyde (50–65%) (see Chapter 8 for details). The species epithet is spelled differently as *burmanni*, *burmannii* and *burmanii*. Here the spelling given by Kostermans (1964) is adopted.

Indian cassia (C. tamala *(Ham.) Th Nees & Eberm.)*

Th. Nees & Ebermayer, Hanb. Med. Pharm. Bot. 2: 426 et 428, 1831; CG Nees, in Wallich, Pl. As. Rar. 2: 75, 1831; in Flora 15(2): 591 et 596, 1831; Syst. Laur. 56 et 666, 1836; Th. Nees, Pl. Office, Suppl. 4:22, 1833; Hayne, Getreue Darstell. Arzneigew, 12:t. 26, 1856, Blume, Rumphia 1(3):t.14, 1833 et 1836; Miquel, Leerb. Artsenijew., 360, 1838; Fl. Ind. Bat.1(1): 892, 1858; Ann. Mus. Bot. Lugd. Bat. 1: 268, 1864; Stendel, Nom. ed. 2, 1: 366, 1840 et 2: 15 et 17, 1841: Meissner, in DC. Prodr.15 (1): 17, 1864; Bentham, fl. Austral. 5: 303. 1870; in Bailey, Queensland. Fl. 4:1309, 1901; Udoy Chand Dutt, Materia Med. Hindus, 224, 1872; rev.ed. 224,320, 1900; Steward and Brandis, For. Fl. N.W. India, 374, 1874; Gamble, List Trees & Shrubs Darjeeling Dist. 63, 1878; ed. 2:64, 1896; Man. Ind. Timb., 306, 1881; ed. 2, 560, 1902; Hooker f., Fl. Brit. India, 5: 128, 1886; Watt, Dict. Econ. Prod. India 2, 319–323; 1889; Pax, in Engl & Pr. Nat. P.fl. fam; 3(2): 114, 1889; Gage, in Rec. bot. Surv. India 1: 355, 1893; 3(1): 98, 1904; Pharmc. J. 12 May 1894, 941; Kanjilal, For. Fl. School circle, N.W. Pror. 275, 1901; For. Fl. Sivalik 326, 1911; Prain, Bengal Pl. 2:899, 1903; in Rec. bot. Survey India 3(2): 270, 1905; Staub, Geschichte gen. Cinnam. 21, 39, 42, t. 17, 1905; Brandis, Ind. Trees, 533, 1906; Lace, List of Trees, shrubs Burma 109, 1912, ed. 2:137, 1922; Duthie, Fl. Upper Gangetic plain 3(1): 57, 1915; Haines, Bot. Bihar & Orissa 797, 1924; Burkill, in Rec. bot, Sur. Ind. 10(2): 351, 1925; Dic. Eco. Prod. Malay Penin, 1: 543 et 556, 1935; Osmanton, For. Fl. Kumaon, 443, 1927; Fischer in Rec. Bot. Sur. India 12(2); 128, 1938; Kanjilal *et al.*; Fl. Assam, 4: 56, 1940; Kostermans, Bib. Lau, 354, 1964. (For full nomenclature citation see Kostermans, 1964).

Indian cassia leaves (known as *'Tejpat'* in India) have been used as a spice in the whole of north India from ancient times. It is an evergreen tree, found mainly in the tropical and subtropical Himalayas up to an altitude of 2400 m. *Tejpat* occurs naturally in the Khasi and Jaintia Hills, but is also grown in the homesteads. The main trading centre for *tejpat* leaves is Shillong in Meghalaya. Baruah *et al.* (2000) found that several species (in addition to *C. tamala*) are traded as *tejpat*, and are used by consumers. The other species used as *'tejpat'* (Indian cassia) are *C. impressinervium* Meissn.; *C. bejolghota* (Buch. Ham.) Sweet and *C. sulphuratum* Nees (see Chapters 8 and 15 for more details).

Baruah *et al.* (2000) identified four morphotypes, two of which are more popularly used. Leaves are alternate, sub-opposite or opposite on the same twig, coriaceous, aromatic, glabrous; pink when young. Leaves are ovate to ovate-lanceolate in morphotype I; elliptic-lanceolate in morphotype II; broadly elliptic-lanceolate in morphotype III; small and elliptic to oblong-lanceolate in morphotype IV; apex acute to acuminate, base acute to obtusely acute, triplinerved, lateral veins not reaching the tip; panicle subterminal to auxillary, equal to the leaves or slightly exceeding them. Floral morphology is similar to other species. In all morphotypes leaves are hypostomatic, stomata sunken, epidermal cells highly sinuous on both surfaces, glabrous except in one morphotype; venation acrodromous, areoles tetragonal to polygonal, variable in size.

Camphor (C. camphora (L.) Bercht. & Presl.)

Berchthold & Presl., Priroz. Rostlin 2: 36 at 47–56, t. 8, 1825; Th. Nees, P1. Off. t. 127, 1825; Sweet, Hort. Brit. 344, 1827; London, Hort. Brit. 160, 1830; Wallich, Catal. No. 6347, 1830; Nees & Ebermayer, Handb. Med. Pharm. Bot. 2, 430–434, 1831; CG Nees in Wallich, P1. As. Rar. 2, 72, 1831; Heynold, Nom. Bot. Hort. 197, 1840; Steudel, Nom. Ed. 21: 271 et 366, 1840; Miquel, F1. Ind, Bat. 1(1), 905, 1858;Meissner in DC. Prodr. 15(1), 24 et 504, 1864; Stewart and Brandis, For. F1. N.W. India, 376, 1874; Gamble, Man. Ind. Timb. 305, 1881; Hooker f., F1. British India, 5: 134, 1885; Engler, Syllabus 127, 1903; Kutze, in Engl. Bot. Jahrb, 33: 421, 1903; Prain, Bengal P1. 2. 899, 1903; Brandis, Ind. Trees 534, 1906; Duthie, F1. Upper Gangetic Plain 3(1): 57, 1915; Sawyar & Dan Nyun, Classif. List. Pl. Burma, 32, 1917; Troup, Silvicult. Ind. Trees, 3: 790–795, 1921; Exot. For Trees Brit. Emp. 70, 1932; Howard in Indian Forest Rec. 9(7): 1–34, 1923; Parker, For. F1. Punjab ed. 2: 429, 1924; Haines, Bot. Bihar & Orissa, 797, 1924; Fujita, Y, in Trans. Nat. Hist. Soc. Formosa, 21: 254–258, 1931; 22: 43–45, 1932; 25: 429–432, 1935; in bot. Mag. Tokyo, 65, 245–250, 1952; Kostermans, Natual. Syst. Paris, 8: 120, 1939; in Humbert, F1. Madag. Fam.81: 86, 1950; in J. Sci. Res. Indon. 1: 84, 85, 1952; Commu. For. Res. Inst. Bogor 57: 6, 10, 21, 24, 1957; in Reinwardtia, 4: 198, 202, 213, 216. 1957; Kanjilal, F1. Assam, 4: 60, 1940.

C. camphoratum Bl. Blume, Bijdr, Fl, Ned. Ind. 11, 571, 1826;

C. camphoratum F. Villar, Villar in Blanco Fl. Filip. 1880;

C. camphoratum Merr. Merril, Enum, Philip. Fl. Pl. 1923;

C. camphoriferum St. Lag. St. Lager in Ann. Soc. Bot. Lyon., 7: 122, 1880;

C. camphoroides Hay. Hayata, Icon. Pl, Formos. 3, 158, 1913.

22 *P.N. Ravindran* et al.

Kostermans (1964) listed the following varieties:
var. *cuneata* Blume
var. *glaucescens* A. Braun
var. *linaloolifera* Fujita
var. *nominalis* Hayata
var. *parviflora* Miquel
var. *procera* Blume
var. *quintuplinerve* Miquel
var. *rotunda* Meissner
var. *rotundifolia* Makino
var. *typica* Petyaev

Hirota (1951, 1953) suggested a sub-division of *C. camphora* based on the predominance of certain chemical constituents (see Chapter 9 for details).

C. camphora is a small to medium sized tree cultivated in Japan, Taiwan, China, Vietnam, Thailand, etc. for its essential oil and camphor. It is a native of the South-East Asian region, and in its natural state extends from Japan to the Himalayan region. Hooker (1886) included *C. camphora* in the section "Camphora" based on the bud character, enclosed in scale leaves as against the naked bud situation present in the taxa in the section "Malabatrum". In *C. camphora*, the leaves are alternate and glabrous on both sides. Leaves are small, highly variable when shoots are coppiced for distillation, elliptic, acuminate or ovate elliptic, extipulate with long petioles, having the characteristic smell of camphor. Inflorescence is a lax panicle, terminal or auxillary, few flowered, usually shorter than leaves, bud scaly. Flowers are small, yellowish-white and floral characteristics are similar to those of *C. verum*. Fruit a small berry, perianth lobes wholly cauducous in age. Fruits contain about 50% yellowish-white aromatic fat having high lauric acid content, useful in soap making.

It is a medium sized evergreen tree having a dense canopy. All plant parts are aromatic having the characteristic smell of camphor. Roots contain more of oil – up to about 10% – having a camphor content of about 27%. For camphor oil distillation, either the whole tree is cut and all plant parts are used, including the main roots, or the leaves are harvested periodically and then distilled. The wood, having a strong smell of camphor, polishes well and is useful for cabinet making. The trees last long and can also grow to a great height. Weiss (1997) mentions that the largest camphor tree in Taiwan is about 51 m tall, 5.3 m diametre (at breast-height), and having a volume of 588 m^3 and can yield 7200 kg oil and 2743 kg camphor. This tree is around 1400 years old. The largest tree in Japan is around 1200 years old having a trunk length of 31 m and a diameter 2.3 m (bh). In a camphor tree of 30–40 years age half the weight is contributed by the trunk, 20% by root, stump 14%, branches 13% and leaves 5% (Weiss, 1997).

Botanical Studies

Botanical studies on *Cinnamomum* are rather meagre in spite of the economic importance of the genus and its wide distribution. The available information is summarised here.

Leaf morphology and anatomy

Leaf characteristics are highly variable in the genus *Cinnamomum*, and this variation is seen both at species and subspecies levels. In *C. verum* leaf length varies from 8.7 to

22.7 cm with a mean of 13 cm; leaf breadth 3.3–8.3 cm with a mean of 5.1 cm. Leaf size index varies from 0.29 to 1.7, the mean being 0.7 (Krishnamoorthy *et al.*, 1992). *C. cassia* has larger lanceolate leaves having a mean length of 16.25 cm and a breadth of 3.8 cm; the L/B value being 4.26 versus 2.28 in the case of *C. verum* and 2.05 for *C. verum* from Sri Lanka. The leaf morphological characteristics of cinnamon and some of the related species are given in Table 2.1.

C. camphora is not closely related to cinnamon, as it belongs to the sub genus *camphora*, characterised by buds covered with orbicular, concave, silky, caducous, and imbricating scales, while the species belonging to the subgenus *malabatrum* have naked buds or buds with very small scales. Santos (1930), Shylaja (1984) and Ravindran *et al.* (1993) studied the leaf morphological and anatomical characteristics of cinnamon and cassia. Baruah *et al.* (2000) reported the leaf characteristics of Indian cassia.

Leaf anatomy

In most of the economically valuable species the leaves are triplinerved, except in camphor tree (and other species that belong to the section camphora) where the leaves are penninerved. Santos (1930), Shylaja (1984) and Bakker *et al.* (1992) reported leaf anatomy of *Cinnamomum* spp. The leaf structure is similar to that of a typical dicot leaf. Both the upper and lower epidermes are covered by more or less thick cuticles. Variations, however, are noted among species. Bakker *et al.* (1992) noted cuticle thickness variation from 2 to 17 μm in different species, and they identified four classes based on cuticle thickness. *C. verum* belongs to the first group (<3 μm), *C. cassia* to the second group (3–8 μm).

Palisade consists of a single layer of elongated cells having large chloroplasts. The relative thickness of palisade and spongy parenchyma varies in different species and even within a species. In *C. verum* (India) palisade is thicker than spongy parenchyma,

Table 2.1 Leaf morphological characters of some *Cinnamomum* spp.

Species	LL (mm)	LB (mm)	L/B (mm)	LT (mm)	ET (mm)	PT (mm)	SPT (mm)	PL (mm)
C. verum (Indian)	81.2	35.5	2.28	0.231	0.015	0.101	0.050	12.7
C. verum (Sri Lanka)	99.0	48.2	2.05	0.197	0.013	0.067	0.096	10.9
C. verum (wild)	81.2	33.0	2.46	0.218	0.013	0.075	0.111	10.0
C. cassia	162.5	38.1	4.26	0.171	0.023	0.054	0.062	15.2
C. camphora	66.0	23.6	2.79	0.197	0.026	0.073	0.070	14.4
C. malabatrum*								
1.	172.7	40.6	4.25	0.218	0.013	0.075	0.104	13.9
2.	119.3	45.7	2.61	0.200	0.013	0.059	0.109	15.2
3.	195.5	66.0	2.96	0.236	0.013	0.106	0.098	11.4
4.	83.8	33.0	2.53	0.244	0.013	0.085	0.122	10.1

Source: Shylaja, 1984.

Notes
* This is the most variable species related to *C. verum*.
LL – leaf length; LB – leaf breadth; L/B – leaf length/leaf breadth; LT – leaf thickness; ET – epidermal thickness; PT – palisade thickness; SPT – spongy parenchyma thickness; PL – petiole length.

24   P.N. *Ravindran* et al.

Table 2.2 Leaf anatomical features in a few *Cinnamomum* spp.

Species	Oil		Muc								ad.ha		ab.ha		
	p	s	p	s	t	cu	cp	p	s	pa	f	l	f	l	v
C. cassia	p	p	1	4	2	2	++	1c	–	+	–		2	2s	t
C. bejolghota	2	–	2	2	3	2	+	1t	±	++	–		–		t
C. burmannii	2	3	3	3	1	2	+	1c	–±	–±	–		–		pl
C. camphora (1)	p	p	p	p	1	1	–	–	–	±	–		–		pl
C. camphora (2)	3	4	2	2	1	1	–	1–	–	±	–		–		pl
C. camphora (3)	4	3	–	–	1	1	–	1–	–	±	–		–		pl
C. citriodora	3	4	2	3	2	2	+	1c*	±	–	–		1	2s	pl
C. culitlawan	2	3	4	4	1	2	+	1c*	±	–	–		–		t
C. keralense	2	4	4	4	2	2	+	1c*	±	–	–		–		t
C. loureirii	2	3	2	2	1	2	+	1tc	+	–	–		–3	3t	t
C. malabathrum	1	2	4	3	2	2	+	1t	±	±	–		–3	2s	t
C. macrocarpum	4	5	2	4	2	2	+	1t	±	–	2	li	4	2s	t
C. sulphuratum	3	4	2	2	2	2	+	1t	–	+	–		2	2s	t
C. tamala	2	4	1	2	1	2	+	1c*	+	–	–		–	3i	t
C. verum	1	4	3	3	1	1	+	1c*	–	±	–		–2	1t	t

Source: Bakker *et al.*, 1992.

Notes

Oil = oil cells; muc = mucilage cells; p = palisade parenchyma; s = spongy parenchyma; – = absent; 1 = ≤0.1 cell/mm leaf width; 2 = 0.1–1 cell/mm; 3 = 1–2 cells/mm; 4 = 2–5 cells/mm; 5 = 5–10 cells/mm; p – present.

t = lamina thickness; 1 = 100–200 μm; 2 = 200–300 μm; 3 = >300 μm; cu = adaxial cuticle thickness; 1 = <3 μm; 2 = 3–8 μm; cp = adaxial epidermis thickness; – = not sclerified; ± = slightly sclerified; + = moderately, distinctly sclerified; ++ = strongly sclerified.

p = palisade parenchyma; 1 = unilayered; – = not sclerified; c = upper layer with sclerified outer periclinal wall; c* = (upper layer) with thickened outer periclinal and anticlinal walls, sclerified; t = (upper layer) totally sclerified.

s = spongy parenchyma; – = not sclerified; ± = weakly sclerified; + = sclerified; pa = papillate abaxial epidermis; – = flat; ± = lowly domed outer periclinal walls; + = dome shaped papillae.

ad.ha = adaxial hairs; ab.ha = abaxial hairs; f = frequency; – = absent; 1 = ≤ 0.1 hair/mm leaf width; 2 = 0.1–1 hair/mm; 3 = 1–2 hairs/mm; 4 = 2–5 hairs/mm; 1 = length; 1 = ≤ 50 μm; 2 = 50–100 μm; 3 = 100–200 μm; s = thick walled (solid); t = thin walled; i = intermediate wall thickness.

v = venation; p = penninerved; 1 = main vein distinct; t = triplinerved.

while in *C. verum* (Sri Lanka) spongy tissue is thicker. A collection of wild *C. verum* has the highest spongy tissue thickness (0.111 mm) (Shylaja, 1984).

Bakker *et al.* (1992) carried out a detailed study on the comparative anatomical features of a large number of *Cinnamomum* spp. (Table 2.2). The palisade is typically one layered. Sclerification of palisade cells is noted in certain species including *C. verum*, in which the outer periclinal and anticlinal walls are thickened and sclerified. *C. verum* (and other species) possess lysigenous cavities in the leaf tissue. In *C. verum*, *C. cassia* and *C. camphora* they are numerous, while in other species they are much less. Lysigenous cavities occur in both palisade and spongy tissues, though in *C. verum* they are found mostly in spongy tissue and rarely in palisade.

Oil and mucilage containing idioblasts are always present in palisade and spongy parenchyma (Bakker *et al.*, 1992). Among species they vary in size, shape, stainability and number. Most of the species show minor anatomical variations among them. The above workers carried out a cluster analysis based on leaf anatomical characteristics and showed that oil and mucilage cells played a significant role in the grouping of the species. Bakker and Gerritsen (1989) studied the ultrastructure of developing and mature mucilage cells in the leaves and shoot apices of *C. verum* and *C. burmannii*. In all mucilage cells a suberised layer is formed in the outer cellulosic cell wall at a very early stage in development. Oil cells are also known to have suberised cell walls. Many features can distinguish these two types of secretory cells. A typical oil cell has a three-layered cell wall – an outer cellulosic layer, a middle suberised layer and an inner cellulosic layer. An oil drop is often attached to a protuberance of the wall, called the cupule, and is enveloped by the plasmalemma (Maron and Fahn, 1979). Oil is usually synthesised in the plastids (Cheniclet and Carde, 1985). Mucilage cells usually lack a suberised layer, *C. verum* is an exception to this. Mucilage is produced in the Golgi apparatus, from which vesicles filled with polysaccharides move towards the plasmalemma and fuse with it. Mucilage accumulates between the plasmalemma and cell wall (Trachtenbeg and Fahn, 1981). The two types of secretory cells can be distinguished by the staining reaction with Sudan IV, Chrysoidin and Alcian Blue (Bakker *et al.*, 1992).

Development of oil and mucilage cells

Bakker *et al.* (1991) reported the development of oil and mucilage cells in the young leaf and shoot apex of *Cinnamomum burmannii*, on which the following discussion is based. Oil and mucilage cells occur in relatively large numbers in a zone underlying epidermis, i.e. in the outer cortex of the shoot and in the mesophyll of developing leaves. Three arbitrary developmental stages can be distinguished in both types of idioblasts; the first two are the same for both oil and mucilage cells (Fig. 2.3).

Stage 1: The young idioblast, possessing a large central vacuole, is recognisable by the absence of osmiophilic deposits in vacuoles (which occur in the vacuoles in the adjoining cells in plenty). There is a parietal layer of cytoplasm that is slightly more electron dense than that of the surrounding cells. The plastids are distinctly small with reduced thylakoids.

Stage 2: At this stage a suberised cell wall is deposited in the cells. This layer consists mostly of two to three discontinuous lamellae and sometimes up to six lamellae. Subsequently, additional wall material gets deposited. The cytoplasm at this stage contains many mitochondria, small plastids, and a slightly increased amount of dictyosomes.

Stage 3 (Oil cells): Development during this stage differs in oil cells and mucilage cells. In the oil forming idioblasts a distinct layer of inner-wall material gets deposited against the suberised layer. Thickness of this inner wall layer increases significantly as development progresses from 44 nm to 162 nm. Based on thickness of the inner cell-wall layer and the composition of the cytoplasm, stage 3 can be subdivided into three intergrading developmental stages – a, b and c. In 3a, the cell contains a suberised cell wall and an inner wall layer. The large central vacuole disappears and is replaced with a few smaller ones having wavy outlines. The cytoplasm is more compact and contains a small oil cavity. In stage 3b, the inner wall thickness increases and the oil cavity increases in size. A cup-shaped cupule that is attached to a thick-ened part of the inner wall layer (cupule base) is formed. At first the oil cavity is distinctly enclosed by plasmalemma, which later disintegrates. Plastids disintegrate, paving the way to the formation of many small vacuoles, and they later fuse with the oil cavity. In stage 3c, the cell reaches maturity, the suberised layer is about 36 ± 10 nm thick and appears as a translucent layer. The inner-wall layer (about 162 ± 69 nm

26 P.N. Ravindran et al.

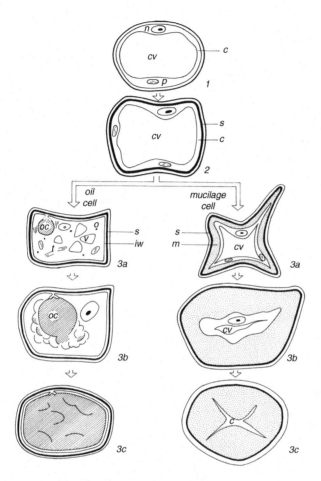

Figure 2.3 Schematic representation showing the developmental stages of oil and mucilage cells in *Cinnamomum burmannii*. Stage 1: Young cells with typical plastids and a central vacuole devoid of deposits. Stage 2: Idioblasts with a suberised layer. *Oil cell*: Stage 3a: An inner wall layer is present, a small oil cavity is attached to a cupule, oil droplets and bundles of tubular ER appear in the cytoplasm. Stage 3b: An enlarged oil cavity, enveloped by a partly disintegrated plasmolemma is in contact with the vacuoles. Stage 3c: Near mature cells with a large cavity filled with oil and no longer enclosed by the plasmolemma. *Mucilage cell*: Stage 3a: Collapsed cell with mucilage deposited against the suberised layer. Stage 3b: Cytoplasm moves inward due to mucilage deposition, the central vacuole disappears. Stage 3c: Near mature cells filled with mucilage surrounding degenerating cytoplasm. c – cytoplasm; cb – cupule base; cv – central vacuole; d – dictyosome; g – starch granule; iw – inner wall layer; m – mucilage; mi – mitochondrion; n – nucleus; o – oil droplet; oc – oil cavity; p – plastid; r – ribosomes; s – suberised layer; t – tubular endoplasmic reticulum; v – vacuole; ve – vesicle; w – initial cell wall. (Source: Bakker *et al.*, 1991.)

thick) is very electron dense. Cytoplasm degenerates completely and oil fills the entire vacuole.

Stage 3 (Mucilage cells): At stage 3 the mucilage idioblasts develop through three stages – a, b, c. In 3a, the cell has a collapsed appearance, and there is an abundance of hypertrophied dictyosomes budding off the vesicles. Mucilage has just started to be deposited as a layer between the suberised layer and the cytoplasm. The cell membrane is pushed inside hard the inner vacuole is still intact. In stage 3b, the cytoplasm consists mainly of dictyosome vesicles full of mucilage and they finally fuse with plasmalemma. The cytoplasm has been forced to move inward by the accumulated mucilage and the central vacuole gets reduced considerably. In stage 3c, the cell is mostly round, the suberised layer is 37 ± 7 nm thick, and the cell is completely filled with mucilage. Cytoplasm gets degenerated and the vacuole disappears completely. Main similarities and differences between the developmental stages of oil and mucilage idioblasts are given in Table 2.3.

Foliar epidermal characters

Ravindran *et al*. (1993) and Baruah and Nath (1997) studied the foliar epidermal characteristics in several species of *Cinnamomum*, including *C. verum*, *C. cassia*, *C. camphora* and *C. tamala*.

Epidermis

Epidermis consists of a single layer of cells covered by smooth cuticle. Cuticle is thick on the upper surface, thin on the lower. Epidermal cells are of two types: (i) In *C. camphora* (and also in species such as *C. parthenoxylon*, *C. cecidodaphne* and *C. glanduliferum*) cell walls are straight or only slightly curved; (ii) In *C. verum*, *C. cassia*, *C. malabatrum*, *C. tamala*, etc. cell walls are sinuous (Fig. 2.4). Epidermal walls of different species show varying degrees of birefringence and autofluorescence of cell walls, which indicate sclerification (Bakker *et al*., 1992).

The epidermis in many species possesses trichomes. Often (as in *C. verum*, *C. cassia*) trichomes are microscopic and occur only on the lower surface, but in species like *C. perrottettii* both leaf surfaces are hirsute and trichomes are visible to the unaided eye. Trichome distribution is sparse in *C. verum*, in certain *C. malabatrum* collections, moderately dense in *C. cassia* and dense in *C. perrottettii* (Ravindran *et al*., 1993). Trichomes are short (less than 0.1 mm), medium (0.1–0.2 mm) or long (above 0.2 mm). Structurally trichomes are identical. They are unicellular, unbranched and nonglandular, thick walled, and enclose a narrow lumen in the centre. In *C. bejolghota* the trichomes are papillate (Baruah and Nath, 1997).

Christophel *et al*. (1996) used leaf cuticular features in relation to taxonomic delimitation in Lauraceae. They found that South American species of *Cinnamomum* have distinctive cuticular signatures and that they are clearly different from the species occurring in Australia. Interestingly the American species of *Cinnamomum* studied by the above authors had been earlier placed in the genus *Phoebe*. Asian species of the genus, including the type species, form a third distinctive group. Further study may well lead to the conclusion that the group is not natural (and is perhaps polyphyletic) as it is currently defined (Christophel *et al*., 1996).

Venation pattern

Kim and Kim (1984) studied the venation pattern in some Lauraceae using soft X-ray analysis and reported acrodromous venation in *C. camphora*, *C. japonicum* and *C. loureirii*.

Table 2.3 Main similarities and differences in the developmental stages 3a–c of oil cells and mucilage cells in the shoot apex of *Cinnamomum burmannii*. For comparison, the characteristics of ground tissue cells are included

	Oil cells	Mucilage cells	Ground tissue cells
Cell size at maturity	Spherical 25 × 18 μm	Spherical 21 × 20 μm	Variable shapes, mostly smaller
Cell wall at maturity	Typically 3-layered with suberised layer and inner-wall layer	Typically 2-layered with suberised layer	No additional wall layers
	Cupule present	First deposited mucilage resembles inner wall layer	Many normal plasmodesmata
	Few specific plasmodesmata	Sometimes a cupule present	
	Local breakdown	Few specific plasmodesmata	
		Local breakdown	
Central vacuole	Absent at stage 3a; small vacuoles, devoid of deposits, disappear at stage 3c	Present at stage 3a; devoid of deposits. Disappear at stage 3b	Present with deposits
Plastids	Typical small plastids with reduced thylakoids, disappear at stage 3b	Typical small plastids with reduced thylakoids, lose their definition at stage 3c	Larger plastids with distinct thylakoids.
Accumulation	Oil is stored in the oil cavity between cupule and plasmalemma	Mucilage is deposited between the wall and the plamalemma	Osmiophilic granules in the central vacuole
Main organelles involved in secretion	Oil-formation in the plastids, migration guided by tubular ER Final fusion with oil cavity membrane: the plasmalemma	Mucilage secretion by hypertrophied dictyosomes Vesicles finally fuse with the plasmalemma	Not applicable
Cytoplasm	Formation of a parietal layer at stage 3c and final degeneration	The parietal layer is forced inward during development and finally degenerates	Includes distinct organelles and groundplasm

Source: Bakker *et al.*, 1991.

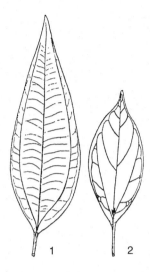

Figure 2.4 Sketches showing two types of venation patterns. 1. A representative leaf of cinnamon showing the typical acrodromous venation pattern. 2. A leaf of camphor tree showing the acrodromous and pinnate-brochydodromous type of venation.

Shylaja (1984) also reported acrodromous venation pattern in south Indian species of *Cinnamomum*. Baruah and Nath (1998) recognised three venation patterns in the *Cinnamomum* species they have studied. In the majority of taxa venation pattern is acrodromous. Here leaves are triplinerved, running in a convergent arch along with two or rarely four lateral nerves. The lateral nerves in most taxa do not reach the apex. In certain species (*C. glanduliferum* and *C. glaucacens*) venation is pinnate-camptodromous (brachidodromous). In *C. camphora* and *C. parthenoxylon* venation is intermediate between the above two patterns (Fig. 2.4).

The 1° veins are moderately stout and straight in *C. malabatrum*, *C. cassia* and *C. tamala*. Lateral 1° veins are generally basal or supra basal. In the case of *C. camphora* and *C. parthenoxylon* 1° veins are anastomosing with the secondaries. In most cases the 2° veins arise from both sides of the 1° veins in an opposite or sub-opposite manner and they are upturned and gradually disappear at apical and basal margins. In *C. camphora*, *C. glanduliferum* and *C. parthenoxylon* 2° veins arise on both sides of 1° veins in an alternate or sub-opposite manner and extend towards the margins and join to form a series of prominent arches (Baruah and Nath, 1998).

The number of 2° veins varies from leaves to leaves in a single taxon, as does the angle between 1° and 2° veins. In *C. cassia*, *C. glanduliferum*, *C. glaucacens*, and *C. parthenoxylon* the sub-adjacent 2° arches are enclosed by 3°–4° arches, while these 2° arches are simple in other taxa studied by Baruah and Nath (1998).

Minor veins of the 4° and 5° order, which originate from 3° veins, constitute the areoles or vein islets. The areoles are generally tetragonal or polygonal in shape. The size and number of areoles are variable in apical, middle and basal portions of lamina. Veinlet endings are generally simple (linear or curved).

30 P.N. Ravindran et al.

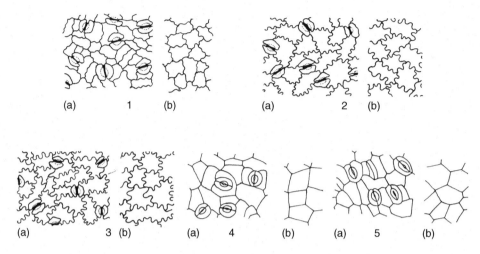

Figure 2.5 Foliar epidermal characteristics of some *Cinnamomum* spp. 1. *C. tamala*; 2. *C. verum*; 3. *C. cassia*; 4. *C. camphora*; 5. *C. panthenxylon*. a-lower epidermis; b-upper epidermis. In 1, 2 and 3 the cell walls are sinuous and stomata anomocytic; in 4 and 5 the cells are straight and stomata paracytic.

Stomata

Cinnamon leaves are hypostomatic, with stomata confined to the lower surface of leaves. Stomata are anomocytic, surrounded by a variable number of cells that are indistinguishable in size or form from the rest of the epidermal cells (Fig. 2.5). In *C. camphora*, stomata are of paracytic type, i.e. they are accompanied on either side by one or more subsidiary cells parallel to the long axis of the pore and guard cells. The guard cells may be equal or unequal in size and are attached to each other at the margins of the concave side with the aperture lying in between the walls. Rare occurrence of other types of stomata was also reported (Baruah and Nath, 1997) (Fig. 2.6).

Guard cells are unevenly thickened, and thickening is heavy along the aperture. Stomatal characters are given in Table 2.4. Stomatal frequency is found to be lower in

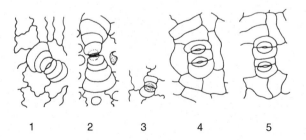

Figure 2.6 Rare stomatal types observed in *Cinnamomum* spp. 1. Paracytic (amphibrachial) stoma in *C. cassia*. 2. Sunken stoma in *C. bejolghota*, flanked by eight subsidiary cells. 3. Paracytic stoma in *C. sulphuratum*. 4, 5. Adjoining paracytic stoma with three or four subsidiary cells in *C. cecidodaphne*. (Source: Baruah and Nath, 1997.)

Table 2.4 Stomatal characteristics of four species of *Cinnamomum*

Species	Frequency/mm^2	Guard cell length (mm)	Guard cell breadth (mm)
C. verum (India)	609	0.018	0.016
C. verum (Sri Lanka)	637	0.017	0.015
C. verum (wild)	609	0.019	0.018
C. cassia	425	0.020	0.017
C. camphora	354	0.023	0.021
C. malabatrum (1)	637	0.016	0.017
C. malabatrum (2)	651	0.018	0.018
C. malabatrum (3)	623	0.018	0.017
C. malabatrum (4)	680	0.016	0.016

Source: Shylaja, 1984.

C. camphora and in *C. cassia* while *C. verum* and *C. malabatrum* have a higher frequency. Guard cell dimensions do not show much variation among species except in *C. camphora*, in which the dimensions are higher than those in other species.

Stomatogenesis

Metcalfe and Chalk (1950) and Kasapligal (1951) were of the opinion that stomata in Lauraceae were paracytic (Rubiaceous), i.e. that they remain surrounded by two subsidiary cells placed parallel to the pore. Pal (1974) reported that in *C. camphora* and *C. zeylanicum* (*C. verum*) stomata were anomocytic, i.e. stomata were surrounded by a number of epidermal cells that were not differentiated into subsidiary cells. Kostermans (1957) and Avita and Inamdar (1981) were of the opinion that stomata in *Cinnamomum* were paracytic. Shylaja (1984) confirmed anomocytic stomata in *C. verum*, *C. cassia*, *C. malabatrum*, *C. perrottettii*, *C. riparium*, *C. macrocarpum* and paracytic stomata in *C. camphora*.

The process of anomocytic stomatogenesis in *C. verum* was outlined by Pal (1978) and Shylaja (1984). In young leaves, epidermal cells are regularly arranged, straight walled and polygonal. Stomata differentiation initiates when leaves attain a length of 11–15 mm and a width of 5–8 mm. The development of stoma begins when an epidermal cell differentiates into a protoderm cell by the formation of a septum that produces two unequal cells (Fig. 2.7). The smaller cell is lenticular in shape, having a prominent nucleus. This cell functions as the meristemoid, while the larger cell becomes a normal epidermal cell. The meristemoid increases in size, becomes almost circular and divides vertically to form two equal or sometimes unequal guard cells. These guard cells enlarge and become kidney shaped, leaving a pore in the centre. As the leaf enlarges the epidermal cell walls become sinuous in *C. verum* and in many other species. At this stage four to five neighbouring cells of various shapes and sizes surround the guard cells. The meristemoid directly acts as guard cell mother cell, and the neighbouring cells are derived from other protoderm cells, hence the development is perigenous type.

Anomocytic stoma is generally regarded as the primitive type and occurs in the primitive land plants such as Psilophytales. The paracytic type is also primitive, occurring in Equisetales, but it is derived from the anomocytic type. In this genus different types of stomata have been reported. Such developmental variations are common in any genus

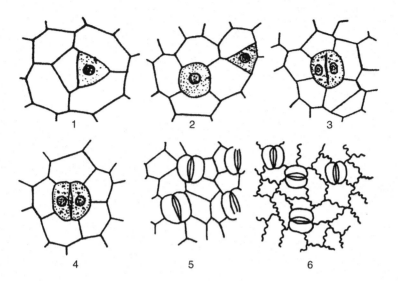

Figure 2.7 Stomatal development in *C. verum*. 1. A meristomoid (m) cutting off from the protoderm cell. 2. Enlarged and spherical meristomoid. 3 and 4. Vertical divisions of the meristomoid producing two guard cells. 5. Mature stomata in *C. camphora*. 6. Mature stomata in *C. verum*. (Source: Shylaja, 1984.)

that is still evolving, and some species may be evolutionarily more advanced. The conflicting reports on the type of stomata lead one to doubt whether both types exist in cinnamon.

Cell inclusions in the leaf tissue

Leaf tissues of *C. verum* and other species contain cellular inclusions such as mucilage, oil globules and calcium oxalate crystals. Tannic substances, appearing as deep brown or black deposits, are found in the epidermal and mesophyll cells of most species. Yellow or golden yellow deposits of mucilage occur in mesophyll cells. Such deposits are abundant in *C. verum*, *C. cassia*, *C. camphora* and less in *C. malabatrum*. Lysigenous cavities containing oil occur within mesophyll. Oil globules are found extensively distributed in leaf and petiolar cells.

Raphides are present in leaf tissue except in *C. camphora*. They are usually needle shaped, and occur either isolated or in groups. Raphides are more common in the cells adjacent to the vascular bundles and in spongy parenchyma cells. In addition, crystalline inclusions are common.

Petiole anatomy

A comparison of the petiolar structure of *C. verum* and some of the related species is given in Fig. 2.8. Balasubramanian *et al.* (1993) and Shylaja (unpublished) studied the anatomical features of petioles. In transections, petiole is somewhat flat at the adaxial side and strongly convex at the abaxial side in *C. camphora*, *C. macrocarpum*, *C. cassia* and *C. verum*. In *C. tamala* the adaxial side is concave and the abaxial side strongly convex.

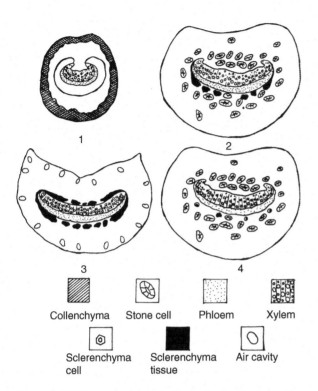

Figure 2.8 Petiole structure of *Cinnamomum*. 1. *C. camphora* petiole t.s, showing hypodermal collenchyma and crescent-shaped vascular strands ensheathed by a continuous layer of abaxial sclerenchyma. 2. *C. macrocarpum* t.s showing arc-shaped vascular strand with abaxial sclerenchyma patches and stone cells. 3. *C. tamala* t.s. showing arc-shaped vascular strand ensheathed by discontinuous patches of sclerenchyma. 4. *C. verum* t.s showing arc-shaped vascular strand, sporadic sclerenchyma cells and stone cells. (Source: Balasubramanian *et al.*, 1993.)

The epidermis is uniseriate and cutinised. Hypodermis is parenchymatous in all the species, in *C. camphora* it is chlorenchymatous and consists of four to five layers of cells. In *C. tamala* air cavities are present in the hypodermis at regular intervals. In *C. cassia*, *C. camphora*, *C. macrocarpum*, and *C. tamala* oil, mucilage and tannin cells are present in the ground tissue. In *C. verum* and *C. cassia* tannin cells are comparatively fewer than in *C. macrocarpum*, *C. riparium* and *C. perottettii*.

The shape of vascular strand and distribution of associated sclerenchyma vary in different species. In *C. verum*, the vascular bundle is crescent shaped and collateral, while phloem is in the abaxial side and xylem in the adaxial side. Sclereids lie scattered around the vascular bundles. Resin canals and cells containing mucilage and tannin are distributed in ground tissue. Cortical cells contain starch grains. In *C. cassia* the shape of the vascular bundle is quite characteristic. The two ends of the vascular bundle curve towards the centre (incurving), thereby presenting a wide arc-shaped appearance. Sclereids lie scattered around the vascular strand. Rows of xylem cells are surrounded by a thick continuous zone of sclerenchyma at the abaxial side. In *C. macrocarpum* the vascular strand is arc shaped with incurving edges. Isolated patches of sclerenchyma

34   *P.N. Ravindran* et al.

are distributed at the abaxial side of the strand in *C. macrocarpum* while in *C. verum* individual sclerenchymatous cells are found scattered in the tissue. Stone cells are distributed in the ground tissue of *C. macrocarpum* and *C. verum*. In *C. tamala* the vascular strand is arc shaped and patches of sclerenchyma occur around the strand. Balasubramanian *et al.* (1993) are of the opinion that petiolar anatomy is useful in taxonomical delimitation of species.

Bark anatomy

Cinnamon bark is of great economic importance, but very little work has gone into its structure and development. Earlier studies on bark anatomy were conducted by Birnstiel (1922), Santos (1930), while more recent studies have been carried out by Chaudhuri and Kayal (1971), Bamber and Summerville (1979), Shylaja (1984) and Shylaja and Manilal (1992). These studies have shown differences in bark structure in cinnamon, especially with regard to the distribution of bast fibres, sclerified tissue, stone cells and secretory cells. Shylaja (1984) and Shylaja and Manilal (1992) reported the anatomy of fresh bark of *C. verum*, *C. cassia*, *C. camphora* and *C. malabatrum*. Externally the bark is grey in colour. Bark tissues are characterised by secretion cells, containing mucilage or essential oil droplets, but the frequency and distribution of such cells vary in the four species. Bark is characterised by islands of sclerenchyma in the pericyclic region, which are connected by a continuous band of stone cells in three of the species (*C. cassia*, *C. malabatrum* and *C. verum*), while in *C. camphora* such a continuous band of stone cells is absent. This band is a demarcating zone between the extrapericyclic region (comprising phellem and phelloderm) and the secondary phloem. The characteristic features of barks of the four species are given in Table 2.5.

Phloem ray frequency per unit length (1 mm) is highest in *C. camphora*, followed by *C. verum* and lowest in *C. malabatrum*. Ray cells of both old and young barks are filled with a dark, golden or brown substance in *C. verum* and *C. cassia*, while such contents are rarely seen in *C. camphora* bark. In *C. malabatrum* such deposits are absent in young bark while present in older. Islands of sclerenchyma in pericyclic regions are

Table 2.5 Bark characteristics (in mm) of four *Cinnamomum* spp.

Species	1	2	3	4	5	6	7	8	9	10	11
C. verum	0.95	0.64	0.30	0.92	0.03	0.29	+	+++	+	++	++
C. cassia	0.97	0.60	0.36	0.65	0.03	0.30	+	+++	+	++	++
C. camphora	0.59	0.38	0.28	0.92	0.03	0.62	++	+	−	+	−
C. malabatrum	0.73	0.38	0.36	0.46	0.06	0.12	+++	−	+++	+	+

Source: Shylaja, 1984.

Notes
1. Bark thickness.
2. Thickness from epidermis to pericycle.
3. Phloem thickness.
4. Phloem ray frequency (per mm).
5. Phloem ray width.
6. Distance between sclerenchyma islands.
7. Phloem fibres.
8. Oil globules.
9. Raphides.
10. Other crystalline inclusions.
11. Tannin in phloem rays.
+: sparse; ++: frequent; +++: very frequent; −: absent.

characteristic features of *Cinnamomum* barks. The distance between such groups is more in *C. camphora* (mean 0.621 mm), while it is lesser in *C. malabatrum* (mean 0.137 mm). The distribution of bast fibres is another characteristic feature of the barks. In *C. cassia* and *C. verum* such fibres are rare and sparsely distributed. In *C. camphora* they are frequent, while in *C. malabatrum* they are very frequent.

The cortical region of young bark and the phellem and phelloderm regions of older bark contain a large number of cells with brown tannic deposits. Such cells are frequent in *C. cassia* and *C. verum*, and less frequent in *C. camphora* and *C. malabatrum*. Oil globules are abundant in the outer bark tissues of *C. cassia* and *C. verum*, but are very few in *C. camphora* and almost absent in *C. malabatrum*.

Crystalline inclusions, mostly in the form of raphides or prismatic crystals, occur in the bark of all the species except in *C. camphora*. Raphides are distributed sparsely in *C. cassia* and *C. verum*. They are abundant in *C. malabatrum*. Raphides are needle-shaped or spindle-shaped, occurring alone or in groups in the phloem tissues.

Chaudhuri and Kayal (1971) made a detailed study on the barks of four species of *Cinnamomum* (*C. verum*, *C. camphora*, *C. tamala* and *C. iners*). The following discussion is based on the above authors.

In *C. verum* the bark is yellowish brown to brown, the outer surface either smooth or rough, uneven, irregularly fissured and longitudinally striate showing circular or irregular brownish patches occasionally with perforations that indicate the positions of nodes. In commercial samples the outer surface is devoid of the suberous coat and often some parts of the middle region. The middle region is granular due to the presence of groups of stone cells; the inner surface is brownish to dark brownish in colour, smooth and soft with faint striae.

Phellem or cork consists of five to ten layers of tangentially elongated, more or less thin-walled, slightly suberised cells. In commercial samples these layers may be completely absent. The cork cells adjacent to the cortex are thin-walled with the inner tangential walls sclerosed.

The phellogen and phelloderm layers are not very distinct and the cortex is not sharply defined from the pericyclic region. The cortex consists of thin-walled parenchymatous cells, which are tangentially elongated and 10–16 layered in thickness. Cortical parenchyma cells contain tannin. Many secretory cells containing oil are intermingled with cortical parenchyma cells. Secretory cells measure 19–35 μ × 35–90 μ in T.S. Pericycle consists of a continuous ring of stone cells, three to four cells in width, which are elongated tangentially, thick-walled and pitted. Inner cells are thicker than the outer ones thereby giving a characteristic appearance. Pericycle fibres occur at intervals and are lignified, elongated and taper at both ends. These fibres measure 300–650 μ in length and 18–35 μ in breadth. The inner region of the stone cell ring consists of six to ten layers of parenchymatous, thin-walled, tangentially elongated cells, intermingled with oil cells and cells containing mucilage. Phloem constitutes about 50% of the thickness of the bark and consists of sieve tubes, phloem parenchyma, phloem fibres and medullary rays. Sieve tubes are arranged in tangential bands, which are completely collapsed in the outer layer. Sieve plates are on the transverse walls. Phloem parenchyma consists of sub-rectangular or rounded cells, which are tangentially elongated and filled with starch grains and small acicular crystals of calcium oxalate. Starch grains are simple as well as compound. Phloem parenchyma cells contain tannin. Associated with the phloem parenchyma are the phloem fibres, which are elongated tangentially, lignified, thick walled and with very narrow lumen. Phloem fibres are single or present in groups of three to four, radially arranged, measuring 250–750 μ in length and 16–30 μ in breadth. Secretory cells containing oil and mucilage are present. Medullary ray cells are usually two cells wide, thin-walled and radially arranged. They are wide near the pericycle and narrow towards the periphery, filled with starch grains and calcium oxalate crystals. The differences in bark structure of the four species studied by Chaudhuri and Kayal (1971) are given in Table 2.6.

Based on the bark anatomical characteristics Chaudhuri and Kayal (1971) proposed the following key for the identification of barks of the four species: (1) Stone cells present, occurring in the cortical region, half stone cells absent – *C. camphora*; (2) Stone cells abundant, forming a continuous ring in the pericycle – *C. verum*; (3) Stone cells occurring in groups in the outer region of the cortex – *C. tamala*; (4) Stone cells abundant, but not forming a continuous ring – *C. iners*.

Namba *et al.* (1987) carried out anatomical and chemical studies on commercial grades of Sri Lankan cinnamon barks (such as Alba, Continental 5 special, Continental 5, Continental 4, Mexican 5, Mexican 4, Hamburg 1 and Hamburg 2). In general, superior grades of cinnamon have less mechanical tissue and more essential oil cells (Fig. 2.9a and b). They also found that in superior grades the percentage of cinnamylacetate is higher and cinamaldehyde is lower than in the lower grades. Mikage *et al.* (1987) subjected cinnamon bark to x-ray microradiogram and observed that the lower grade barks had broad linear bands due to calcium oxalate crystals in the phloem rays, while such bands were absent in high-grade barks. Comparative analysis also showed that higher-grade barks came from slender shoots, while the lower grade ones corresponded to thicker shoots and stem.

Parry (1969) briefly discussed the histology of cassia bark. Unscraped bark of Chinese cassia consists of (a) cuticle, (b) epidermis, (c) cork, (d) cortex, (e) pericycle, and (f) phloem. In scraped cassia bark epidermis, cuticle, epidermis, cork and a portion of the cortex tissue are absent. The cuticle is well-developed and varies in thickness. The epidermis consists of cells that are tangentially oblong-rectangular and that have dark contents and vary in size from 10 μ to 25 μ (long axis). Cork tissue consists of stone cells, thin-walled cells and arch cells. Thickening of the stone cell is more or less uniform; that of the arch cells occurs on the outer wall, causing the arch like appearance. Thin-walled cells of the cork layer are disorganised. Parenchyma cells of the cortex are irregular in shape and vary in size. Stone cells are numerous in the cortex, and occur both singly and in groups. They vary in shape and in size. Starch granules occur in the parenchyma cells. The pericyclic region consists of numerous stone cells occurring in groups, but they do not form an unbroken ring. Walls of stone cells are mostly evenly thickened. The fibres are thick. Phloem constitutes the inner bark. Phloem fibres are not numerous, and they vary in size ranging from 440 μ to 690 μ in length and 27–35 μ in width. Sieve tubes in CS appear irregularly compressed, parenchyma cells are irregular in shape. Phloem ray consists of radial rows of rectangular parenchymatous cells containing small, prismatic crystals of calcium oxalate. The histological nature of the Indonesian cassia is similar to Chinese cassia.

Jackson and Snowdon (1990) studied the microscopical nature of bark powders of cinnamon and cassia. Cinnamon bark powder is characterised by: (i) abundant sclereids, occurring singly or in small groups, more or less isodiametric; walls moderately thickened, pits numerous and conspicuous; (ii) fibres occur in abundance, occur singly,

Table 2.6 Microscopic characteristics of barks of four *Cinnamomum* spp.

	1 Microscopic characters	2 C. verum	3 C. camphora	4 C. tamala	5 C. iners
(a)	Phellem	5–10 layers – thin-walled, tangentially flattened, suberised cork cells	5–20 layers – tangentially elongated cells, slightly suberised	10–20 layers – two kinds of cork cells alternating each other which are either suberised or sclerosed	10–15 layers – slightly suberised; contain reddish-brown substance
(b)	Phellogen	Not distinct	Not distinct	More or less distinct due to presence of brownish content	Somewhat distinct
(c)	Phelloderm	Not sharply defined from pericyclic region. 10–16 layers of tangentially elongated cells that contain either starch grains or filled with a brownish substance	Several layers of cells which contain either starch grains that are sparingly distributed; contain acicular crystals of calcium oxalate	16–20 layers of tangentially elongated cells that either contain starch grains or acicular crystals of calcium oxalate	Several layers of tangentially elongated cells that contain starch grains, acicular crystals and a few prismatic crystals of calcium oxalate
(d)	Secretory cells	Cells containing oil are intermingled with cortical parenchyma cells and some containing oil and large cells containing mucilage are scattered in the phloem region	Distributed in the cortical and phloem region which contain oil	Distributed in the cortical and phloem region. These cells contain oil, mucilage and tannin	Few cells containing essential oils and mucilage in the cortical and pericycle regions
(e)	Stone cells	Middle part of the bark contains tangentially elongated, sometimes thick-walled sclerenchymatous stone cells with reduced cavities. Inner walls thicker than outer	Distributed in the cortical region, isolated or in groups, associated with fibres, generally thick walled with their walls striated	Distributed in the outer region of the cortex thick walled	Stone cells are separated by cortical parenchyma cells with pitted walls and slightly thickened at one side. A few are occasionally found in the bast region
(f)	Half stone cells	Present	Absent	Present	Present

Table 2.6 (Continued)

	1 Microscopic characters	2 C. verum	3 C. camphora	4 C. tamala	5 C. iners
(g)	Cortical fibres	Pericyclic fibres lignified, elongated and tapering at both ends	Absent	Same as *C. verum*	With pointed ends and large lumen
(h)	Phloem parenchyma	Tangentially elongated, rectangular or rounded, contain minute starch grains, small calcium oxalate crystals in raphides or clinorhombic forms	Polygonal or rectangular, contain starch grains and crystals of calcium oxalate in raphides	Tangentially elongated, polygonal, contain starch grains and crystals of calcium oxalate in raphides	Thin-walled, either round or rectangular, contain starch grains, crystals of calcium oxalate in raphides or prismatic forms
(i)	Phloem fibres	Lignified, thick walled with very narrow lumen, isolated or in small radially arranged groups of 3–4; 250–750 μ long and 16–30 μ broad	Lignified, thick walled with narrow lumen and with their walls striated; 250–700 μ long and 16–35 μ broad	Lignified with narrow lumen and with distinct striations; 280–900 μ long and 21–35 μ broad	Distributed throughout the phloem region, either isolated or in radial or tangential groups with distinct striations and comparatively narrow lumen. 400 μ – 1 mm long and 16–46 μ broad
(j)	Oil secretion cells	Secretory cells containing oil and large cells containing mucilage are scattered throughout	Comparatively smaller secretory cells containing oil and mucilaginous cavities	Large secretory cells containing mucilage and smaller ones containing oil are distributed in the lower region of the secondary phloem either single or in groups	Secretory cells are somewhat smaller and fewer in number than *C. verum*

(k)	Medullary ray cells	2 cells wide, radially arranged, wider near the pericycle and narrow towards the periphery, filled with starch grains and raphides of calcium oxalate	1–3 cells wide, radially elongated	1–3 cells wide, radially elongated, contain numerous raphides of calcium oxalate	2–4 cells wide, radially elongated, filled with starch grains, raphides of calcium oxalate and a brownish substance
(l)	Powder	Light brown or yellowish brown in colour, odour fragrant, taste warm, sweet, pungent and aromatic. Characterised by the presence of cortical cells with starch grains and small acicular crystals of calcium oxalate, pericyclic and phloem fibres, acicular and a few prismatic crystals of calcium oxalate	Greyish to light brownish with camphoraceous odour. Taste aromatic and camphoraceous. Characterised by the presence of stone cells, cortical and phloem fibres, oil and mucilaginous cells, cortical cells and acicular crystals of calcium oxalate	Dark brown to chocolate brown, odour fragrant, taste aromatic and mucilaginous. Characterised by the presence of cork cells, cortical cells, secretory cells containing oil, medullary ray cells with or without acicular crystals of calcium oxalate	Deep brown to chocolate brown, odour aromatic, taste astringent and slightly pungent. Characterised by the presence of cork cells, cortical cells with or without starch grains and acicular and few prismatic crystals of calcium oxalate. Normal as well as half stone cells, mucilaginous cells, cortical and pholem fibres

Source: Chaudhuri and Kayal, 1971.

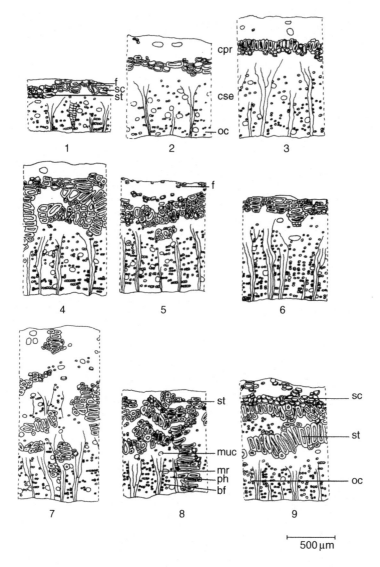

Figure 2.9a Cinnamon bark – diagrams illustrating the transverse sections of commercial grades of Sri Lankan Cinnamon. 1 – Alba; 2 – Continental 5 – special; 3 – Continental 5; 4 – Continental 4; 5 – Mexican 5; 6 – Mexican 4; 7 – Hamburg 1 (thin bark); 8 – Hamburg 1 (thick); 9 – Hamburg 2. bf – phloem fibre, cpr – primary cortex, cse – secondary cortex, f – fibre, mr – medullary ray, muc – mucilage cell, oc – oil cell, ph – phloem, sc – selereid, st – stone cell. (Source: Namba *et al.*, 1987.)

thick walled and lignified with small uneven lumen, having inconspicuous slit shaped pits; (iii) starch grains present abundantly, found scattered, small, simple or compound; (iv) thin-walled oil cells, large ovoid and usually occur singly; (v) thin-walled parenchyma and medullary rays of the phloem, the medullary ray cells contain acicular crystals of calcium oxalate; and (vi) the presence of cork cells, thin-walled and polygonal in surface

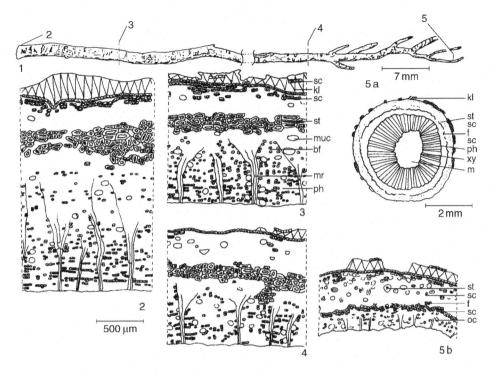

Figure 2.9b Nature of cinnamon bark. 1. A cinnamon stem (branch). 2–4. Diagrammatic sketches of transverse sections of the bark from three regions: 2. base of the branch, 3. middle of the branch, 4. top of the branch. 5a. T.S. of stem tip at region indicated. 5b. T.S. of bark from the region indicated. bf – phloem fibre; f – fibre; kl – cork layer; m – pith; mr – medullary ray; muc – mucilage cell; oc – oil cell; ph – phloem; sc – sclereid; st – stone cell; xy – xylem. (Source: Namba *et al.*, 1987.)

view (Fig. 2.10). The characteristic features of cassia bark powder are very similar to cinnamon, distinguishable by the larger size of starch granules, the greater diameter of fibres and the abundance of cork fragments. In the case of cinnamon, starch granules are rarely over 10 mμ in diameter, fibres up to 30 mμ in diameter. In the case of cassia single starch granules are more than 10 mμ in diameter and fibres are up to 40 mμ in diameter (Fig. 2.11). Brief descriptions on bark and bark powder characteristics are also given by Parry (1969) and Wallis (1967).

Bark regeneration

Cassia bark is extracted from standing trees (in Vietnam and some parts of China) and fast bark regeneration is very important. No effort has so far been made to improve the bark regeneration after bark removal. Hong and Yetong (1997) have shown that application of a bark regeneration fluid followed by wrapping with plastic film resulted in 80% regeneration of bark and such bark is thicker with higher oil and cinnamaldehyde contents. The composition of the regeneration fluid is not known.

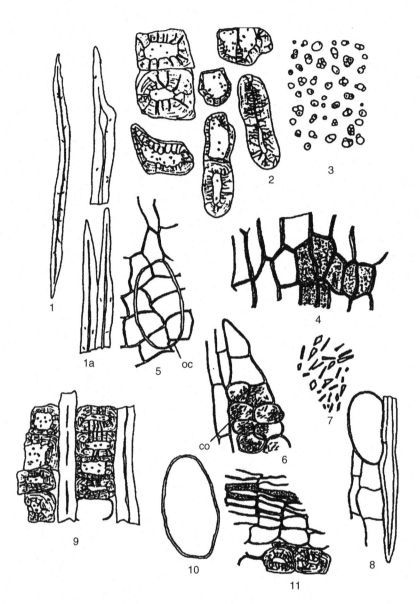

Figure 2.10 Microscopical features of bark powder of *C. verum*. 1, 1a. Fibres. 2. Selereids. 3. Starch granules. 4. Cork in surface view. 5. Phloem parenchyma and oil cells (oc). 6. Parts of medullary ray with some of the cells containing acicular crystals of calcium oxalate (co) and associated phloem parenchyma in tangential longitudinal section. 7. Calcium oxalate crystals. 8. Part of the fibre with an associated oil cell and phloem parenchyma. 9. Part of a group of fibres and selereids from the pericycle. 10. A single oil cell. 11. Part of the cork and cortex in sectional view. (Source: Jackson and Snowdon, 1990.)

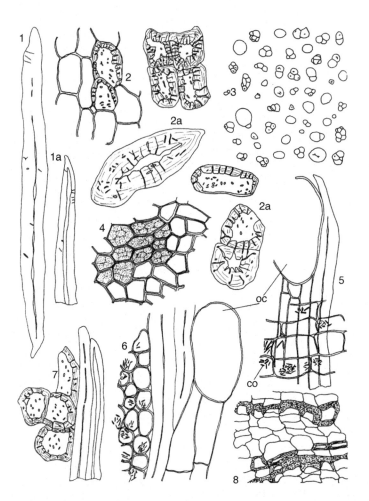

Figure 2.11 Microscopical features of bark powder of *C. cassia*. 1, 1a. A single fibre and part of a fibre. 2. Selereids with associated parenchymatous cells. 2a. Different types of selereids. 3. Starch granules. 4. Cork in surface view. 5. Part of phloem in radial longitudinal section showing a fibre, part of an oil cell (oc), parenchyma, and a medullary ray with some of the cells containing acicular crystals of calcium oxalate (co). 6. Part of the phloem in tangential longitudinal section showing an oil cell (oc) with associated fibres, parenchyma, and part of a medullary ray with some of the cells containing acicular crystals of calcium oxalate. 7. Selereids and fibre of the pericycle. 8. Part of the cork in sectional view showing the alternating layers of thin-walled and thicker walled cells. (Source: Jackson and Snowdon, 1990.)

Wood anatomy

Pearson and Brown (1932), Dadswell and Eckersley (1940) and Stern (1954) carried out wood anatomical studies on *Cinnamomum* species. Stern (1954) studied the wood anatomical features of *C. camphora* and *C. porrectum* and gave the following description of xylem features of *C. camphora*: growth rings distinct, pores 17–35 in a square mm, angular to circular, tending to angularity, vessel wall thickness 2–4.5 µ, tyloses

44   *P.N. Ravindran* et al.

Table 2.7 Xylem anatomical features of *C. camphora* and *C. porrectum*

Features	C. camphora	C. porrectum
Pore distribution (% per group)		
Solitary	51	30
Multiple	40	63
Cluster	9	7
Perforation	Scalariform	Simple + scalariform
Pore diameter (range μ.)	41–95	68–136
Vessel element length (range μ.)	150–462	204–639
Fibres	Fibre-tracheid	Septate, fibriform wood fibres
Parenchyma	Terminal, vasicentric	Vasicentric

Source: Stern, 1954.

thin-walled, vessel element end wall inclination 30°–50°, mostly 45°, intervascular pits circular to oval. Pitting alternate with some opposite, vascular rays heterogeneous, one to three cells wide, uniseriate rays uncommon; one to six but mostly two to three cells high, multiseriate rays 4–17 mostly 7–14 cells high; secretory cells in rays, axial parenchyma present and all inclusions occur as idioblasts; sclerotic pith flecks present. Xylem anatomical features of *C. camphora* in comparison with *C. porrectum* are given in Table 2.7.

Floral morphology, biology and breeding behaviour

Flowers of cinnamon are produced in lax terminal or axillary panicles that are almost equal to or slightly longer than the leaves. The flowers are abundant with long greenish white peduncles. The flowers are softly hairy, bracteate, actinomorphic, bisexual, trimerous, perigynous, perianth six in two whorls of three each, free, stamens 9 + 3 in four whorls of three each; outer two whorls introrse and glandless, third whorl extrorse and flanked by two prominent glands, fourth whorl is represented by glandless, sagittate, stalked staminodes. Fertile stamens show valvular dehiscence. The filaments of stamens and staminodes are provided with minute hairs. Ovary superior, unilocular with a single pendulous anatropous ovule. Style long, ending in a stigma which belongs to dry, papillate class (Heslop-Harrison and Shivanna, 1977) (see Fig. 2.2).

Joseph (1981) and Kubitzsky and Kurz (1984) and Mohanakumar *et al.* (1985) have studied the floral biology but information is still incomplete. Flowering starts by November and lasts until the early part of March. On average, flower development takes 14 days from the stage of its visible initiation (Fig. 2.12a). The female and male phases are separated temporally (protogynous dichogamy) thereby ensuring outcrossing. Protogynous dichogamy leads to the maturing of the female phase first, and the male phase later. The two phases are separated by almost a day, during which time the flower closes completely.

Joseph (1981) reported that every flower opens twice in two stages. In stage one on the first day, when a flower opens, its stigma is whitish and fresh and appears to be receptive. There is no anther dehiscence, and the stamens of the first whorl and those of the third whorl appear fused. After about five hours the flower closes. In stage two, the next day, the flower opens again. The stigma by then appears shrivelled and not

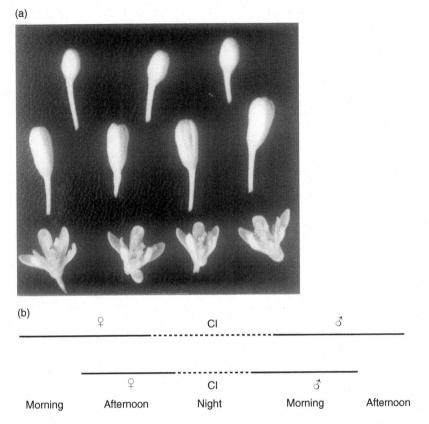

Figure 2.12 Flower biology of cinnamon. (a) Flowers of cinnamon at different stages. (b) Flowering behaviour in cinnamon (synchronised dichogamy).

receptive. The stamens of the third whorl now occur separated from those of the first and they now adhere to the pistil. The anthers dehisce 30–60 minutes after the flower opening. After about five hours the flower again closes and will not open again.

Variations were noticed in the time of occurrence and duration of the first and second stages. Joseph (1981) noted that out of 55 plants studied, in 33 the first stage occurred in the morning, with the second stage occurring at about noon next day. In 22 plants the first stage occurred at about noon with the second stage occurring the next morning. On any day both stage one and two flowers occurred randomly in a population of trees, and functionally female and functionally male flowers (female and male phases) would appear both in the morning and at noon. Thus the flower of cinnamon is highly adapted for cross-pollination.

A similar situation is present in another Lauraceae plant avocado, *Persea americana* (Sedgley and Griffin, 1989). Here all flowers show protogynous dichogamy, the female and male stages are separated by a distinct time period, during which the flower closes completely. During the female phase the pistil is conspicuous and receptive, anthers reflexed against the petals and not dehisced. In the male phase the pistil is obscured by the anthers, which are now dehisced. Two flowering types occur: (a) female phase in the morning and male phase next day after noon; and (b) female phase in the afternoon and male phase on the next morning. Interplanting of (a) and (b) types can therefore ensure outcrossing in avocado. The situation is very similar in *Cinnamomum* as shown by the studies by Joseph (1981) and Kubitzski and Kurz (1984). The latter workers investigated *C. verum* and *C. camphora* and confirmed the observations detailed above. They used the term synchronised dichogamy for the situation existing in the genus (Fig. 2.12b).

The type (a) plant is essentially female in the FN of the first day and male in the AN of the following day. Type (b) plant is functionally female in the N/AN of the first day, and male in the FN of the following day. Both (a) and (b) plants occur mixed in any natural population and hence large numbers of functional male and female flowers are available for any given day. Because of the temporal separation, within-flower pollination (autogamy) as well as between-flower pollination (allogamy) are prevented, leaving between-tree pollination as the only alternative. According to Kubitzski and Kurz (1984) this is the basic floral behaviour in the family. This view is supported by the fact that this phenomenon occurs in all tribes and in both new and old world species. Protogynous (or synchronised) dichogamy is evolved as an adaptation for outcrossing and is the forerunner of dioecy, the obligatory outbreeding mechanism (present in some taxa in Lauraceae).

No research has so far gone into the physiology of dichogamy in cinnamon. However, the studies by Sedgley (1977) on avocado (Lauraceae) shed some light. Pollination of the female phase flower results in fertilisation and fruit set. Pollination of the male stage flower results in some pollen tube growth, but the growth is arrested in the lower style and the tube does not reach the ovary or fertilise the ovule. This inhibition is correlated with an increase in callose (a β-1,3-glucan cell wall component). The thickening of the cell walls of the stylar tissue may prevent or reduce the availability of nutrients to the pollen tube, thereby preventing its sustained growth. The existence of two flowering types within the same species may have a complicated physiology and may be genetically determined. Sedgley (1985) has shown that in avocado the opening of flowers is related to light and dark periods.

Mohanakumar *et al.* (1985) in a preliminary study reported that the anthesis commenced at around 9:00 am, continued until 1:00 pm and reached a peak at between 11:00 am and 12:00 noon. After 1:00 pm anthesis came to an end. Anther dehiscence began by 11:00 am and continued up to 2:00 pm. Pollen grains are spherical, the equatorial diameter is about 23.4 μm. Stainability is around 96.8% (in 0.5% acetocarmine). The highest germination as well as tube growth is in 35% sucrose solution (53.7% and tube length 46.8–78 μm). Pollen grains are binucleate (Erdtman, 1952; Brewbaker, 1967). Cinnamon flowers are insect pollinated. Mohanakumar *et al.* (1985) reported 13 insect pollinators on cinnamon, but nothing is known about their relative efficiency. Honey bees (*Apis cerana*, *A. dorsata* and *A. corea*) are the most active as they are noted most often on blooming cinnamon trees.

Floral anatomy

The floral anatomy of *Cinnamomum iners* was studied by Sastri (1952). Shylaja (1984) studied the floral anatomy of *C. verum* and found that the floral anatomical features are very similar to those of *C. iners*. In *C. verum* the vascular supply of the pedicel consists of a closed ring of vascular bundles, which show secondary thickening to some extent. At its base the closed ring is triangular in shape from which six prominent traces

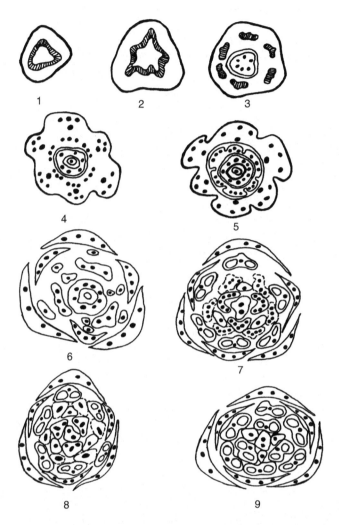

Figure 2.13 Floral anatomy of *C. verum*. 1 and 2. Pedicel with a closed ring of vascular bundles. Initiation (projections) of six traces from the central stale. 3. The six peripheral bundles to tepals and stamens and the central bundle to the ovary. 4 and 5. Differentiation of dorsal and ventral bundles. The perianth cup with the outer series of bundles for the outer two whorls of stamens and an inner series for the inner two whorls of stamens. In 5, tepals and stamens starts to differentiate. 6 and 7. Further differentiation of tepals and stamens. The third whorl of stamens contains three bundles each, the latter ones supplying the glandular outgrowths. 8 and 9. Final stages indicating three whorls of bilocular anthers of three each and the staminodal fourth whorl. In the centre there is the vascular supply to the ovary. (Source: Shylaja, 1984.)

diverge at a slightly higher level (Fig. 2.13, 1–3). These are the vascular traces for perianth and stamens. The remaining bundles in the centre serve as the vascular supply to the ovary. The peripherals divide tangentially to form both the outer series of bundles which supply the tepals and the outer two whorls of stamens, and the inner

48 P.N. Ravindran et al.

series which supply the inner whorls of stamens. The outer bundles again divide tangentially and form the bundles for stamens towards the inner side and for the tepals towards the outer side (Fig. 2.13, 4–5).

The six tepals are differentiated in two whorls of three each. All tepals are supplied with three bundles each. Of the four whorls of stamens, the outer two have one bundle each, and the inner two three bundles each (Fig. 2.13, 6–7). The third whorl of stamens has two lateral outgrowths (glands). Each of these stamens is supplied with a single vascular bundle at its base. However, this bundle gives off one trace each laterally towards its two sides, which diverge out to supply the flattened lateral outgrowths of stamens of this whorl. Higher up this whorl has only one central bundle. The glands are traversed by a number of minute bundles, which are formed by the splitting of the primary bundles. At this level staminodes appear curved in T.S. and the single vascular bundle in each of them divides laterally to form two branches on either side so that the staminodes come to possess a number of bundles (Fig. 2.13, 7–8). These bundles traverse upwards for a short distance and then disappear, so that at their upper parts the staminodes also show only one bundle.

Vascular supply to the ovules consists of two prominent bundles, one dorsal and one compound ventral, and several smaller secondary marginal bundles. The single ovular supply arises from the compound ventral bundle. The secondary marginals disappear at various levels of the ovary, so that the upper part of the carpel is supplied with only the dorsal and the compound ventral, both of which enter the style and the reduced stigma (Fig. 2.13, 8–9).

From the floral anatomical point of view the trimerous androecium is the most interesting part of the cinnamon flower. Saunders (1939) claims that stamens in Lauraceae resulted from chorosis. Recee (1939), on the other hand, considered them to be derived from an original branched type by the process of reduction. Sastri (1952) reported the presence of glandular outgrowths for third and fourth whorls (in *C. iners*), though the fourth whorl is staminodal in nature. Shylaja (1984) found that in all the South Indian species studied only the third whorl of stamens is flanked by glands, except in the case of *C. macrocarpum*. In *C. verum* Shylaja (1984) observed that even though the staminodes are not provided with glands, the vascular traces entering each of them divide to form three bundles, of which the two laterals disappear quickly so that at a higher level only the median trace is found, which then split into a number of minute strands. The lateral splitting of the vascular trace of staminodes possibly indicates the existence of lateral glands (as in a few other species) that might have disappeared in *C. verum*, though the three-trace condition of the vasculature still persists. The fasciculated nature of the stamens and the transformation of the inner whorl of stamens into staminodes are primitive characteristics (Eames, 1961).

The ovary was regarded tricarpellary by Eames and multicarpellary by Sastri (1952), though it appeared to be monocarpellary. In *C. iners*, Sastri (1952) reported more than two bundles, though only two (dorsal and ventral) continued their course into style. Sastri (1952) also noted that the vascular supply to the ovary consisted of a dorsal and a ventral trace only, indicating fusion of the two ventral traces.

Embryology

The earlier studies were those of Tackholm and Soderberg (1917) in *C. seiboldi*, who recorded a polygonum type of embryosac development. Giviliani (1928) reported the same type of embryo sac development in *C. camphora*. A more detailed embryological study was carried out by Sastri (1958) on *C. zeylanicum* (*C. verum*) and *C. iners*. The following discussion is mainly based on the above studies.

Ontogeny of the carpel

Carpellary primordium arises as a small conical protuberance on the thalamus, which becomes cup shaped due to differential growth. One side of the cup continues to grow and form the style and the stigma, from the other side the ovule primordium arises in the subterminal position as a lateral protuberance.

Ovule

The ovule soon becomes bent as a result of the funiculus and comes to lie in a position facing away from the style at the megaspore mother cell stage. The curvature of the ovule continues further, the rate of growth varying from species to species. In *C. verum* the curvature is arrested early, and at the mature embryosac stage the ovule takes a transverse position in relation to the micropyle at right angles to the style, pointing towards the lateral wall of the gynoecium. In *C. iners* at the megaspore tetrad stage or even earlier, the tip of the ovule is at a right angle to the style. At the four nucleate stage the curvature is complete, and the micropyle points to the style. The ovule is anatropous, pendulous, crassinucellate and bitegmic (Fig. 2.14). Integument initials are seen in the young ovule at the primary archesporium stage. Integuments are not very well developed at the megaspore mother cell stage, but are fully formed at the tetrad stage. The inner integument is three to four layered, while the outer integument is four to five layered.

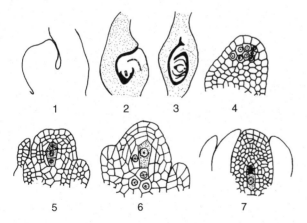

Figure 2.14 Ovule development in *Cinnamomum*. 1–3. L.S. of carpel showing successive stages in the development of ovule. 4. L.S. of nucellus showing multi-cellular archesporium. 5. L.S. of ovule showing megaspore mother cell. 6. L.S. of young ovule showing linear tetrad of megaspores. 7. L.S. of ovule showing linear tetrad of megaspores of which the micropylar three are degenerating. (Source: Sastri, 1958.)

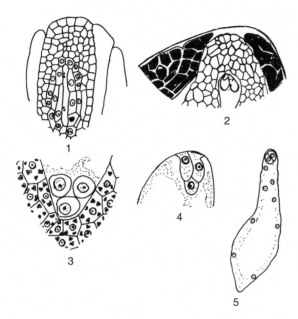

Figure 2.15 Stages in the development of embryosac. 1. L.S. of ovule at 4 – nucleate embryosac stage showing nucellar cells with prominent nuclei. 2. L.S. of apical region of ovule showing synergids 3. L.S. of chalazal portion of nucellus showing antipodal cells. 4. Apical portion of embryosac showing egg apparatus. 5. L.S of embryosac showing endosperm nuclei and two-celled proembryo (1–3, 5: *C. iners*; 4: *C. verum*). (After Sastri, 1958.)

Embryosac

A group of hypodermal cells and one or two cells of the nucellus below them get differentiated into the primary archesporial cells in the young ovule. Only one of them is functional. The nucleus of this cell undergoes meiotic division, giving rise to a linear tetrad of megaspores, of which the chalazal one is functional while the other three degenerate. This nucleus undergoes three mitotic divisions, giving rise to an eight-nucleated embryosac with the 3 + 2 + 3 arrangement. The mature embryosac is narrow, elongated and develops numerous small vacuoles (Fig. 2.15). The synergids persist, while the antipodals degenerate before fertilisation (Sastri, 1958).

Embryo

Sastri (1958) studied the embryo development in *C. iners*, and the process is believed to be the same in other species, including cinnamon. The zygote divides transversely producing two superposed cells, the apical cell (ac) and the basal cell (bc) (Fig. 2.16). They then divide by vertical walls resulting in a four-celled proembryo. One of the daughter cells derived from ac divides by means of an oblique wall and a triangular cell (the epiphyseal initial, e) is formed and this cell undergoes further division only at a much later stage. The remaining cells undergo many divisions giving rise to the many-celled proembryo (Fig. 2.16). The subepiphyseal cells undergo transverse

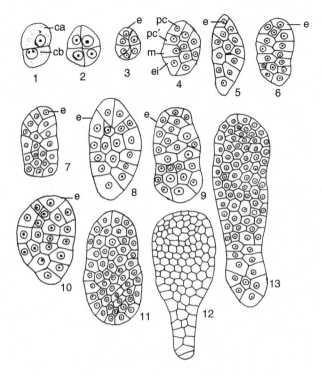

Figure 2.16 Stages in the development of embryo (for explanations see text). ca – apical cell; cb – basal cell; e – epiphyseal initial; pc, pc' – tiers of cells derived from subepiphyseal cells; m – hypophyseal initial, that divides and forms the hypophyseal region; ci – suspensor initial that eventually gives rise to the suspensor. (Source: Sastri, 1958.)

divisions producing two superposed tiers of cells (*pc* and *pc'*). They undergo further divisions and finally lead to the formation of the central cylinder of the stem and cotyledons and the hypocotyledonary region, respectively. The two daughter cells of the basal cell (bc) divide transversely giving rise to two superposed tiers of cells. The upper tier divides further in all planes and forms the hypophyseal region. The lower tier gives rise to the suspensor. The epiphyseal cell undergoes further divisions, ultimately forming the stem apex.

Since the derivatives of the apical cell of the two-celled proembryo alone contribute to the formation of the embryo proper, while the suspensor and hypophyseal regions are derived from the basal cell, the embryo development is of Onagraceae type.

Endosperm

The primary endosperm nucleus divides earlier than the fertilised egg. Endosperm development is of the nuclear type (Sastri, 1958). By the time the zygote undergoes the first division, eight to ten free nuclei derived from the division of the endosperm nucleus are seen distributed along the periphery of the embryosac. These nuclei are large with one to three nucleoli. The endosperm nuclei have glistening crystalline bodies at the centre. The cell wall formation begins from the micropylar end and, at the globular stage of the embryo, the entire endosperm becomes cellular.

Fruit and seed

The fruit is an oblong or cylindrical berry. Testa 10–12 cells thick; outer epidermis is composed of short radially elongated cells; mesophyll unspecialised, composed of longitudinally elongated cells with unlignified spiral thickening on the radial walls. Vascular bundle of the raphe divides into two post chalazal branches ascending to the micropyle. Pericarp consists of five layers: (i) outer epidermis that is short celled; (ii) outer hypodermis, four to five cells thick as stone cells; (iii) pulpy mesophyll four to five cells thick, thin walled; (iv) fibrous endomesophyll (inner hypodermis), four-cells thick, the fibres tangentially elongate; and (v) inner epidermis composed of columnar cells in one layer (Choudhury and Mitra, 1953; Sastri, 1958). Fruit disposal is conducted mostly by birds (Sedgley and Griffin, 1989) (Fig. 2.17).

Mature seed is exalbuminous. The adult embryo fills the entire seed and has two massive cotyledons, enclosing the radicle and plumule, the latter shows leaf primordia.

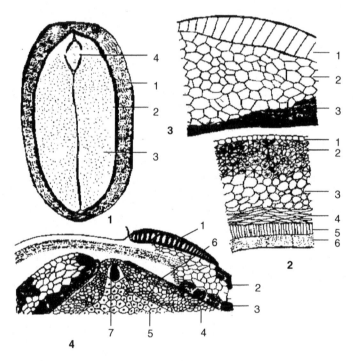

Figure 2.17 Seed structure of cinnamon. 1. L.S. of mature fruit. 1. Fruit wall; 2. Seed coat; 3. Cotyledons; 4. Plumule. 2. L.S. of mature fruit wall and seed coat. 1. Outer epidermis; 2. Stone cells; 3. Parenchymatous cells; 4. Fibrous cells; 5. Columnar stone cells; 6. Seed coat. 3. L.S. of mature seed coat. 1. Outer epidermis; 2. Parenchymatous cells; 3. Inner epiderminal cells with helical thickening. 4. L.S. of apical portion of young seed of *C. iners*. 1. Stone cells; 2. Outer integument; 3. Cells with helical thickening; 4. Nucellus; 5. Endosperm; 6. Inner integument; 7. Embryo.

Seed shows differentiation of procambial vascular supply. In the fertilised ovule, the outer integument is many layered while the inner is two-to-four layered. After fertilisation the inner integument gradually disintegrates and is absent in the mature seed. The outer integument gradually increases in thickness and the outer epidermal cells become filled with tannin. The cells of the inner epidermis acquire band shaped helical thickening. In between them there are eight to ten layers of parenchymatous cells (Sastri, 1958).

Cassia buds

Cassia bud is an item of commerce and consists of dried immature fruits of *C. cassia*. They vary from about 6 to 14 mm in length, the width of the cup varying from about 4–6 mm. The immature fruit is enclosed within the calyx cup. Dried cassia buds have a sweet, warm, pungent taste similar to that of cassia bark.

Parry (1969) has provided a detailed description of the histology of the cassia bud, on which the following discussion is based (Fig. 2.18). The calyx lobes consists of: (i) an outer epidermis of small cells, rectangular or nearly so, with thick outer walls; (ii) a cortex of parenchyma cells, secretion cells and occasional stone cells; and (iii) fibrovascular tissue consisting of short and long parenchyma. The short fibres are porous with numerous pits. The long fibres are pitted and striated. Phloem cells are numerous. Xylem vessels are small and consist of annular, spiral and reticulate types. There is an inner epidermis of small, rectangular cells, similar to that of the outer epidermis.

Calyx tube: The calyx tube consists of the following parts: (i) an epidermis of small cells, rectangular or nearly so, with thick outer walls; (ii) a cortex of parenchyma cells, isodiametric that are about 45 μ in diameter, secretion cells about 66 μ in diameter; and occasional stone cells. The cell contents are light to dark brown in colour; (iii) a pericycle marked by a ring of fibres, the ring is broken by the parenchymatous tissue of pith rays. The fibres are thick-walled and up to about 66 μ in diameter; (iv) fibrovascular tissue similar to that of the calyx lobes; and (v) pith of isodiametric, parenchymatous cells and occasional stone cells.

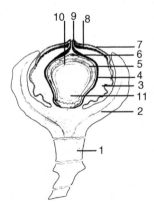

Figure 2.18 Longitudinal section of "cassia bud" (young fruit). 1. Pedicel; 2. Calyx; 3. Mesocarp; 4. Endocarp; 5. Reticulate cells; 6. Pigmented parenchyma; 7. Exocarp; 8. Pericarp; 9. Compressed parenchyma; 10. Cotyledon. (Source: Parry, 1969.)

54   *P.N. Ravindran* et al.

Pericarp: The pericarp consists of the following tissues: (i) a sclerenchymatous epicarp the cells of which have thick outer walls and thin inner walls; (ii) a mesocarp of stone cells, parenchyma cells, secretion cells and vascular tissue. Stone cells vary in size and shape and they underlie the epicarp. Parenchyma cells occupy the major portion of the mesocarp. Cell contents vary in colour from light to dark brown. Secretion cells are numerous, calcium oxalate crystals present; (iii) an endocarp of sclerenchymatous tissue except in the basal region of the fruit, where it is parenchymatous.

Seed coat: The seed coat consists of the following regions: (i) an outer tissue of four to six layers of deeply pigmented (dark brown) parenchyma cells. These cells are tangentially elongated to about 90 μ; (ii) an almost complete tissue of reticulated cells. In the upper part of the seed these cells are arranged in palisade fashion, the innermost cells being the longest (about 100 μ). Reticulated cells are not evident in the basal region; and (iii) an incomplete parenchymatous tissue underlying the reticulate cells, five to six layers and irregular in shape. Calcium oxalate crystals are present in the cells.

Pedicel: In CS the pedicel consists of the following parts: (i) epidermis – epidermal cells are small and rectangular cells with thick outer walls; (ii) cortex – consists of isodiametric, thin walled parenchyma cells with dark contents, secretion cells and occasional stone cells; (iii) pericycle – marked by an almost continuous ring of thick walled fibres; (iv) an almost complete ring of fibrovascular tissue; and (v) a pith of isodiametric parenchyma cells.

Biosystematics and inter-relationships

Shylaja (1984), Shylaja and Manilal (1998) and Ravindran *et al.* (1991, 1996) carried out numerical taxonomic studies on *C. verum* and some of the related taxa. Cluster analysis of characters led to the clustering of the following characters: (i) bud type, stomatal type, phyllotaxy and epidermal thickness; (ii) hairiness, hair frequency and hair length; (iii) inflorescence length, nature and type; (iv) thickness of leaf, palisade and spongy parenchyma; and (v) leaf length and leaf breadth. The first three sets of characteristics are important for taxonomic delimitation of taxa. Cluster analysis of taxa using centroid linkage led to the grouping of *C. verum* with some collections of *C. malabatrum*, while *C. cassia* and *C. camphora* formed unique groups. Much variability exists in *C. malabatrum*, in which some of the collections clustered with *C. verum*, while others formed eight distinct groups. This result led to a re-examination of the *C. malabatrum* complex from which one new species and one variety were identified (Shylaja, 1984; Manilal and Shylaja, 1986). In fact, the *malabatrum* complex requires a closer study. Principal component (PC) analysis (Shylaja, 1984; Ravindran *et al.*, 1996) was used to analyse the characteristics responsible for species divergence. The dispersion of OTUs between the PCs shows that *C. camphora* gets diverged from other species due to the first PC (bud type, stomatal type, phyllotaxy, epidermal thickness). The second PC (hairiness, hair size, and hair frequency) is responsible for the divergence of *C. perrottettii*, *C. riparium* and *C. macrocarpum*. *C. cassia* diverges from the rest of the species by the third PC (inflorescence length, nature and type). The fourth PC (thickness of leaf, palisade and spongy parenchyma) and fifth PC (leaf length and leaf breadth) are responsible for the divergence of various *C. malabatrum* collections. Species such as *C. camphora*, *C. cassia*, *C. riparium*, *C. macrocarpum* and *C. perrottettii* remain independent entities with respect to the PCs studied, thereby indicating their taxonomic independence. *C. verum* and *C. malabatrum* show considerable closeness, indicating their taxonomic affinity.

Infraspecific variability in *C. malabatrum* is evident from the divergence of some of the *C. malabatrum* collections from the rest.

Bakker *et al.* (1992) carried out a phenetic analysis of a large number of *Cinnamomum* spp. using leaf anatomical characteristics. They applied cluster analysis and the phenogram constructed based on Manhatten distances indicated ten clusters. According to this indication *C. burmannii* and *C. verum* are in cluster 1, together with many other species. Within this cluster *C. burmannii* occupies an independent sub-cluster position, while *C. verum* is in another sub-cluster together with 21 other species (including *C. tamala* (Indian cassia), *C. culitlawan*, *C. keralaense*, *C. sintok*, etc.). *C. cassia* is in cluster 2. Different accessions of *C. camphora* are included in clusters 4, 9 and 10, thereby indicating the infraspecific variability present in this taxon. In clusters 1 and 2 the idioblast distribution shows the largest variation. Cluster 3 always possesses mucilage cells in both mesophyll layers and oil cells in spongy parenchyma. The species in cluster 4 can be recognised by the combination of the following features: presence of both idioblastic types in both layers; non-sclerified epidermal cells; and, a weakly or non-sclerified spongy parenchyma. Cluster 5 does not possess a discriminating leaf anatomical characteristic distribution. Cluster 6 lacks mucilage cells in the spongy parenchyma and also possesses non-sclerified epidermal cells and spongy parenchyma, a two-layered sclerified palisade parenchyma and penninerved leaves. Cluster 7 is characterised by the presence of oil and mucilage cells in both palisade and spongy layers, a thick lamina and cuticle, a two-layered sclerified palisade parenchyma and triplinerved leaves. Cluster 8 lacks oil cells and clusters 9 and 10 are characterised by the absence of mucilage cells. In these last three clusters the idioblast distribution pattern is the discriminating factor.

Meissner (1864) has subdivided *Cinnamomum* into two sections, *Malabathrum* and *Camphora*, which was adopted by later florists like Hooker (1886). Further infrageneric studies have not been made so far (Kostermans, 1986). Bakker *et al.* (1992) tried to correlate leaf anatomical characteristics of the species with the existing classification. They found that most of the species under the section *Malabathrum* were in groups 1–3. All these species possess sclerified epidermal cells and/or palisade cells. Those species in the section *Camphora*, which fall in groups 4, 9 and 10, lack sclerified epidermal and palisade cells. Most neotropical species included in the study by Bakker *et al.* (1992) were transferred from other genera (such as *Phoebe*). These species occupy clusters 6 and 9. Nine of these species have penninerved leaves, and they all possess non-sclerified epidermal cells.

Chemotaxonomy

Shylaja (1984) and Ravindran *et al.* (1992) carried out chemotaxonomical studies on *C. verum* and some of its related taxa occurring in the Western Ghats of Kerala, India. They have analysed flavonoids, terpenoids and steroids and found much variation among species. *C. verum*, *C. camphora*, *C. cassia* and *C. riparium* are chemically very distinct among themselves and from other species. Much infraspecific chemical variability was noticed in *C. malabatrum*. They also found that Sri Lankan and Indian accessions of *C. verum* were chemically identical. The above workers also carried out a centroid clustering analysis of the flavonoid-triterpenoid data, which resulted in independent clustering of *C. verum*, *C. riparium*, *C. cassia* and *C. camphora*. Close clustering resulted between *C. perrottettii* and *C. macrocarpum* as well as between

56   *P.N. Ravindran* et al.

C. malabatrum and *C. nicolsonianum*. Flavonoid complexity is greater in *C. verum* and *C. camphora*. The complexity in flavonoid pattern in *Cinnamomum* is found to be the result of o-methylation, which is considered an advanced characteristic in flavonoid evolution. It was also noticed that flavones have replaced the simpler flavonols which again is an advanced character in the evolutionary history of flavonoids (Crawford, 1978).

Infraspecific chemical variability exists in many species of *Cinnamomum* (Table 2.8). Fujita (1967) attempted a sub-specific classification of many species based on volatile oil composition. Such infraspecific variability might have evolved as a result of inter-breeding, and forces such as segregation, chance mutations, and isolation mechanisms acting on the populations. Tetenyi (1970) used the term "polychemism" to denote the simultaneous existence of more than one chemically distinct form in a species. This aspect is well studied in *C. camphora*. The Formosan camphor tree (*C. camphora* ssp. *formosana*) is known to have at least seven distinct chemovarieties (Tetenyi, 1970):

chvar. Borneol
chvar. Camphor
chvar. Safrole
chvar. Sesquiterpene alcohol
chvar. Cineol
chvar. Linalool
chvar. Sesquiterpene

Table 2.8 Infraspecific chemical variability in *Cinnamomum* species (compiled from published literature)

Species and references	Chemovar	Major component
C. camphora Hirota, 1951 Hirota and Hiroi, 1967	A	Camphor type (Camphor >80%)
Shi, 1989 Wan-Yang *et al.*, 1989	B	Linalool type (Linalool 80% + mono terpenes 10%)
Shi, 1989	C	Cineole type (Cineole 76% + α-terpinol 20%)
Wan-Yang *et al.*, 1989	D	Nirolidol type (nerolidol 40–60% + sesquiterpenoids 20% + monoterpenoids 10%)
	E	Isonerolidol type (>57% isonerolidol)
	F	Safrole type (safrole 80% + monoterpenoids 10%)
	G	Borneol type (Borneol >80%)
C. bejolghota Baruah and Nath, 1997	A	Linalool + α-terpineole
	B	Linalool + α-phellandrene
	C	Linalool
C. cecidodaphne Birch, 1963	A	Methyleugenol
	B	Methyleugenol 45% + Safrole 20%
	C	Safrole (major) + elemicin + myristicin (minor)
C. culitlawan Spoon-Spruit, 1956	A	Safrole 70–80%
	B	Safrole (35–53%) + methyl eugenol (41–50%)
	C	Eugenol (80–100%)

C. glanduliferum	A	Safrole (major) + myristicin + elemicine
Gildemeister and	B	Methyl eugenol
Hoffmann, 1959		
Birch, 1963	C	Methyl eugenol 45% + Safrole 25%
C. kiamis	A	Cinnamaldehyde
Birch, 1963	B	Cinnamaldehyde (45–62%) + eugenol (10%)
C. loureirii	A	Cinnamaldehyde (major) + eugenol (little)
Gildmeister and		
Hoffmann, 1959		
Birch, 1963		
	B	Eugenol (major)
C. parthenoxylon	A	1,8-cineole
Baruah and Nath, 2000		
	B	Linalool
C. pedunculatum	A	Safrole (60%) + eugenol
Birch, 1963		
	B	Eugenol + methyl eugenol
C. sintok	A	Eugenol
Gildmeister and		
Hoffmann, 1959		
Birch, 1963		
	B	Eugenol + methyl eugenol
	C	Methyl eugenol + safrole
	D	Safrole
C. sulphuratum	A	Linalool
Baruah *et al.*, 1999		
	B	Citral
	C	Cinnamaldehyde
	D	Methyl cinnamate
C. tamala	A	Linalool
Baruah *et al.*, 2000		
	B	Citral
	C	Cinnamaldehyde
	D	Methyl cinnamate
C. verum	A	Cinnamic aldehyde (65–76%)
Gildmeister and		
Hoffmann, 1959		
Birch, 1963	B	Cinnamic aldehyde + eugenol
	C	Cinnamic aldehyde + safrole

Chvar. Linalool (designated by later taxonomists as *C. camphora* var *linaloolifera* (Fujita, 1967) consists of two further chemoforms : chforma 86% linalool and chforma 71% linalool (terminology of Tetenyi, 1970).

Hirota and Hiroi (1967) and Wan-Yang *et al.* (1989) recognised the following chemical races among Chinese camphor trees:

(i) linalool type that yields 80% linalool and 10% monoterpenes;
(ii) cineole type having 76% cineole and 20% α-terpene + 1, α-terpenol;
(iii) sesquiterpene type having 40–60% nerolidol and 20% each of sesquiterpenoids and monoterpenes;

(iv) safrole type with 80% safrole and 10% monoterpenoids; hard
(v) camphor type that contains mainly camphor.

Khein *et al.* (1998) recognised eight chemovarieties among camphor trees in Vietnam. They are: camphor type; camphor + sesquiterpene type; sesquiterpene type; cineole type; linalool type; p-cymene + α-phellandrene types; p-cymene + β-phellandrene type and p-cymene + α + β-phellandrene type. They also reported an interesting study of the chemical segregation of camphor trees in two chemovarieties. They identified four chemoforms among the progenies of a camphor tree rich in camphor.

Chemoform   1 – Camphor type (62–93%)
                    2 – 1,8-cineole (30–57%) + α-terpineol (14–25%)
                    3 – E-nerolidol + 9-oxonerolidol (47–84%)
                    4 – E-nerolidol + 9-oxonerolidol (22–36%) and safrole (30–70%)

The progenies of another camphor tree (Linalool type) segregated as follows:

Chemoform   1 – Linalool (72–96%)
                    2 – Camphor (70–86%)
                    3 – E-nerolidol + oxonerolidol (61–91%)
                    4 – Camphor (19–71%) and linalool (14–72%)
                    5 – 1,8-cineole (54–56%), β-pinene (16–24%) and α-terpineol (9–11%)
                    6 – Linalool (38%), 1,8-cineole (36%), β-pinene (11%).

Many other studies also indicated the existence of such chemoforms in camphor tree (see Chapter 9).

Guang-Fu and Yang (1988) investigated the *Cinnamomum* species occurring in the Hubei province of China and did a cluster analysis using chemical characteristics. They observed good correlation among morphological and chemical characteristics. The species studied by them clustered in three groups:

(1) *C. appelianum*, *C. pauciflorum*, *C. wilsonii*
(2) *C. bodinieri*, *C. septentrionale*, *C. platiphyllum*
(3) *C. parthenoxylon*, *C. longepaniculatum*.

Based on a Q clustering analysis the above workers also proposed the probable relationships among the species (mainly in the light of chemical relationships). The chemical evolution indicated by them is given below:

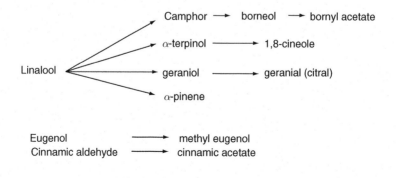

Botany and Crop Improvement of Cinnamon and Cassia 59

Based on the chemical evolution Guang-Fu and Yang (1988) have suggested the probable interrelationships among these species.

Cytology

The earliest cytological study was that of Tackholm and Solderberg (1917), who reported the somatic number as 2n = 24 in *C. sieboldii*. Sugiura (1936) and later Chuang *et al.* (1963) reported the same number in *C. camphora*, *C. japonicum*, *C. linearifolium*, *C. obtusifolium*, *C. sieboldii* and *C. zeylanicum*. Sharma and Bhattacharya (1959), in a detailed study on four species (*C. camphora*, *C. verum*, *C. tamala* and *C. iners*), found 2n = 24 consistently in all of them. Mehra and Bawa (1968, 1969) found the same number in *C. camphora*, *C. caudatum*, *C. cecidodaphne*, *C. impressinervium*, *C. obtusifolium* and *C. tamala*. Okada (1975) and Okada and Tanaka (1975) carried out cytological studies on Japanese Lauraceae, including five species of *Cinnamomum*, and confirmed the same chromosome number in all of them. These authors could establish clear cytological differences based on the nature of the interphase nuclei, which lends support to the subdivision of Laureaceae into the subfamilies Lauroideae and Cassythoideae.

In *C. camphora*, as well as in other species studied by Okada (1975), chromosomes can be distinguished by their size, position of centromere as well as satellite and location of heterochromatic segments (Fig. 2.19, a–c). In interphase nuclei, chromatin forms about 20 condensed bodies, which are stained darkly and are round or rod shaped. Species-specific chromatin distribution is also noticed. In *C. camphora* the 5th pair (chr. 9, 10, satellite chromosomes) shows the following chromatin distribution: the satellites and the long arms are euchromatic (Fig. 2.19, a). In the 7th pair (chr. 13, 14) there are heterochromatic regions in the proximal regions of short arm and in the long term. The distal regions of the short arms consist of late condensing chromatin. The twelth pair is composed of a heterochromatic segment in the proximal region of the short arm, of euchromatin in the long arm and of late condensing chromatic segments in the distal region of short arms. In *C. daphnoides* the morphological characteristics of the chromosomes at interphase, prometaphase and metaphase are more or less the same as that of *C. camphora*. The satellite chromosome pair (9, 10) showed a different chromatin pattern from those of *C. camphora*. All of the satellite and proximal regions of both arms consisted of heterochromatin and the distal region of the long arm was euchromatin. In *C. sieboldii* the distribution of chromatin was also found to resemble that in *C. camphora* but differed in the two pairs of satellite chromosomes. The eighth pair (15, 16) showed the distribution pattern as follows: euchromatin located in the distal region of long arm and in all of the satellite. Heterochromatin was found in the proximal regions of both arms. The other satellite chromosomes (23, 24) possessed euchromatic segments in all of the satellite in the proximal region of the short arm and the distal region of the long arm, and a hetrochromatic segment in the proximal region of the long arms. At metaphase the secondary constructions of these chromosomes were invisible probably due to heavy condensation (Okada, 1975).

Propagation

Cinnamon and cassia can be propagated either through seeds or clonally by cuttage. Seed is recalcitrant and loses viability quickly when stored. When sown immediately after harvest, seeds give 90–94% germination, while on storage for five weeks the

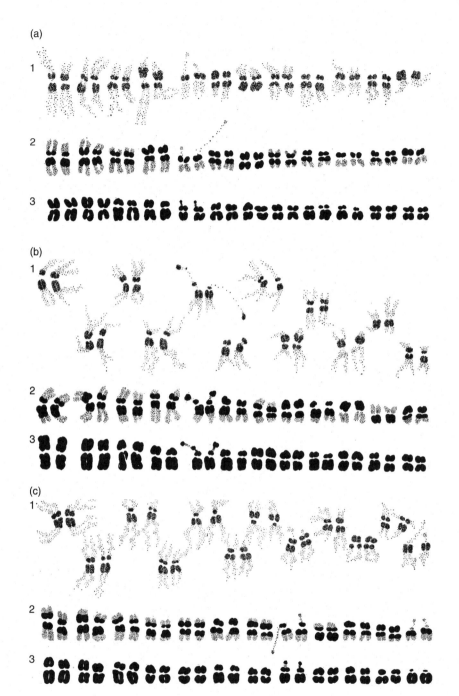

Figure 2.19 Karyotypical representation of somatic chromosomes of three species of *Cinnamomum*. a. *C. camphora*, b. *C. daphnoides*, c. *C. sieboldii*. 1. Late prophase; 2. Prometaphase; 3. Metaphase. In all species 2n = 24. One pair of nucleolar chromosomes can be seen in all the three species. (Source: Okada, 1975.)

Botany and Crop Improvement of Cinnamon and Cassia 61

viability is completely lost (Kannan and Balakrishnan, 1967). Lin (1996) found that in *C. subavenium* (and also in certain other genera in Lauraceae) the storage behaviour of seeds deviated from that of the orthodox and recalcitrant types. The seeds of this species exhibited partial desiccation tolerance and freezing sensitivity deviating from the definitions of either orthodox or recalcitrant storage behaviour.

Ripe fruits are collected from selected mother trees having the following desirable charcteristics: (i) erect stem with smooth bark; (ii) vigorous growth; (iii) easiness of peeling stem bark; (iv) free from pests and diseases; and (v) good quality character-istics – sweetness, pungency and flavour. This can be judged by the 'bite test' of petiole or bark. Fully ripe berries are collected and left in heaps for two to four days in shade to soften and rot. Seeds have to be washed and cleaned and then good seeds are sown without delay in shaded seed beds or in poly bags. Germination starts in about 20 days. Germination is epigeal. For field planting one-year old seedlings are used.

Sebastian *et al.* (1995) and Bhandari (1996) studied the effect of phytohormones on seed germination. Soaking of seeds in GA_3 (150 ppm) or thiourea (1500 ppm) resulted in significantly high germination (98%). GA_3 at 50 ppm reduced the number of days taken for commencement of germination (13 days) when compared to the control (22 days). The subsequent seedling growth (root length, root number, dry weight of shoot, seedling vigour) was more in seedlings raised from seeds treated with 1500 ppm thiourea. GA_3 at 300 ppm resulted in more leaves and greater shoot length. Bhandari (1996) found that GA and kinetin showed stimulating activity in breaking seed dormancy, while IAA and IBA had no such effect. ABA retarded germination.

Vegetative propagation

Both cinnamon and cassia are cross-pollinated species and wide variability has been observed in yield (Ponnuswami *et al.*, 1982; Krishnamoorthy *et al.*, 1992), quality of produce and oil content (Paul and Sahoo, 1993) and other morphological characteristics. Hence vegetative propagation is necessary to produce uniformly high yielding plantations, and also for propagating elite lines. In Chinese cassia vegetatively propa-gated plants are not used for commercial planting, as such plants are known to give poor quality stem and bark, less vigorous growth and regeneration (Dao, this volume). However, clonal propagation is recommended in cinnamon (Weiss, 1997).

Cinnamon can be propagated through cuttings. Single node cuttings with leaves can be rooted in a month's time under high humidity conditions (CPCRI, 1985). Application of IBA or IAA at 2000 ppm enhanced the rooting to 73 and 65%, respectively. Rema and Krishnamoorthy (1993) noted much variability in the rooting response of various cinnamon accessions (genotypes) (Table 2.9). Variation in rooting during different seasons has also been reported and this has been interpreted to be due to the differences in the endogenous levels of auxins, reducing and nonreducing sugars, nitrogen and C:N ratio (IISR, 1996). In a study involving nine genotypes the rooting percentage ranges from 3.3 to 60.5 and genotypes differ in their ability to root.

Nageswari *et al.* (1999) studied the effect of biofertilisers on rooting of cinnamon and found that phosphobacteria (soil application + dipping the cutting in phosphobacteria containing slurry) application gave a significantly higher rooting percentage (63.3%), longer roots (8.2 cm) and a greater number of roots (2.3) per cutting. Nageswari *et al.*

62 *P.N. Ravindran* et al.

Table 2.9 Variability in rooting response among cinnamon genotypes

Elite line	Rooting (%)		
	I year	II year	Mean
SL-5	58.4	62.6	60.5
	(49.8)	(52.3)	
SL-44	18.6	32.5	25.5
	(25.5)	(34.8)	
SL-53	3.4	10.4	6.9
	(10.5)	(18.7)	
SL-63	7.8	18.6	13.2
	(16.2)	(25.6)	
SL-65	1.7	8.6	5.2
	(7.5)	(17.1)	
In-189	14.9	26.5	20.7
	(22.6)	(30.9)	
In-203	10.8	21.8	16.4
	(19.2)	(27.8)	
In-310	12.0	16.3	14.2
	(20.2)	(23.8)	
In-312	0.0	6.6	3.3
	(0.6)	(14.9)	
CD at 5%	18.4	13.5	

Source: Rema and Krishnamoorthy, 1993.

Note
Figures within parenthesis are transformed values.

(2000) also reported that the use of IAA (100 and 500 ppm) gave better rooting (50%) in hardwood cuttings.

Air layering

Air layering is also a successful method of vegetative propagation. Semi-hardwood shoots are suitable for this purpose. Banerjee *et al.* (1982) found that application of 100 ppm gallic acid during air layering resulted in 80% rooting. Hegde *et al.* (1989) obtained rooting in non-girdled shoots treated with 2500 ppm NAA. IBA 3000 ppm when used in semi-hardwood cutting resulted in 70% rooting (NRCS, 1990). Ranaware *et al.* (1995) tried different rooting media and sphagnum moss was found to be the best. They also reported seasonal variations in rooting of air layers; 80% in July, 65% in June and no rooting during January and February (Ranawere *et al.* 1995).

Cassia

Cassia is propagated through seeds in the major growing countries like China and Vietnam, and as a result wide variability exists in the progenies. Trials were conducted at IISR, Calicut, to evolve a successful clonal method for propagating a few trees in India, which were introduced many years ago. Two methods have been standardised. Rooting of semi-hardwood cuttings of cassia was achieved by treating with IBA 500,

1000, 2000 ppm. Much genotypic variability was noted in the rooting response (IISR, 1996; Rema *et al.*, 1997). Air layering was also successful and 88% and 50% rooting have been reported during July and November, respectively (Krishnamoorthy and Rema, 1994).

Micropropagation of cinnamon and cassia

Micropropagation is useful for clonal multiplication of selected elite lines of cinnamon to augment the conventional vegetative propagation methods. It is also helpful to circumvent seedling variability. The recent status of cinnamon micropropagation has been reviewed by Nirmal Babu *et al.* (2000). Rai and Jagdishchandra (1987) obtained production of multiple shoots using seedling explants. The protocol developed by these workers started with seeds. Multiple shoots were induced from seeds on MS medium supplemented with Benzyladenine (BA) or kinetin. The shoots were multiplied by subculturing hypocotyl and internode segments on a second medium supplemented with BA or kinetin in combination with NAA. The *in vitro* generated shoots were rooted in Whites solution containing a combination of IAA, IBA and IPA. Successful micropropagation of Chinese cassia was reported by Inomoto and Kitani (1989) who used nodal explants from seedlings and MS medium containing low concentrations of BA and NAA, transferring the explants to medium containing BA alone, for shoot elongation. The elongated shoots were rooted in a third medium supplemented with only NAA. Mathai *et al.* (1997) reported the production of multiple shoots from mature tree explants of cinnamon. They found that MS medium or WP medium fortified with 0.5 mg/l kinetin and 2% sucrose was ideal for the initial establishment of the cultures. This initial establishment was followed by transferring the culture to WPM supplemented with 3 mg/l BAP and 1 mg/l kinetin (Fig. 2.20). The regenerated shoots were transferred to WPM supplemented with NAA, IBA and activated charcoal for rooting. WPM medium supplemented with 2 mg/l activated charcoal and without any growth regulators was reported to be the best for rooting (Mathai *et al.*, 1997). The addition of auxins induced callus formation.

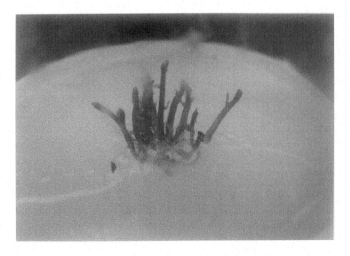

Figure 2.20 Micropropagation in *Cinnamomum verum*.

Figure 2.21 Somatic embryogenesis in *Cinnamomum verum*.

When young seeds are inoculated on WPM supplemented with 0.5 mg/l kinetin, about 20–30% cultures showed production of somatic embryos as well as embryogenic calli from cotyledonary axis. Somatic embryos mature in the same medium and about 5% of them develop into plantlets (Fig. 2.21) (Mathai *et al.*, 1997). Sheeja *et al.* (2000) reported micropropagation from nodal explants of cinnamon in WPM supplemented with benzyl adenine (2 mg/l) and kinetin (0.5 mg/l), followed by multiple shoot induction in WPM supplemented with BA (3 mg/l) and kinetin (1 mg/l). The excised healthy shoots were given a pulse treatment of IBA (3000 ppm) for 15 seconds and then cultured in WPM containing IAA (0.5 mg/l) and IBA (0.5 mg/l) for rooting. Nirmal Babu *et al.* (1997) have provided the protocol for micoporpagation of cinnamon from mature trees.

Micropropagation protocols for *C. camphora* were developed by Huang *et al.* (1998) and Nirmal Babu *et al.* (1997). Nirmal Babu *et al.* (1997) got multiple shoots from shoot tips and nodal explants of camphor tree on Woody Plant Medium (WPM) supplemented with BAP and kinetin. Nodal segments developed from shoots *in vitro* could be induced to produce a large number of harvestable shoots in the same medium. Harvested shoots are rooted *in vitro* in WPM supplemented with activated charcoal and IBA. Plantlets are subsequently hardened and planted out in the nursery with 90% survival rate. Huang *et al.* (1998) have evolved a protocol for camphor micropropagation. They used MS medium supplemented with different concentrations of BA and thidiazuron (TDZ). They also used NAA for rooting and a commercial formulation (EM_2) for the prevention of hyperhydracity. BA stimulated shoot formation and callus development, whereas TDZ promoted only callus development. Rooting of shoots occurred best when supplemented with 0.5 μM NAA. Micropropagation of *C. cassia* was also reported by Nirmal Babu *et al.* (1997).

Other tissue culture studies

Yazaki and Okuda (1993) discussed the production of condensed tannins in tissue culture of *C. cassia*. They induced callus from petiole explants cultured in Linsmaier and Skoog's (LS) agar medium containing several combinations of growth regulators (1965). Callus tissues induced from petiole segments were further grown on LS medium containing 10^{-5} m IAA and 10^{-5} m 6-benzyl adenine as well as on a medium containing 10^{-4} m NAA and 10^{-6} m BA. Callus was further subcultured with 0.2% gellan gum at 25 °C in the dark. The clumps of callus tissues developed on HPLC analysis were found to contain epicatechin, procyanin B_2, procyanidin B_4, and procyanidin C_1, which are precursors of condensed tannins. The vanillin-HCL test showed the occurrence of a large amount of high-molecular condensed tannin in the callus cultures.

Physiological studies

Little is known about the physiology of cinnamon. Pathiratna *et al.* (1998) studied the shade effects on photosynthesis of cinnamon plants grown under four levels of growth irradiances ranging from 12% to 100% daylight. Photosynthetic rate at light saturation derived from the data fitted to a quadratic model showed an increase from 5.14 to 7.25 μmol m^{-2}s^{-1}. Light required for 50% light saturation of photosynthesis also increased with the increase of growth irradiance. The above authors found that light compensation point and dark respiration derived by fitting the data from 0 to 81 μmol m^{-2}s^{-1} photon flux density to a linear model, and dark respiration measured directly increased with the increase of growth irradiance. The plants under 12% daylight had the highest quantum efficiency. The percentage increase of photosynthetic rate from growth irradiance to that at light saturation was greatest at 12% daylight. The size of leaves, the leaf area per unit, fresh and dry weight of leaves and leaf area per unit dry weight of plants decreased with the increase in growth irradiance. These results also indicated the C_3 nature of photosynthesis and the characteristics of shade adapted photosynthesis in cinnamon.

In another study Pathiratna and Perera (1998) reported the growth and bark yield of cinnamon grown under the shade of rubber trees, having an incident solar radiation of 1170 μmol m^{-2}s^{-1}. The major yield components of bark yield (number of shoots per bush and mature shoot length) were highest in cinnamon bushes under 21% daylight. Bark thickness and average shoot diametre were higher under 100% and 53% daylight but significantly lower under 21% and 5% daylight. Bark yield per bush in plants under 21% daylight was significantly higher than under other light levels. Those under 5% daylight gave the lowest bark yield per bush and were on par with 100% and 53% daylight.

Bregvadze (1975) reported that frost resistance in *Cinnamomum* species is related to the degree of growth inhibitor activity in leaves; the greater the activity, the greater the resistance to frost. Thus selection for a high degree of growth inhibitor activity to increase frost resistance is possible.

Menon *et al.* (1993) used linear measurements for calculating leaf area. The model $Y = aL \times B$ (0.72 L $\times$ B) had given an R^2 value of 0.98, indicating that 98% of the variance in leaf area could be explained by this model. The computed leaf area was in close agreement with the actual area based on measurement.

Crop Improvement

Genetic resources, varieties

Based on leaf morphology, bark pungency, grittiness of the bark and leaves, eight different types of cinnamon are recognised by growers in Sri Lanka (Wijesekera *et al.*, 1975; Anon, 1996). They are (in Sinhalese):

> *Pani Kurundu* or *Pat kurundu* or *Mapat kurundu*, *Naga kurundu*, *Pani Miriskurundu*, *Weli kurundu*, *Sewala (sevel) Kurundu*, *Kahata kurundu*, *Penirasa kurundu*, *Pieris kurundu*.

In addition, there are 19 selections, identified after screening 210 accessions by the Department of Export Agriculture of Sri Lanka. These selections are being popularised. The Indian Institute of Spices Research (IISR) at Calicut (Kerala, India) maintains 300 accessions of cinnamon and related taxa. This collection comprises lines from the cinnamon estates of Ancharakandi (Kannur District, Kerala; one of the oldest cinnamon estates in Asia, established by the British in nineteenth century and raised from seeds brought from Ceylon) the Mangalamcarp estate of Wynad (Kerala) and plants raised from selected trees of a few other collections existing in Kerala. Twelve lines introduced from Sri Lanka during the 1970s also form part of the collection of IISR. Apart from cinnamon, 35 lines of Chinese cassia (*C. cassia*) developed from the open pollinated progenies of cassia trees introduced during the early 1950s from China and maintained at the Srikundra Estate (under Brook Bond Tea Ltd.) at Valparai (Tamil Nadu), are also being conserved. At the Aromatic and Medicinal Plants Research Station of Kerala Agricultural University, Odakkali, Kerala, 236 lines of cinnamon are being maintained. These lines are mainly derived from seed progenies of the material maintained at Ancharakandi. The collections at IISR and KAU (as well as other collections derived from these sources) have been analysed for genetic variability with a view to develop better lines. Flavonoid analysis (Shylaja, 1984; Ravindran *et al.*, 1992) indicated absolute similarity between *C. verum* collections cultivated in India and those introduced from Sri Lanka, illustrating the common origin of both.

In India, at IISR, 291 lines of cinnamon (all originally raised from open pollinated seeds from selected mother plants) were evaluated for quality characteristics such as bark oil, oleoresin and leaf oil and nine elite lines were identified. Clonal progenies of some of these lines with high quality parameters were evaluated in replicated trials

Table 2.10 Yield and quality characteristics of elite cinnamon lines

Lines	Shoot production/ year	Fresh bark yield (gm)	Dry bark yield (gm)	Bark recovery %	Bark oil %	Bark oleoresin %	Leaf oil %
SL 53	15.60	256.05	100.0	33.0	2.8	10.0	3.0
SL 63 (*Navashree*)	25.5	489.0	201.0	40.0	2.7	8.0	2.8
SL 65	23.0	469.0	187.0	37.0	1.0	8.6	2.7
IN 189 (*Nithyashree*)	18.9	511.0	195.0	31.7	2.7	10.0	3.0
IN 203	18.2	314.0	123.0	31.0	2.9	9.0	1.7

Source: Krishnamoorthy *et al.*, 1996.

Table 2.11 Bark and leaf oil constituents of elite cinnamon lines

Lines	Bark oil			Leaf oil			Bark oleoresin
	%	CA%	Eg	%	Eg%	CA	%
SL 53	2.8	68.0	6.5	3.0	75.0	15.0	10.0
SL 63 (*Navashree*)	2.7	73.0	6.0	2.8	62.0	15.0	8.0
SL 65	1.0	NA*	NA	2.8	NA	NA	9.0
IN 189 (*Nithyashree*)	2.7	58	5.0	3.0	78.0	14.0	10.0
IN 203	2.9	NA	NA	1.7	NA	NA	9.0

Source: Krishnamoorthy *et al.*, 1996.

Notes
CA – Cinnamaldehyde; Eg – Eugenol; * not analysed.

(Tables 2.10, 2.11). Four of the lines were poor in establishment and subsequently discarded. SL 63 and IN 189 were finally selected based on regeneration capacity, fresh bark yield, dry bark yield, leaf oil, percentage eugenol in leaf oil and cinnamaldehyde in bark oil, etc. (Krishnamoorthy *et al.*, 1996). These lines, named as *Navashree* (SL 63) and *Nithyashree* (IN 189), were released for cultivation and are being popularised in India. The yield and quality characteristics of these selections are given in Tables 2.10 and 2.11.

Haldankar *et al.* (1994) screened 300 seedlings of cinnamon collected from IISR, Calicut, for isolating promising genotypes and four promising selections from these were further tested in field trials for yield and quality characteristics. One selection, B-iv, exhibited the highest yield of fresh and dried bark (289.7 g and 84.5 g, respectively), bark oil (3.2%) with a good percentage of cinnamaldehyde (70.2%). Leaf oil content was 2.28%, having 75.5% eugenol. This was released as *Konkan Tej* for the Konkan region in Maharashtra areas of India.

Pugalendhi *et al.* (1997) reported a cinnamon selection named YCD-1 from open pollinated seedlings. This selection is reported to be suitable for cultivation in the hill regions of Tamil Nadu at an altitude of 500–1000 m above msl. Its yield potential is 360 kg/ha dry bark, with a yearly regeneration capacity of 19.2 harvestable shoots. The bark recovery is 35.0% and the volatile oil contents of bark and leaves is 2.8 and 3%, respectively. At the Regional Research Laboratory, Bhubaneshwar, a population of 2500 plants (seed-propagated progenies) were analysed for quality characteristics and 20 selected elite plants were propagated vegetatively and field evaluated. The oil content in six-year old plants of these selections ranged from 0.2% to 1.2% (mean 0.78%), while in seven-year old plants the range was 0.60–1.4% (mean 0.90%). Leaf eugenol ranged from 73.63% to 92.19% (sixth year) and 90.8–95.7% (seventh year). Based on the evaluation results, RRL (B) C-6 was selected as the most promising (having 94% eugenol in leaf oil and 83% cinnamic aldehyde in the bark oil). This line was released for commercial cultivation (Paul *et al.*, 1996; Paul and Sahoo, 1993; Sahoo *et al.*, 2000).

Joy *et al.* (1998) evaluated 234 accessions of cinnamon maintained at the Aromatic and Medicinal Plants Research Station, Odakkali, based on growth, yield and quality parameters. They identified three superior accessions (ODC-130, ODC-10 and

68   *P.N. Ravindran* et al.

ODC-67) as the most promising. The best accession (ODC-130) had given 18.34 kg fresh leaf per tree per year, 294.7 ml leaf oil per tree per year, an oil recovery of 1.6% (fresh weight basis −3.73% on dry weight basis) and a eugenol content of 93.7% in the oil. The eugenol yield per tree per year was 275.1 ml. This line was released under the name *Sugandhani*, exclusively for leaf oil production purpose.

Variability and association

Joy *et al.* (1998) investigated the genetic variability among 234 accessions of cinnamon, maintained at the Aromatic and Medicinal Plants Research Station (Odakkali). Based on quality and yield parameters 50 superior genotypes (trees) were selected and evaluated further. The extent of variability for various characteristics can be understood from Tables 2.12 and 2.13. The large variability existing in the germplasm can be further exploited for developing superior genotypes. Among the cinnamon population investigated, 14% had deep purple flushes, 72% medium coloured flushes, and 14% light coloured or green flushes. Leaves were small to medium in 46%, medium to large in 22%, and small to large in 32%. Tree canopy was compact in 64%, semicompact in 24%, loose in 12%; canopy shape was spherical in 82%, semispherical in 12%, and linear in 6%. In a study of 239 cinnamon plants, Krishnamoorthy *et al.* (1988) and Gopalam (1997) reported that about 55% of trees had green flushes while in the rest the flushes had various degrees of purple colouration (such as purple dominated with green, green dominated with purple, deep purple, etc.). These workers also noted a

Table 2.12 Growth and yield parameters of the cinnamon accessions

Parameter	Range	Mean ± SE	CV (%)
Plant height cm	108.2–373.7	254.4 ± 3.312	19.92
Canopy spread cm	78.8–311.3	194.5 ± 3.055	24.03
Purple colour 0–9 score	1.5–8.5	4.37 ± 0.087	30.48
Fresh leaf yield kg/tree/yr.	0.425–18.335	7.587 ± 0.268	54.06
Dry leaf yield kg/tree/yr.	1.28–7.88	4.77 ± 0.224	33.21
Leaf oil yield ml/tree/yr.	36.45–294.69	154.12 ± 7.90	36.23
Oil recovery %, FWB	0.95–1.84	1.40 ± 0.030	15.42
Oil recovery %, DWB	2.20–4.29	3.24 ± 0.071	15.46
Eugenol yield ml/tree/yr.	80.53–95.03	90.29 ± 0.411	3.22
Eugenol yield ml/tree/yr.	31.75–275.11	138.53 ± 7.107	36.27
Yield index %	37.25–91.00	63.22 ± 1.543	17.26

Source: Joy *et al.*, 1998.

Table 2.13 Growth and yield parameters of the selected cinnamon accessions at AMPRS, Odakkali (Pooled mean of four years from 1992 to 1995)

Accession no.	Plant height (cm)	Canopy spread (cm)	Dry leaf yield (kg/tree)	Leaf oil yield (ml/tree/year)	Oil recovery FWB (%)	Oil recovery DWB (%)	Eugenol content (%)	Eugenol yield (ml/tree)	Cinnamon yield index (%)
ODC-10	313.8	255.0	7.40	234.53	1.36	3.18	91.36	231.87	80.25
ODC-24	325.8	262.5	5.59	187.04	1.33	3.40	90.18	169.39	69.50
ODC-28	260.0	240.0	4.96	168.77	1.46	3.41	93.74	158.03	66.75
ODC-32	338.3	282.5	7.77	229.93	1.27	2.96	87.86	198.21	75.25
ODC-37	300.8	267.5	4.99	168.78	1.49	3.49	92.44	155.37	66.25
ODC-38	318.8	293.8	6.34	231.02	1.64	3.82	88.78	203.46	79.25
ODC-39	373.8	311.3	6.67	184.71	1.24	2.89	85.52	155.43	68.00
ODC-40	299.5	255.0	5.67	200.10	1.47	3.43	92.80	182.33	72.25
ODC-45	301.3	247.0	6.21	215.49	1.46	3.41	89.29	193.19	74.50
ODC-67	286.3	269.8	6.46	239.84	1.56	3.64	87.21	209.71	78.00
ODC-72	342.5	228.0	5.21	182.50	1.52	3.53	88.73	163.03	69.50
ODC-74	288.3	241.3	5.54	178.37	1.41	3.29	91.62	162.45	69.00
ODC-91	297.5	239.0	6.02	173.96	1.24	2.88	89.34	155.49	66.75
ODC-99	287.8	227.8	4.68	159.07	1.48	3.45	90.48	144.10	65.75
ODC-101	365.0	265.5	6.14	216.66	1.50	3.49	90.72	197.59	74.75
ODC-111	323.8	237.5	5.77	195.03	1.46	3.40	87.21	170.20	72.50
ODC-128	242.8	219.5	4.23	159.68	1.66	3.88	95.03	151.67	68.75
ODC-130	319.0	291.3	7.88	294.69	1.60	3.73	93.67	275.11	91.00
ODC-133	362.0	266.5	7.33	232.21	1.34	3.14	89.11	206.38	77.00
ODC-147	306.3	278.3	7.25	204.96	1.19	2.78	91.33	186.49	72.25
ODC-183	306.0	247.8	6.11	205.24	1.57	3.68	88.52	181.44	72.25
ODC-197	295.0	257.5	4.73	176.20	1.55	3.61	90.35	158.70	66.50
ODC-209	299.8	245.0	6.41	212.70	1.44	3.36	63.88	199.95	76.50
ODC-224	288.5	263.8	5.15	205.24	1.67	3.89	80.53	162.50	69.75
Mean	292.8	238.1	4.77	154.12	1.39	3.24	90.25	138.53	63.20
CD 0.05	29.53	29.73	1.50	60.64	0.31	0.72	4.10	53.60	10.26

Source: Joy *et al.*, 1998.

Note
AMPRS: Aromatic and Medicinal Plants Research Station.

correlation between flush colour and quality – the purple coloured plants having more bark oil (about 29% more).

Correlation and path analysis studies conducted by Joy et al. (1998) indicated that the economic yield characteristics (fresh leaf yield, leaf oil yield and eugenol yield) were highly correlated among themselves, the values ranging from 0.92 to 0.99. The component characteristics, plant height and canopy spread, were highly and positively correlated between themselves as well as to the three economic characteristics mentioned above. On this basis of high correlation, plant height, canopy spread, fresh leaf yield, leaf oil yield and eugenol yield were used for working out multiple regression and path coefficients. High coefficients of determination existed for all equations, and values ranged from 0.72 to 0.99. However, the R^2 value for leaf oil yield on plant height, canopy spread and fresh leaf yield did not show improvement over the correlation between leaf oil yield and fresh leaf yield ($\gamma = 0.93$). Similar was the case with the multiple regression equation of eugenol yield on plant height and leaf oil yield ($\gamma = 0.99$). The above workers suggested that variability in eugenol yield was determined more by the indirect effects rather than by the direct effect of plant height, canopy spread or fresh leaf yield. Similarly, the variability in leaf oil yield was determined more by the indirect effects than by the direct effect of plant height or canopy spread. The path diagram suggested by the above workers (Joy et al., 1998) is given below. The characteristic, that is ultimately important is the eugenol yield. There is a unidirectional relationship among fresh leaf yield → leaf oil yield → eugenol yield as indicated by the high path coefficients. Canopy spread is directly linked to plant height, but is poorly associated with fresh leaf yield.

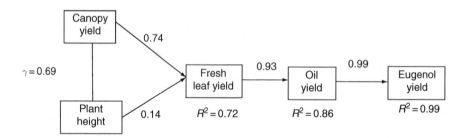

Therefore, selection for improvement in oil or eugenol yield, which requires cumbersome procedures, can be done indirectly using leaf yield, which in turn is related closely to canopy spread.

Krishnamoorthy et al. (1988) observed significant variability for bark oil content in cinnamon germplasm. Krishnamoorthy et al. (1991) also reported significant variation in progeny performance of nine lines for plant height, number of branches per tree, fresh and dry weight of bark and percentage recovery of bark. Krishnamoorthy et al. (1992) also studied the variability and association studies in 71 cinnamon germplasm accessions maintained in the germplasm conservatory of IISR, Calicut (Table 2.14). Among the characteristics, the highest variation was for dry weight of bark followed by fresh weight of bark, bark oleoresin and leaf oil. Moderate variability was noted for bark oil and leaf size index. High association existed between fresh weight of bark and leaf oil with dry weight of bark. Ponnuswamy et al. (1982) also conducted some studies on variability among seedling progenies of cinnamon. In field plantations of

Table 2.14 Mean, range and coefficient of variation for nine characteristics in cinnamon

Character	Mean	Range	W (%)
Leaf length (cm)	13.08	8.75–20.69	17.83
Leaf breadth (cm)	5.06	3.31–8.30	18.74
Leaf size index	0.67	0.29–1.71	35.13
Fresh weight of bark (g)	207.41	30.0–840.0	67.19
Recovery of bark (%)	32.08	10.70–80.0	34.00
Bark oleoresin (%)	8.48	1.32–20.02	57.08
Bark oil (%)	1.81	0.51–3.85	36.28
Leaf oil (%)	1.97	0.72–4.80	48.46
Dry weight of bark (g)	64.70	8.00–305.00	74.73

Source: Krishnamoorthy *et al.*, 1992.

cinnamon in the Orissa state, Paul and Sahoo (1993) recorded wide variations in many characteristics such as plant height (2.17–3.30 m), stem girth (7–16.6 cm), leaf oil (0.38–1.80%), eugenol in leaf oil (traces to 80–98%) and bark oil (0.05–2.18%). They also noticed plants having very low eugenol and high benzyl benzoate content ranging from 2.3% to 66.0%. The presence of benzyl benzoate and eugenol showed a negative relationship (Rao *et al.*, 1988; Paul and Sahoo, 1993; Paul *et al.*, 1996; Sahoo *et al.*, 2000).

Lin *et al.* (1997) reported allozyme variation in *C. kanchirae*, a Taiwanese endangered species. Genetic diversity within and genetic differentiation among four geographic areas were investigated using 164 clones. Seven out of 11 loci examined were polymorphic. The mean, expected heterozygous loci per individual ranged from 13.9% to 21.6%, the number of alleles per locus ranged from 1.7 to 1.9, and the effective number of alleles per locus ranged from 1.34 to 1.54 at the area level. The large seed size and insect pollination impose a barrier to free gene flow, according to the above studies.

Improvement of Chinese cassia

The crop improvement programme in Chinese cassia aims mainly at: (1) selection of clones with high quality (high essential oil content having high cinnamaldehyde); (2) high bark recovery; and (3) high regeneration capacity and growth. With these objectives, cassia germplasm maintained at IISR were evaluated for morphological and quality parameters (Krishnamoorthy *et al.*, 1999, 2001) (Tables 2.15, 2.16).

Table 2.15 Quality characteristics of four elite lines of cassia

Sl. no.	Character	Range	Mean	CV%	CD (P:0.05)
1	Bark oil (%)	1.2–4.9	3.09 ± 0.59	26.79	1.71
2	Leaf oil (%)	0.4–1.6	1.08 ± 0.21	26.79	0.60
3	Bark oleoresin (%)	6.0–10.5	8.40 ± 1.13	18.98	3.29
4	Cinnamaldehyde (%) (bark oil)	61.5–91.0	77.2 ± 6.31	11.5	18.42

Source: Krishnamoorthy *et al.*, 1999.

72 P.N. Ravindran et al.

Table 2.16 Evaluation of chinese cassia accessions for quality

Acc. no.	Oil %	Bark		Oil %	Leaf
		Oleoresin %	Cinnamaldehyde %		cinnamaldehyde %
A_1	3.75	10.20	81.50	1.00	80.50
A_2	2.95	8.30	88.00	0.61	57.00
A_6	2.80	9.75	61.50	1.60	71.00
A_7	2.70	9.80	70.50	1.00	55.00
B_1	2.80	7.55	78.50	1.10	80.60
B_2	2.65	6.00	81.00	0.60	71.60
B_3	1.20	8.75	70.00	0.65	60.00
B_4	2.80	9.00	80.50	1.10	67.40
B_5	3.75	7.70	71.50	1.35	58.00
B_6	3.85	8.30	76.50	0.80	66.50
C_1	3.45	10.50	86.50	0.50	76.50
C_2	3.35	9.00	72.00	0.80	65.00
C_3	2.25	7.90	73.50	1.45	73.00
C_4	3.50	7.70	71.00	1.45	86.00
C_5	3.35	7.50	66.50	1.45	70.50
C_6	3.10	9.35	68.00	1.35	71.50
C_7	2.90	7.45	87.50	1.35	60.00
C_8	3.25	7.15	87.00	1.05	62.20
D_1	4.70	9.30	86.50	0.95	76.00
D_2	2.70	9.70	91.00	1.25	40.70
D_3	4.90	8.65	90.50	1.15	75.00
D_4	2.45	8.10	62.00	0.04	71.50
D_5	4.25	7.00	85.50	1.45	58.00
D_6	2.55	9.20	79.50	1.40	53.50
D_7	1.30	9.15	64.00	0.82	77.50
Mean ± SE	3.09 ± 0.59	8.40 ± 1.13	77.22 ± 6.31	1.08 ± 0.21	67.38 ± 6.94
Range	1.2–4.9	6.0–10.5	61.5–91.0	0.4–1.6	4.07–86.0
CV%	26.79	18.98	11.55	26.74	14.57
CD (P = 0.05)	1.71	3.29	18.42	0.60	20.26

Source: Krishnamoorthy *et al.*, 2001.

Bark oil in various lines ranged from 1.2% to 4.9%, leaf oil 0.4% to 1.6% and bark oleoresin 6.0% to 10.5%. The cinnamaldehyde content of leaf oil ranged from 40% to 86%, and that of bark oil 61% to 91%. The coefficient of variation was high for bark and leaf oil (26.76% and 26.74% respectively). (In this trial cassia plants were maintained and harvested similarly to Sri Lankan cinnamon and hence may not be comparable with the practices prevalent in the producing countries).

Based on quality and other characteristics four promising lines were identified. Two lines (A_1 and C_1) have high bark oleoresin (10.2% and 10.5%, respectively) and two others (D_2 and D_3) have high bark oil (4.7% and 4.9%, respectively) and high cinnamaldehyde in bark oil (91.0% and 90.5%, respectively). Taking the overall yield, chemical and flavour profiles, C_1, D_1, and D_3 were selected and are in pre-release yield evaluation trials (Krishnamoorthy *et al.*, 1999, 2001).

Little crop improvement work has gone into the Chinese cassia in China or Vietnam, where the production is located. This is true of the Indonesian cassia as well. In all these countries cassia cinnamon is treated, and is managed and exploited as a forest tree crop.

References

Allen, C.K. (1939) Cinnamomum In. Cinnamomum and Neocinnamomum, *J. Arnold Arboretum*, **20**, 52–53.

Anonymous (1950) *The Wealth of India*, CSIR, New Delhi Vol. III.

Anonymous (1996) *Cinnamon* Tech. Bull., Dept. of Export Agri., Sri Lanka.

Avita, Sr. and Inamdar, J.A. (1981) Stomatal complex in Lauraceae: structure and ontogeny, *Acta Bot. Indica*, **9**, 50–56.

Bakker, M.E. and Gerritsen, A.F. (1989) A suberized layer in the cell wall of mucilage cells of *Cinnamomum. Ann. Bot.*, **63**, 441–448.

Bakker, M.E., Gerritsen, A.F. and Van Der Schaaf, P.J. (1991) Development of oil and mucilage cells in *Cinnamomum burmannii* – an ultrastructural study. *Acta Bot. Neerl.*, **40**, 339–356.

Bakker, M.E., Gerritsen, A.F. and Van Der Schaaf, P.J. (1992) Leaf anatomy of *Cinnamomum* Schaeffer (Lauraceae) with special reference to oil and mucilage cells. *Blumea*, **37**, 1–30.

Balasubramanian, A., Jacob, T. and Saravanan, S. (1993) Petiolar anatomy as an aid to the identification of *Cinnamomum* species (Lauraceae). *Indian Forester*, **119**, 583–586.

Bamber, R.K. and Summerville, R. (1979) Taxonomic significance of sclerified tissue in the bark of Lauraceae. *IAWE Bull*, No. 4, 69–75.

Banerjee, D.P., Chatterjee, B.K. and Sen, S. (1982) Air layering of Cinnamon (*Cinnamomum zeylanicum* Nees.). *South Indian Hort.*, **30**, 272–273.

Baruah, A. and Nath, S.C. (1997) Foliar epidermal characters in twelve species of *Cinnamomum* Schaeffer (Lauraceae) from northeastern India. *Phytomorphology*, **47**, 127–134.

Baruah, A. and Nath, S.C. (1998) Diversity of *Cinnamomum* species in north-east India a micromorphological study with emphasis to venation pattern. In A.K. Goel, V.K. Jain and A.K. Nayak (eds) *Modern Trends in Biodiversity*; Jaishree Prakashan, Muzaffernagar, pp. 147–167.

Baruah, A., Nath, S.C. and Boissya, C.L. (1999) Taxonomic discrimination amongst certain chemotypes of *Cinnamomum sulphuratum* Nees with emphasis to foliar micromorphology. *J. Swamy Bot. Club*, **16**, 3–7.

Baruah, A., Nath, S.C. and Boissya, C.L. (2000) Systematics and diversities of Cinnamomum species used as Tejpat spice in north-east India. *J. Econ. Tax. Bot.*, **24**, 361–374.

Baruah, A. and Nath, S.C. (2000) Certain chemomorphological variants of *Cinnamomum bejolghota* (Buch Ham) Sweet and their relevance to foliar micromorphology. *J. Swamy Bot. Cl.*, **17**, 19–23.

Bhandari, J. (1996) Effect of phytohormones on seeds and seedlings of *Cinnamomum camphora*. *Indian Forester*, **122**, 767–769.

Birch (1963) cited from Tetenyi (1970).

Birnstiel, W. (1922) cited from Metcalfe and Chalk (1950).

Bourdillon, T.F. (1908) *The Forest Trees of Travencore*, Reprint 1988, Bishensigh Mahendrapalsingh, Dehra Dun.

Bregvadze, M.A. (1975) The activity of endogenous growth regulators in *Cinnamomum* spp. in relation to frost hardiness. *Soob. Akad. Nauk. Gruzinskoi*, 79(1): 153–156 (cited from Weiss 1997).

Brewbaker, J.L. (1967) The distribution and phylogenetic significance of binucleate and trinucleate pollen grains in the angiosperms. *American J. Bot.*, **54**, 1069–1083.

74   *P.N. Ravindran* et al.

Burkill, I.H. (1935) *Dictionary of the Economic Products of the Malay Peninsula*. Crown Agents for the colonies, London.

Chalk, L. (1937) The phylogenetic value of certain anatomic features of dicotyledonous woods. *Ann. Bot.*, 1, 409–427.

Choudhury, J.K. and Mitra, J.N. (1953) Abnormal trichotyledonous embryo and the morphological structure of normal fruit and seed of *Cinnamomum camphora* Nees. *Sci & Cult.*, 19, 159–160.

Chaudhuri, R.H.N. and Kayal, R.N. (1971) Pharmacognostic studies on the stem barks of four species of *Cinnamomum*. *Bull. Bot. Surv. India*, 13, 94–104.

Cheniclet, C. and Carde, J.P. (1985) Presence of leucoplasts in secretory cells and monoterpenes in the essential oil: a correlative study. *Israel J. Bot.*, 34, 219–238.

Christophel, D.C., Kerrigan, R and Rowett, A.I. (1996) The use of cuticular features in the taxonomy of the Lauraceae. *Ann. Missouri Bot. Garden*, 83, 419–432.

Chuang, T.I., Chao, C.Y., Hu, W.W. and Kwan, S.C. (1963) Chromosome numbers of the vascular plants of Taiwan. *Taiwania*, 1, 357–375.

Coppen, J.J.W. (1995) *Flavours and Fragrances of Plant Origin*. FAO, Rome.

Corner, E.J.H. (1976) *The Dicotyledonous Seed*, Vol. I. Cambridge Uni. Press.

CPCRI (1985) Annual Report for 1984, Central Plantation Crops Research Institute, Kasargod, India.

Crawford, D.J. (1978) Flavanoid chemistry and angiosperm evolution. *Bot. Rev.*, 44, 431–456.

Dadswell, H.E. and Eckersley, A.M. (1940) The wood anatomy of some Australian Lauraceae with methods for their identification. CSIR Australia, Division of Forest Products, Tech. P. No. 34.

Dassanayaka, M.D. (1995) *A Revised Handbook to the Flora of Ceylon*. Oxford & IBH, New Delhi (Revised edn.) pp. 112–129.

Eames, A.J. (1961) *Morphology of Angiosperms*, Mc Graw Hill Pub. Co. USA.

Erdtman, O.G.E. (1952) *Pollen Morphology and Plant Taxonomy – Angiosperms*. Almsquist Wicksell, Stockholm.

Fujita, Y. (1967) Classification and phylogeny of the genus *Cinnamomum* in relation to the constituent essential oils. *Bot. Mag. Tokyo*, 80, 261–271.

Gamble, J.S. (1925) *Flora of the Presidency of Madras*, Vol. II. Botanical Survey of India, Calcutta (Reprint).

Gildmeister, E. and Hoffmann, Fr. (1959) cited from Purseglove *et al.* (1981).

Giviliani (1928) Cited from Sastri (1958).

Gopalam, A. (1997) Profile of essential chemical constituents in tree spices. In S.S. Handa and M.K. Kaul (eds) *Supplement to Cultivation and Utilization of Aromatic Plants*, RRL, (CSIR), Jammu; pp. 405–423.

Guang-Fu, T. and Yang, Z. (1988) Study on numerical chemotaxonomy of *Cinnamomum* in Hubei province. *Acta Bot. Sinica*, 26, 409–417.

De Guzman, C.C. and Siemonsma, J.S. (1999) *Plant Resources of South East Asia, No. 13, Spices*. Backhuys Pub., Leiden.

Haldankar, P.M., Nagwekar, D.D., Patil, J.L. and Gunjate, R.T. (1994) Varietal screening for yield and quality in cinnamon. *Indian Cocoa, Arecanut & Spices J.*, 18(3), 79–81.

Hegde, K.R., Sulikeri, G.S. and Hulamani, N.C. (1989) Effect of growth regulators and pre-girdling treatments on rooting of cinnamon (*Cinnamomum verum* Presl.) air layers. *South Indian Hort.*, 37, 329–332.

Heslop-Harrison, Y., and Shivanna, K.R. (1977) The receptive surface of the angiosperm stigma. *Ann. Bot.*, 41, 1233–1258.

Hirota, N. (1951) In Mem. Ehime Univ., Sect. 2, Sci.1(2): 83–106 (cited from Kostermans (1964).

Hirota, N. (1953) In *Perf. Essent. Oil Rec.*, 44, 4–10 (cited from Kostermans 1964).

Hirota, N. and Hiroi, M. (1967) Later studies on the camphor tree, on the leaf oil of each practical form, and its utilization. *Perf. Ess. Oil Rec.*, 58, 364–367.

Botany and Crop Improvement of Cinnamon and Cassia 75

Hooker, J.D. (1886) *Flora of the British India*, Vol. V. Reeve & Co., London (Reprint), pp. 78–95.

Hong, L. and Yetong, C. (1997) Regeneration of bark of *Cinnamomum cassia* Presl. after girdling. *J. Plant Resources and Environment*, 6(3), 1–7.

Huang, L.C., Huang, B.L. and Murashige, T. (1998) A micropropagation protocol for *Cinnamomum camphora*. *In vitro Cellular and Developmental Bio. – Plant*, 34, 141–146.

IISR (1996) Indian Institute of Spices Research, Annual Report for 1995–96, IISR, Calicut, pp. 17–20.

Inomoto, Y. and Kitani, Y. (1989) In vitro propagation of *Cinnamomum cassia*. *Plant Tissue Cul. Lett.*, 6, 25–27.

Jackson, B.P. and Snowdon, D.W. (1990) *Atlas of Microscopy of Medicinal Plants, Culinary Herbs and Spices*. Belhaven Press, UK.

Joseph, J. (1981) Floral biology and variation in cinnamon. In S. Vishveshwara (ed.) *Proc. PLACROSYM IV*, ISPC, CPCRI, Kasaragod, India, pp. 431–434.

Joy, P.P., Thomas, J., Mathew, S. and Ibrahim, K.K. (1998) Growth, leaf oil yield and quality investigations in Cinnamon (*Cinnamomum verum*). *J. Med. Aromatic Plants*, 28, 401–406.

Kannan, K. and Balakrishnan, S. (1967) A note on the viability of cinnamon seeds. *Madras Agric. J.*, 54, 78–79.

Kasapligal, B. (1951) Morphological and ontogenetic studies on *Umbellularia californica* Nutt and *Laurus nobilis*, L. *Univ. California pub. Bot.*, 25, 115–240.

Khein, P.V., Chien, H.T., Duang, N.X., Leclercq, A.X. and Leclercq, P.A. (1998) Chemical segregation of progeny of camphor trees with high camphor c.q. linalool content. *J. Essent. Oil Res.*, 10, 607–612.

Kim, K.S. and Kim, M.H. (1984) Systematic studies on some Korean (R.) woody plants-venation pattern of Lauraceae. *Korean J. Bot.*, 27, 15–24.

Kostermans, A.J.G.H. (1957) Lauraceae. *Reinwardtia*, 4, 193–256.

Kostermans, A.J.G.H. (1961) The New World Species of *Cinnamomum* Trend (Lauraceae). *Reinwardia*, 6, 17–24.

Kostermans, A.J.G.H. (1964) *Bibliographia Lauracearum*, Bogor Bot. Gardens, Djakarta.

Kostermans, A.J.G.H. (1980) A note on two species of *Cinnamomum* (Lauraceae) described in *Hortus Indicus Malabaricus*. In Manilal, K.S. (ed.) *Botany and History of Hortus Malabaricus*, Oxford IBH, New Delhi. pp. 163–167.

Kostermans, A.J.G.H. (1983) The South Indian species of *Cinnamomum* Schaeffer (Lauraceae). *Bull. Bot. Surv. India*, 25, 90–133.

Kostermans, A.J.G.H. (1986) A monograph of the genus *Cinnamomum* Schaeffer (Lauraceae) Part I. *Ginkgoana*, 6, 1–71.

Kostermans, A.J.G.H. (1995) Lauraceae. In M.D. Dasanayake, F.R. Fosberg and W.D. Clayton (eds) *A Revised Handbook of the Flora of Ceylon*, Vol. 9. Amerind Pub. Co; New Delhi, pp. 112–115.

Krishnamoorthy, B., Gopalam, A. and Abraham, J (1988) Quality parameters of cinnamon (*Cinnamomum verum*) in relation to flush colour. *Indian Cocoa, Arecanut and Spices J.* 12, 38.

Krishnamoorthy, B., Rema, J. and Sasikumar, B. (1991) Progeny analysis in cinnamon. *Indian Cocoa, Arecanut and Spices J.*, 14, 124–125.

Krishnamoorthy, B., Sasikumar, B., Rema, J., Gopalam, A. and Abraham, J. (1992) Variability and association studies in cinnamon (*Cinnamomum verum*) *J. Spices and Aromatic Crops*, 1, 148–150.

Krishnamoorthy, B. and Rema, J. (1994) Air layering in cassia cinnamon. (*Cinnamomum aromaticum* Nees) *J. Spices and Aromatic Crops*, 3, 48–49.

Krishnamoorthy, B., Rema, J., Zachariah, T.J., Abraham, J. and Gopalam, A. (1996) Navashree and Nithyashree – two new high yielding and high quality Cinnamon – (*Cinnamomum verum* Bercht & Presl.) selections. *J. Spices and Aromatic Crops*, 5, 28–33.

Krishnamoorthy, B., Sasikumar, B., Rema, J., George, G.K. and Peter, K.V. (1997) Genetic resources of tree spices and their conservation in India. *Plant Gen. Res. News Lett.*, III, 53–58.

Krishnamoorthy, B., Zachariah, T.J., Rema, J. and Mathews, P.A. (1999) Evaluation of selected Chinese cassia, *Cinnamomum cassia* accessions for chemical quality. *J. Spices and Aromatic Crops*, **8**, 215–217.

Krishnamoorthy, B., Zachariah, T.J., Rema, J. and Mathews, P.A. (2001) High quality cassia selections from IISR, Calicut. *Spice India*, 14(6), 2–4.

Kubitzki, K. and Kurz, H. (1984) Synchronised dichogamy and dioecy in neotropical Lauraceae. *Plant Syst. Evol.*, 147, 253–266.

Lang-Yang, S., Wei, H. and Guang-yu, W. (1989) Study on chemical constituents of the essential oil and classification of types from *Cinnamomum camphora*. *Acta Bot. Sinica*, **31**, 209–214.

Lin, T.P. (1996) Seed storage behaviour deviating from the orthodox and recalcitrant type. In Proc. International Seed Testing Association, ISTA Secretariat, Switzerland.

Lin, T.P., Cheng, Y.P. and Huang, S.G. (1997) Allozyme variation in four geographic areas of *Cinnamomum kanehirae*. *The J. Heredity* (USA), **88**, 433–438.

Linsmaier, E.M. and Skoog, F. (1965) Organic growth factor requirements of tobacco tissue cultures. *Physiol. Plant.*, **18**, 100–127.

Manilal, K.S. and Shylaja, M. (1986) A new species of *Cinnamomum* Schaeffer (Lauraceae) from Malabar. *Bull. Bot. Surv. India*, **28**, 111–113.

Maron, R. and Fahn, A. (1979) Ultrastructure and development of oil cells in *Laurus nobilis* leaves. *Bot. J. Linn-Soc.*, **78**, 31–40.

Mathai, M.P., Zachariah, J.C., Samsudeen, K., Rema, J., Nirmal Babu, K. and Ravindran, P.N. (1997) Micropropagation of *Cinnamomum verum* (Bercht & Presl.) In S. Edison, K.V. Ramana, B. Sasikumar, K. Nirmal Babu and S.J. Eapen (eds) *Biotechnology of Spices, Medicinal and Aromatic Plants*, ISS, IISR, Calicut, pp. 35–38.

Mehra, P.N. and Bawa, K.S. (1968) B-Chromosomes in some Himalayan hardwoods. *Chromosoma*, **25**, 90–95.

Mehra, P.N. and Bawa, K.S. (1969) Chromosomal evolution in tropical hardwoods. *Evolution*, **23**, 466–481.

Meissner (1864) In *DC Prodr.* 15(1), 1–260. Cited from Kostermans (1957).

Menon, R., Shylaja, M.R. and Mercy, K.A. (1993) Leaf area estimation in cinnamon (*Cinnamomum zeylanicum* L.). *South Indian Hort.*, **41**, 239–241.

Merrill (1920) Cited from Allen (1939).

Metcalfe, C.R. and Chalk, L. (1950) *Anatomy of Dicotyledons*. Clarenden Press, Oxford.

Mikage, M., Komatsu, K. and Namba, T. (1987) Fundamental studies on the evaluation of crude drugs. IX. Soft X-ray analysis of crude drugs. (1) Evaluation of the quality. *Yakugaku Zasshi*, **107**, 192–198.

Mohanakumar, G.N., Mokashi, A.N., Narayana Swamy, P., Prabhakar, N., Devar, K.V., Reddy, D.N.R. and Nalawadi, U.G. (1985) Studies on the floral biology of Cinnamon. *Indian Cocoa, Arecanut & Spices J.*, **8**(4), 100–102.

Nageswari, K., Pugalendhi, L. and Balakrishnamurthy, G. (1999) Effect of biofertilizers on rooting of Cinnamon (*Cinnamomum verum* Presl.). *Spice India*, Nov. 1999, 9–10.

Nageswari, K., Pugalendhi, L. and Azhakiamanavalan, R.S. (2000) Propagation studies in cinnamon (*Cinnamomum zeylanicum*). *Spice India*, Feb. 2000, 11–12.

Namba, T., Kikuchi, T., Mikage, M., Kadota, S., Komatsu, K., Shimizu, M. and Tamimori, T. (1987) Studies on the natural resources from Sri Lanka. (1) On anatomical and chemical differences among each grade of 'cinnamomi veri cortex'. *Shoyakugaku zasshi*, **41**, 35–42.

Nees, Th. (1836) *Systema Laurinarum* (cited from Kostermans, 1957).

Nirmal Babu, K., Ravindran, P.N. and Peter, K.V. (2000) Biotechnology of Spices. In K.L. Chadha, P.N. Ravindran and Leela Sahijram (ed.) Biotechnology in Horticultural and Plantation Crops, Malhotra Pub., New Delhi, pp. 487–527.

Nirmal Babu, K., Ravindran, P.N. and Peter, K.V. (1997) Protocols for micropropagation of Spices and Aromatic Crops. Indian Institute of Spices Research, Calicut, Kerala, p. 35.

NRCS (1990) National Research Centre for Spices, Ann. Report for 1989, NRCS, Calicut.

Okada, H. (1975) Karyomorphological studies on woody polycarpicae. *J. Sci. Hiroshima Uni.*, B(2), 15(2), 115–200.

Okada, H. and Tanaka, R. (1975) Karyological studies in some species of Lauraceae. *Taxon*, 24, 271–280.

Pal, S. (1974) Stomatogenesis in *Cinnamomum. Acta Bot. Indica*, (Suppl.) 171–173.

Pal, S. (1976) Studies in Lauraceae. Node in Lauraceae. *J. Indian Bot. Soc.*, 55, 226–229.

Pal, S. (1978) Epidermal studies in some Indian Lauraceae and their taxonomic significance. *Acta Bot. Indica*, 6, 68–73.

Parry, J.W. (1969) *Spices*, Vol. II. Chemical Pub. Co., New York.

Pathiratna, L.S.S. and Perera, M.K.D. (1998) The effect of shade on the bark yield components of Cinnamon (*Cinnamomum verum* J. Presl) inter-cropped with rubber (*Hevea brasiliensis* Muell Arg.). *J. Plantation Crops*, 26, 70–74.

Pathiratna, L.S.S., Nugawela, A. and Samarasekera, R.K. (1998) Shade effects on photosynthesis of cinnamon (*Cinnamomum verum* J. Presl). *J. Rubber Res. Inst. Sri Lanka*, 81, 29–37.

Paul, S.C. and Sahoo, S. (1993) Selection of elite cinnamon plants for quality bark production. *J. Eco. Tax. Bot.*, 17, 353–355.

Paul, S.C., Sahoo, S. and Patra, P. (1996) Elite *Cinnamomum zeylanicum* genotypes isolated from germplasm of India and Sri Lanka. *J. Plant Gen. Resources*, 9, 307–308.

Pearson, R.S. and Brown, H.P. (1932) *Commercial Timbers in India*, Vol. 2, Govt. of India, Central Pub. Branch, Calcutta.

Ponnuswamy, V., Irulappan, I., Annadurai, S. and Vadivel, E. (1982) Variability studies in cinnamon (*C. zeylanicum* Breyn). *South Indian Hort.*, 30, 159–160.

Pugalendhi, L., Nageswari, S.K. and Azhakiamanavalan, R.S. (1997) YCD-1 Cinnamon. *Spice India*, 10(4), 4.

Prance, G.T. and Prance, A.E. (1993) *Bark*, Timber Press, USA, pp. 84–88.

Rai, R. and Jagadishchandra, K.S. (1987) Clonal propagation of *Cinnamomum zeylanicum* Breyn. by tissue culture. *Plant Cell Tissue & Organ Culture*, 9, 81–88.

Ranaware, V.S., Nawale, R.N. and Khandekar, R.G. (1994) Effect of rooting media and wood maturity on rooting in air layering of cinnamon (*Cinnamomum zeylanicum* Blume). *Spice India*, 7(6), 19–21.

Ranaware, V.S., Nawale, R.N., Khandekar, R.G. and Magdun, M.B. (1995) Effect of season on air layering of cinnamon (*Cinnamomum zeylanicum* Blume). *Indian Cocoa Arecanut Spices J.*, 19, 81–84.

Rao, M.R. (1914) *Flowering Plants of Travancore*. Bishensingh Mahendrapal Singh, Dehra Dun. (Reprint).

Rao, Y.R., Paul, S.C. and Dutta, P.K. (1988) Major constituents of essential oils of *Cinnamomum zeylanicum. Indian Perfumer*, 32, 86–89.

Ravindran, S., Balakrishnan, R., Manilal, K.S. and Ravindran, P.N. (1991) A cluster analysis study on *Cinnamomum* from Kerala, India. *Feddes Repertorium*, 102, 167–175.

Ravindran, S., Krishnaswamy, N.R., Manilal, K.S. and Ravindran, P.N. (1992) Chemotaxonomy of *Cinnamomum* Schaeffer occurring in Western Ghats. *J. Indian Bot. Soc.*, 71, 37–41.

Ravindran, S., Manilal, K.S. and Ravindran P.N. (1993) Dermal morphology of *Cinnamomum* spp. and its taxonomic significance. *J. Pl. Anatomy and Morph.*, 6, 42–50.

Ravindran, S., Manilal, K.S. and Ravindran, P.N. (1996) Numerical taxonomy of *Cinnamomum* II. Principal component analysis of major species from Kerala, India. In K.S. Manilal and A.K. Pandey (eds) *Taxonomy and Plant Conservation*, IBH Pub. Co. New Delhi, pp. 321–332.

Reece, P.C. (1939) Floral anatomy of avocado. *Amer. J. Bot.*, 26, 429–432.

Rema, J. and Krishnamoorthy, B. (1993) Rooting response of elite cinnamon (*Cinnamomum verum* Bercht & Presl.) lines. *J. Spices and Aromatic Crops*, 2, 21–25.

78 *P.N. Ravindran* et al.

Rema, J., Krishnamoorthy, B. and Mathew, P.A. (1997) Vegetative propagation of major spices – a review. *J. Spices and Aromatic Crops*, 6, 87–105.

Rheede, H. von. (1685) *Hortus Indicus Malabaricus* Vol. I, Ametelodami, Holland.

Richards, A.J. (1986) *Plant Breeding Systems*. George Allen & Unwin, London.

Sahoo, S., Paul, S.C. and Patra, P. (2000) Quality cinnamon production in India. *J. Med. Aromatic Plant Sci.*, 22, 361–365.

Santos, J.K. (1930) Leaf and bark structures of some cinnamon trees with special reference to Philippine trees. *Philippine J. Sci.*, 43, 305–365.

Sastri, R.L.N. (1952) Studies in Lauraceae. 1. Floral anatomy of *Cinnamomum iners* Reinw. and *Cassytha filiformis* Linn. *J. Indian Bot. Soc.*, 31, 240–246.

Sastri, R.L.N. (1958) Studies in Lauraceae. II Embryology of *Cinnamomum* and *Litsea J. Bot. Soc.*, 37, 266–278.

Sastri, R.L.N. (1963) Studies in Lauraceae. IV Comparative embryology and phylogeny. *Ann. Bot.*, 27, 425–433.

Sastri, R.L.N. (1965) Studies in Lauraceae. V. Comparative morphology of flower. *Ann. Bot.*, 29, 39–44.

Saunders, E.R. (1939) *Floral Morphology*, Vol. II, Cambridge.

Sebastian, S., Farooqi, A. and Subbaiah, T. (1995) Effect of growth regulators on germination and seedling growth of Cinnamon (*Cinnamomum zeylanicum* Breyn.) *Indian Perfumer*, 39, 127–130.

Sedgley, M. (1977) Reduced pollen tube growth and the presence of the callose in the pistil of the male floral stage of avocado. *Sci. Hort.*, 7, 27–36.

Sedgley, M. (1985) Some effects of daylength and flower manipulation on the floral cycle of two cultivars of avocado (*Persea americana* Mill. Lauraceae) a species showing protogynous dichogamy. *J. Exp. Bot.*, 36, 823–832.

Sedgley, M. and Griffin, A.R. (1989) *Sexual Reproduction of Tree Crops*. Academic Press, New York.

Sharma, A.K. and Bhattacharya, N.K. (1959) Chromosome studies on four different species of *Cinnamomum*. *Jap. J. Bot.*, 17, 43–54.

Sheeja, G., Jisha, K.G., Seetha, K., Joseph, I. and Nair, R.V. (2000) Tissue culture of cinnamon, *Cinnamomum verum* (Presl.). In N. Muralidharan and R. Raj Kumar (eds) *Recent Advances in Plantation Crops Research*, Allied Pub., New Delhi, pp. 13–15.

Shi, W.Y. (1989) Study on the chemical constituents of essential oil and classification of types from *C. camphora*. *Acta Bot. Sinica*, 31, 209–214.

Shylaja, M. (1984) *Studies on Indian Cinnamomum*. Ph.D thesis, University of Calicut.

Shylaja, M. and Manilal, K.S. (1992) Bark anatomy of four species of *Cinnamomum* from Kerala. *J. Spices and Aromatic Crops*, 1, 84–87.

Shylaja, M. and Manilal, K.S. (1998) Biosystematics of the Genus *Cinnamomum*. In K.S. Manilal and M.S. Muktesh Kumar (eds) *A Handbook on Taxonomy Training*. Dept. Sci. & Technology, New Delhi, pp. 200–206.

Spoon-Spruit (1956) Cited from Tetenyi (1970).

Stern, W.L. (1954) Comparative anatomy of xylem and phylogeny of Lauraceae. *Tropical Woods*, 100, 7–72.

Sugiura, T. (1936) Studies on the chromosome numbers in higher plants, with special reference to cytokinesis. *Cytologia*, 7, 544–595.

Tackholm, G. and Soderberg, E. (1917) Cited from Sastri (1958).

Tetenyi, P. (1970) *Infraspecific Chemical Taxa of Medicinal Plants*. Chem. Pub. Co., New York.

Trachtenberg, S. and Fahn, A. (1981) The mucilage cells of *Opuntia fieus-indica* (L) mill – Development, ultrastructure, and mucilage secretion. *Bot. Gaz.*, 142, 206–213.

Wallis, T.E. (1967) *Textbook of Pharmacognosy*. J&A Churchill Ltd., London, pp. 78–82.

Wan-Yang, S., Wei, H., Guang-yu, W., De-xuan, G., Guang-yuan, L. and Yin-gou, L. (1989) Study on chemical constituents of the essential oil and classification of types from *Cinnamomum camphora*. *Acta Bot. Sinica*, 31, 209–214.

Weiss, E.A. (1997) *Essential Oil Crops*. CAB International, U.K.

Wijesekera, R.O.B., Ponnuchamy, S. and Jayewardene, A.L. (1975) *Cinnamon*. Ceylon Institute of Scientific and Industrial Research, Colombo, Sri Lanka.

Willis, J.K. (1973) *A Dictionary of the Flowering Plants and Ferns*. 8th edn., Cambridge Uni. Press, Cambridge.

Yazaki, K. and Okuda, T. (1993) *Cinnamomum cassia* Blume (Cinnamon): In vitro culture and the production of condensed tannins. In Y.P.S. Bajaj (ed.) *Biotechnology in Agriculture and Forestry 24. Medicinal and Aromatic Plants*. V. Springer-Verlag, pp. 122–131.

3 Chemistry of Cinnamon and Cassia

U.M. Senanayake and R.O.B. Wijesekera

Introduction

The "true" cinnamon or spice cinnamon is the dried inner stem-bark of *Cinnamomum verum* (Syn. *C. zeylanicum*). This species is mostly cultivated in Sri Lanka, Malagasy Republic, and Seychelles. Sri Lanka produces about three-quarters of the total world production of cinnamon. Although the Sri Lankan grown spice has a special appeal because of its organoleptic properties, there are, however, significant compositional variations even within the plantations in Sri Lanka. The presence of chemical cultivars has been recorded previously (Wijesekera, 1978; Wijesekera *et al.*, 1975). Even within the commercially cultivated species such chemical cultivars have been noted. These cultivars are recognised by sensory evaluation by the planters and denoted by such nomenclature as: "sweet", "honey", "camphoraceous" and also "mucilaginous", "wild" and "bloom". Only the "sweet" and "honey" varieties are extensively cultivated (Senanayake, 1977). There has also been a variety recently termed "Pieris" cinnamon, probably first identified by a planter by the name of Pieris. This variety is present within the commercial plots and has been investigated by the authors recently (unpublished).

"Chinese cinnamon" or cassia was initially produced from *C. cassia*, grown in the south-eastern provinces of China and Vietnam. Chinese cassia was marketed through Canton and Hong Kong. Cassia trees are grown on hillsides, about 100–300 m above sea level. Peeling of the bark is done after six years of growing (Brown, 1955, 1956). Fairly large quantities are exported from China and through Hong Kong, mainly to the USA (Manning, 1970).

C. burmannii is the source of another commercially accepted quality of bark known as Indonesian cassia or Batavia or Korintgi cinnamon. These plantations are found in the Padang area of the Indonesian island of Sumatra. Over 80% of US imports come from this source. At present the quantity remains at around 10,000 t/annum.

The nomenclature attached to commercially produced barks is somewhat confusing. It is recognised in trade that barks (and oils distilled) from *C. verum* in Sri Lanka are of "superior" quality. Lawrence (1967) has proposed that barks from *Cinnamomum* species be known as cinnamon – except *C. cassia* which will be known as cassia – and the country of origin indicates the quality. This proposal is not acceptable to all concerned. The chemical differences between each variety coupled with organoleptic considerations dictate that a combination of geographic and botanical nomenclatures should be derived.

Though the camphor tree (*C. camphora*) belongs to the genus *Cinnamomum*, it is not used as a spice. Instead, the camphor tree produces the bulk of the world's supply of

0-415-31755-X/04/$0.00 + $1.50
© 2004 by CRC Press LLC

natural camphor. From time immemorial the camphor tree has been growing in China, Japan and Taiwan (Guenther, 1950). Through Marco Polos thirteenth-century journal camphor became widely known as a curious and valuable product of the orient. From that time, camphor has occupied an important position in commerce. Due to increase in demand, production of camphor, which originated in China, later spread to Japan and Taiwan. True cinnamon as well as cassia are considered as spices on account of their distinct chemistry. However, *C. camphora* has camphor as the main chemical constituent. The root bark oil of *C. verum* also has camphor as a main constituent. This makes the biosynthesis of volatiles of *Cinnamomum* quite interesting (Senanayake, 1977).

Analytical Methodology

Classical methods

The methods in vogue several decades ago, despite their limitations, determined the standard sample of a cinnamon product. These methods could be termed classical methods for the assessment of quality. The development of these methods, and the formulation of standards for cinnamon and cinnamon oils, took place in the era preceding the development of modern analytical methods based on instrumentation. Both classical and modern studies of *Cinnamomum* volatiles have been reviewed by Senanayake (1977), Wijesekera (1978), Jayewardene (1987), Senanayake and Wijesekera (1989).

The advent of instrumental techniques

Gas liquid chromatography

The introduction of analytical methods based on gas liquid chromatography (GLC) created a new dimension in the study of volatiles. Relatively minor compounds, which could not be detected by classical methods, could be detected by GLC. As a result, the number of compounds detected in cinnamon oils, as in the case of all essential oils, increased several-fold. Even though the quantity of purified fractions that could be obtained by GLC was quite small, a combination of infrared (IR) spectroscopy and mass spectroscopy (MS) helped in the positive identification of the compounds separated. The improved techniques helped to develop the concept that aroma characteristics could often be considerably modified by the presence or absence of the compounds present in small quantities, and need not necessarily be dependent solely on the major compounds present.

Datta *et al.* (1962) identified benzaldehyde and cinnamic aldehyde in cassias (Saigon, Korintji and Batavia) oils by GLC. They proposed that GLC could be used to identify the geographical origin of spices. Bhramaramba and Sidhu (1963) identified α-pinene, α-phellandrene, p-cymene, caryophyllene, humulene, linalool, and O-methyl eugenol in the non-eugenol fraction of Indian cinnamon leaf oil by GLC. Lawrence (1967) initially proposed that the GLC profile of cinnamon oils could serve as an analytical control of purchased cinnamon oils, and Wijesekera and Jayewardene (1972) showed the characteristic differences between the oils of cinnamon and cassia, where GLC profiles were compared.

82 U.M. Senanayake and R.O.B. Wijesekera

As sensitivity of GLC techniques increased towards the end of the 1960s with the advent of improved electronics, stationary phases and support materials, and the development of capillary columns, many previously unidentified cinnamon volatiles were described. The introduction of capillary columns, particularly support coated open tubular (SCOT) columns, combined with mass spectrometry, permitted the identification of compounds that were present in very minute quantities, and yet could have significant impact on flavour.

Asakawa *et al.* (1971) detected the presence of 19 compounds from the benzene extracts of the root, bark and heartwood of *C. loureirii*. All three extracts had the same array of compounds but in different proportions. Cinnamic aldehyde was the major compound in both root and bark extracts; both extracts contained over 70%. There was no cinnamic aldehyde in the heartwood. Asakawa *et al.* (1971) identified eight monoterpenes and four aromatic compounds by GLC (Table 3.1).

Table 3.1 Volatiles identified in *C. verum* oils by various workers

Compound	Method of identification			Literature reference
	Classical	*TLC**	*GLC**	
α-Pinene	+		+	1,4,5,6,7,9,10,11,12,13
Dipentene	+		+	2,5,9
Camphene			+	10,11,13
β-Pinene			+	10,11,13
Sabinene			+	10
Δ^3-Carene			+	10,13
α-Phellandrene	+		+	1,5,6,9,10,11,13
α-Terpinene			+	10,13
Limonene			+	9,10,13
1:8 Cineole			+	9,10,11
β-Phellandrene	+		+	3,4,5,6,13
γ-Terpinene			+	10,13
p-Cymene	+	+	+	1,5,6,7,9,13
β-Ocimene			+	13
n-Methylamylketone	+			1,5
Nonaldehyde	+		+	1,5
Furfural	+		+	1,5,11
Benzaldehyde	+		+	1,3,5,6,9,11,13
α-Ylangene			+	10
Camphor	+		+	2,9,10,11,13
Linalool	+	+	+	1,3,5,7,9,10,11,13
Borneol	+			3
β-Caryophyllene	+	+	+	1,3,5,6,7,9,13
4-Terpinen-1-ol			+	10
α-Caryophyllene	+		+	3,4,5,11
α-Humulene	+		+	10,13
α-Terpineol	+		+	3,4,9,10,11,13
Piperitone	+		+	3,10
Cuminaldehyde	+		+	1,5,10,11
Geraniol	+		+	3,5,10
Safrole	+		+	3,5,9,10
Cinnamic aldehyde	+	+	+	3,5,6,8,9,10,11,13
Hydrocinnamic aldehyde	+		+	1,5,11
Methyl cinnamate			+	10,13

Ethyl cinnamate			+	10
Cinnamyl acetate			+	9,10,13
Eugenol	+	+	+	1,4,5,6,8,9,10,11,13
Acetyl eugenol	+		+	3,9,10
O-methyl eugenol	+	+	+	6,7,13
Cavacrol			+	9
Cinnamyl alcohol	+		+	3,9,10
Benzyl benzoate	+	+		3,5,7,10,13
2-Phenyl ethanol			+	11
3-Phenyl propanal			+	13
Thymol			+	11

References for Table:
1. Walbaum and Huthis (1902).
2. Pilgrim (1909).
3. Glichitech (1924).
4. Shintare and Rao (1932).
5. Guenther (1950).
6. Bhramaramba and Mahboob (1963).
7. Bhramaramba and Sidhu (1963).
8. Betts (1965).
9. Angmor et al. (1972).
10. Wijesekera et al. (1974).
11. Zurcher et al. (1974).
12. Senanayake and Wijesekara (1989).
13. Chalchat and Valade (2000).

Notes
TLC: Thin layer chromatography.
GLC: Gas liquid chromatography.

Angmor *et al.* (1972), Wijesekera and Jayewardene (1972), and Wijesekera *et al.* (1974) reported systematic compositional analyses of leaf, stem bark and root bark oils of the commercially prominent true cinnamon, *C. verum*, using GLC methods, the former group working with plants growing in Ghana and the latter with those under cultivation in Sri Lanka. *C. verum* leaves contained 1.2% steam volatile oil (Angmor *et al.*, 1972), although the oil content varied between 0.5% and 1.8% depending on the water content of the leaves and the method of distillation (Brown, 1955, 1956; Wijesekera, 1978). The principal constituent of leaf oil is eugenol, and again the reported levels can vary from 65% to 92% in leaves from both sources. The most prominent minor volatiles are β-caryophyllene, linalool, safrole, cinnamic aldehyde, cinnamyl acetate, cinnamyl alcohol and benzyl benzoate (Table 3.1). The presence of benzyl benzoate was reported for the first time in cinnamon by the Wijesekera group (1974) and they later proposed a likely mechanism for its formation in the plant (Wijesekera, 1978). Nath *et al.* (1996) reported a variety of *C. verum* growing in Brahmaputra Valley, India, containing benzyl benzoate as its major constituent in the leaf (65.4%) and bark (84.7%) oils. Chalchat and Valade (2000) analysed the bark oil of cinnamon growing in Madagascar and found 52.2% cinnamaldehyde together with 15.2% camphor. Raina *et al.* (2001) reported the leaf essential oil composition of cinnamon growing in Little Andaman (located in the Andaman and Nicobar group of islands in Indian ocean). Forty-seven constituents have been identified that constitute 99.96% of oil. The main components are eugenol (76.60%), linalool (8.50%), piperitone (3.31%), cinnamyl acetate (2.59%), eugenyl acetate (2.74%) and α-phellandrene (1.19%).

The yield of the stem bark oil can vary from 0.4% to 0.8% and root bark oil from 0.9 to 2.8% (Angmor *et al.*, 1972; Wijesekera, 1978). It is possible that the yield of oil will vary with the type and age of the plant, as well as water content of the material and the method of distillation. Cinnamic aldehyde and camphor are the major

84 U.M. Senanayake and R.O.B. Wijesekera

constituents in stem bark oil and root bark oil, respectively. The cinnamaldehyde content varies from 60 to 80% in the stem bark oil, and camphor comprises about 60% of root bark oil. All three oils possess the same array of monoterpene hydrocarbons, but in different proportions (Table 3.1). The important components are represented in Fig. 3.1. GLC analyses of the steam distilled oil of cinnamon bark were carried out by Zurcher et al. (1974) and they prepared a model mixture using the pure substances in the same ratio as found in the bark oil. They obtained an identical GLC profile and identical 33 compounds in Sri Lanka cinnamon oil (Table 3.1). By comparing the GLC profile of the model mixture with those from oils of cinnamon quills, chips and feathering it was possible to differentiate each type of oil. The same model mixture was used to distinguish commercial Chinese cassia oil from those of commercial cinnamon.

The volatiles of *C. cassia* and *C. camphora* have also been subjected to detailed analysis. Unlike *C. verum*, from which two types of commercial oils are obtained (Wijesekera, 1978), only one type is obtained for both *C. cassia* and *C. camphora*, where leaves and barks are distilled together (Guenther, 1950; Brown, 1956; Baruah et al., 1975; Senanayake, 1977) (Tables 3.2 and 3.3) (see chapter on Chinese/Indonesian cassia and camphor for details).

Table 3.2 Volatile constituents identified and their relative abundance in *C. cassia* oil

No.	Peak retention time (min)	Compound	Relative %		
			Leaf oil A	Bark oil A	Bark oil B
1	8.6	α-Pinene	0.03	0.05	tr
2	9.3	Camphene	0.02	tr	tr
3	10.6	β-Pinene	0.02	tr	tr
4	10.8	Sabinene	0.02	tr	–
5	13.1	1,8-Cineol	0.12	0.7	tr
6	14.8	γ-Terpinene	tr	tr	–
7	15.8	p-Cymene	tr	–	–
8	16.7	Terpinolene	0.03	tr	tr
9	17.4	Unkown	tr	tr	–
10	17.7	Unknown	tr	tr	–
11	18.3	n-Hexanol	0.13	tr	0.03
12	18.9	Unknown	tr	tr	–
13	19.9	3-Hexen-1-ol	1.09	tr	–
14	20.9	Fenchone	0.48	tr	–
15	21.3	Unknown	tr	tr	–
16	23.0	Furfural	tr	tr	–
17	25.7	Unknown	0.02	tr	–
18	26.3	Nonaldehyde	0.08	0.16	0.05
19	27.2	Benzaldehyde	2.68	4.73	1.16
20	28.4	Linalool	0.02	0.13	–
21	31.3	Unknown	tr	tr	tr
22	31.7	Terpinen-4-ol	0.07	0.04	0.07
23	32.2	Unknown	tr	0.05	tr
24	33.1	Methylbenzoate	0.04	tr	tr
25	33.8	Guaiacol	0.13	0.45	0.01
26	34.2	Unknown	tr	tr	tr
27	35.4	α-Humulene	tr	tr	tr

28	36.5	α-Terpineol	0.17	tr	0.11
29	37.6	β-Selanene	tr	tr	–
30	38.3	Geranial	tr	tr	tr
31	39.8	Unknown	tr	0.03	tr
32	40.1	Cuminaldehyde	0.09	0.35	tr
33	40.7	Hydrocinnamic aldehyde	0.97	1.95	tr
34	40.8	Unknown	0.08	tr	–
35	41.1	Phenylethyl acetate	tr	tr	0.24
36	44.2	Geraniol	tr	tr	–
37	45.8	Safrole	tr	tr	–
38	46.4	2-Phenylethyl alcohol	3.57	2.50	0.15
39	47.6	Unknown	tr	tr	–
40	49.7	4-Ethylguaiacol	2.2	0.5	0.80
41	51.6	Unknown	0.25	–	–
42	52.0	Phenol	0.60	0.07	tr
43	53.0	Methyl eugenol	tr	–	–
44	54.0	Cinnamic aldehyde	69.6	87.0	77.2
45	55.6	Methyl cinnamate	1.52	tr	–
46	56.9	Methyl isoeugenol	0.08	–	–
47	57.6	Unknown	0.52	0.02	–
48	58.3	Unknown	0.04	0.09	–
49	58.8	Ethyl cinnamate	1.24	0.42	0.37
50	59.8	Cinnamyl acetate	0.92	0.08	3.55
51	60.6	Eugenol	0.09	0.06	3.55
52	61.2	Unknown	0.01	0.06	tr
53	61.9	Chavicol	0.70	0.33	0.02
54	62.6	Unknown	0.16	0.13	tr
55	63.6	Acetyl eugenol	1.12	0.04	0.09
56	65.4	Cinnamyl alcohol	0.14	tr	tr
57	67.1	Farnesol	0.08	tr	tr
58	68.4	Isoeugenol	1.83	tr	tr
59	69.6	2-Vinylphenol	0.45	–	tr
60	71.3	Coumarin	8.06	0.28	15.3
61	74.2	Unknown	0.02	–	0.47
62	79.6	Benzyl benzoate	0.37	0.14	0.07
tr	:	trace ≤0.01%			
–	:	not detected			

Notes

A Laboratory distilled leaf and bark oils.

B Commercial cassia bark oil analysed by GLC on SCOT column. Values are means of three analyses.

Montes (1963) applied GLC for the analysis of cassia oil constituents and found a number of compounds, including eugenol and salicylaldehyde. Ter Heide (1972) analysed cassia oil (C. cassia) by GLC and identified 35 compounds in the oil. This analysis was conducted after the cinnamic aldehyde had been removed with sodium bisulphite ($NaHSO_3$) solution. The cinnamic aldehyde free fraction was then separated into noncarbonyls, phenolic and acid fractions. Each fraction was separately analysed for the constituents by GLC. Those compounds identified (Senanayake, 1977) are given in Table 3.4.

Lawrence and Hogg (1974) reported the composition of the bark oil of two uncommon Cinnamomum species growing in the Philippines. The bark oil from C. mercadoi contained 17 constituents, and bark oil from C. mindanaense 29 constituents. The major compounds of C. mercadoi were safrole (30.3%), 1,8-cineol (29.4%) and eugenol

Table 3.3 Volatile constituents identified and their relative abundance in commercial *C. camphora* oil analysed by GLC on SCOT column. (Values are means of three analyses)

Peak no.	Retention time (min)	Compound	Relative %
1	8.6	α-Pinene	3.76
2	9.4	Camphene	1.64
3	10.4	β-Pinene	1.26
4	10.6	Unknown	1.53
5	11.3	Unknown	tr*
6	11.7	Sabinene	1.47
7	12.0	α-Phellandrene	0.17
8	12.4	α-Terpinene	0.05
9	13.1	Limonene	2.71
10	13.5	1,8-Cineole	4.75
11	14.1	cis-Ocimene	0.04
12	14.7	γ-Terpinene	0.24
13	15.7	p-Cymene	0.14
14	16.3	Terpinolene	0.30
15	21.5	Fenchone	tr
16	24.9	Furfural	0.16
17	26.5	α-Ylanglene	0.02
18	27.2	Benzaldehyde	tr
19	28.7	Camphor	51.5
20	29.4	Linalool	0.68
21	30.4	Linalylacetate	tr
22	31.6	Bornyl acetate	0.02
23	32.5	Terpinen-4-ol	0.57
24	32.9	β-Caryophyllene	1.49
25	33.4	Phenyl acetaldehyde	0.05
26	34.3	Unknown	tr
27	35.1	Borneol	0.02
28	36.2	Methyl chavicol	0.05
29	36.8	α-Humulene	2.19
30	37.6	α-Terpineol	1.62
31	38.2	β-Selanene	tr
32	39.1	Piperitone	2.41
33	39.7	Geranyl acetate	0.22
34	40.2	Unknown	0.76
35	41.6	Cuminaldehyde	0.15
36	42.7	Unknown	tr
37	43.3	Nerol	tr
38	45.2	Geraniol	0.63
39	45.7	Unknown	tr
40	47.4	Safrole	13.4
41	51.0	3-phenylpropylacetate	tr
42	52.8	Phenol	tr
43	53.6	Caryophyllene oxide	tr
44	50.0	Cinnamic aldehyde	0.08
45	55.4	Methyl cinnamate	0.08
46	55.9	Unknown	0.07
47	57.2	Isoeugenol	tr
51	59.4	Eugenol	0.12
59	64.4	Acetyl eugenol	tr
60	65.3	Cinnamyl alcohol	0.18
65	70.2	Unknown	tr

Note
* tr: ≤0.01%.

Table 3.4 Volatile constituents identified in commercial cinnamon leaf, stem bark oils and root bark oil of 2½-year old plants by GLC on SCOT column. (Values are means of three analyses)

Peak no.	Retention time (min)	Compound	Relative %		
			Leaf oil	Stem bark oil	Root bark oil
1a		Solvent			
1b		Do			
1	9.0	α-Pinene	0.99	0.59	4.28
2	9.4	Unknown	tr	tr	0.14
3	9.8	Camphene	0.38	0.18	1.94
4	10.7	β-Pinene	0.28	0.14	2.23
5	11.7	Sabinene	tr	0.02	0.77
6	11.8	Δ^3-Carene	0.07	0.03	tr
7	12.0	Myrcene	0.08	0.05	2.21
8	12.3	α-Phellandrene	0.88	0.63	0.19
9	12.7	α-Terpinene	0.12	0.42	0.17
10	13.3	Limonene	0.36	0.46	3.08
11	13.7	1,8-Cineole	0.62	2.00	11.7
12*	14.3	cis-Ocimene	0.02	0.03	0.57
13	15.0	γ-Terpinene	0.05	0.03	0.23
14*	15.1	trans-Ocimene	0.02	tr	0.20
15	15.9	p-Cymene	1.15	1.12	1.41
16*	16.4	Terpinolene	0.13	0.11	0.28
17*	18.3	n-Hexanol	tr	tr	–
18*	21.6	Fenchone	tr	tr	0.02
19*	24.3	Furfural	0.02	0.03	0.02
20*	25.0	t-Linalool oxide	0.05	0.05	0.05
22	26.8	α-Ylangene	0.99	0.31	–
23*	27.0	Nonaldehyde	tr	–	–
24*	27.4	Benzaldehyde	0.16	0.26	–
25	27.8	Camphor	tr	tr	56.2
26	28.7	Linalool	3.44	2.39	0.87
27*	29.4	Linalool acetate	tr	tr	0.31
28*	30.7	Bornyl acetate	tr	0.10	0.03
29	31.7	Terpinen-4-ol	0.12	0.36	2.16
30	32.3	β-Caryophyllene	5.78	3.28	0.50
31*	33.4	2-Phenylacetaldehyde	tr	tr	0.04
32*	34.6	Borneol	tr	0.02	0.08
33*	35.0	Methyl chavicol	tr	tr	0.37
34	35.8	α-Humulene	0.91	0.56	0.17
35	36.4	α-Terpineol	0.42	0.66	6.85
36*	37.2	β-Selinene	tr	tr	–
38*	37.9	γ-Cadenene	0.01	–	0.50
39*	38.1	Geranial	0.02	tr	–
40	38.5	Piperitone	0.04	0.07	0.94
41*	38.7	Geranyl acetate	–	tr	–
42	39.0	Cuminaldehyde	0.06	0.04	0.08
43*	39.6	Hydrocinnamic aldehyde	0.17	0.40	tr
45*	40.3	Phenyl ethyl acetate	0.03	0.07	tr
46*	41.4	Nerol	tr	tr	–
47	44.1	Geraniol	0.04	0.06	0.7
48*	44.8	Benzyl alcohol	tr	tr	tr
49	45.7	Safrole	2.28	tr	0.31
50*	46.8	2-Phenyl ethyl alcohol	tr	0.41	tr
51	48.1	Unknown	–	–	tr
52*	49.2	3-Phenyl propyl acetate	1.00	0.13	tr

88 U.M. Senanayake and R.O.B. Wijesekera

Table 3.4 (Continued)

Peak no.	Retention time (min)	Compound	Relative %		
			Leaf oil	Stem bark oil	Root bark oil
53	50.4	Unknown	tr	tr	–
54*	51.8	Phenol	0.02	tr	tr
55*	52.5	Caryophyllene oxide	0.52	tr	0.22
56*	52.9	Methyl eugenol	0.01	tr	–
57	53.5	Cinnamic aldehyde	1.98	75.0	0.67
58	54.5	Methyl cinnamate	0.03	tr	tr
59	55.1	Unknown	tr	–	tr
60*	56.3	Methyl isoeugenol	tr	–	tr
61	56.9	Unknown	–	–	–
62	58.8	Ethyl cinnamate	0.02	–	tr
63	60.1	Cinnamyl acetate	1.65	5.01	tr
64	60.6	Eugenol	70.1	2.20	0.45
65	62.9	Unknown	tr	–	–
66	63.8	Acetyl eugenol	2.51	0.16	0.10
67	64.5	Unknown	tr	tr	–
68	65.2	Cinnamyl alcohol	0.38	0.26	0.08
69*	67.1	Farnesol	0.12	0.03	–
70*	68.8	Ioseugenol	0.13	0.02	–
71*	69.2	2-Vinylphenol	tr	0.03	–
72*	71.3	Coumarin	tr	0.66	–
73*	72.8	Vanillin	–	tr	–
74	79.2	Benzylbenzoate	3.5	0.66	0.25
75*	83.1	2-Phenyl ethyl benzoate	tr	tr	–

Source: Senanayake, 1977.

Notes
– Not detected; * New compound; tr: trace.

(15.4%), whereas *C. mindanaense* contained eugenol (39.2%), linalool (19.4%), and safrole (15.0%). The minor compounds were essentially the same as those from *C. verum* (Table 3.1). Wijesekera and Jayewardene (1974) studied the volatile oil obtained from the bark oil of a rare species of cinnamon, *C. capparu-coronde* (Kostermanns), which is found in central Sri Lanka and is claimed to have medicinal properties. *Capparu-coronde* is a variation of the Sinhala name for the plant *Kapuru kurundu* that means literally "*camphoraceous odour*" of the plant (the species differed from the cultivated species in the comparatively low quantities of eugenol and cinnamaldehyde). The major constituents of the oil were linalool (29%), eugenol (23%) and 1,8-cineole (16%); nineteen minor constituents were established, most of which were similar to those found in *C. verum*. *C. capparu-coronde* is not cultivated, but wild growing varieties are found in the mid-country rain forests of Sri Lanka. The oil has promising organoleptic characteristics (see Chapter 16 for more details on related species of *Cinnamomum*).

High performance liquid chromatography (HPLC) of cinnamon oil

Ross (1976) analysed the cinnamon bark and leaf oils for eugenol and cinnamic aldehyde using the technique of high performance liquid chromatography (HPLC). By comparing the relative abundance of each compound in authentic bark oil, it was possible to check the commercial cinnamon oils for any adulterants. He proposed that HPLC could be gainfully used as a quality control check for the commercial cinnamon oils.

Use of infrared spectroscopy (IR) in the analysis of cinnamon oils

Infrared spectroscopy (IR) is generally considered a tool which is complimentary to GLC, TLC and CC in quality assessment. Shankaranarayana *et al.* (1975) reviewed its applicability to the analysis of flavours. IR has been used for routine analysis of essential oils (Carrol and Price, 1964). Wijesekera and Fonseka (1974) demonstrated the use of IR both as a method to establish the genuineness of Sri Lankan cinnamon oils, and also to quantitatively estimate the main constituents in them in rapid fashion, once the standardisation of the procedure had been accomplished. Compounds such as eugenol, cinnamic aldehyde, acetyl eugenol, cinnamyl acetate and benzyl benzoate in the leaf oil; cinnamic aldehehyde, eugenol and cinnamyl acetate in the stem bark oil; camphor, 1,8-cineol and cinnamic aldehyde in the root bark oil can be estimated rapidly by the technique developed by these workers. The results obtained are in close agreement with those obtained by GLC. By comparing the IR of authentic cinnamon oil with those of commercial oil, it is possible to detect any adulteration, or to establish the deficiency or excess of any particular compound due to reasons like faulty distillation.

A detailed study of the chemical composition of the volatiles of cinnamon is that by Senanayake (1977). In this study cinnamon leaf, stem bark and root bark oils from Sri Lankan sources were analysed using GLC techniques employing a glass SCOT column (65 m × 1.6 mm OD). Prior to detailed analysis the major constituent of the leaf oil, eugenol, was removed by means of KOH and the hydrocarbon and oxygenated fractions were separated on silica gel columns (Table 3.4, Fig. 3.1).

Similarly in the study of stem bark oil, cinnamic aldehyde was removed using $NaHSO_3$ and hydrocarbon and oxygenated fractions were separated by column chromatography. Since root bark oil does not contain appreciable quantities of either eugenol or cinnamic aldehyde, it was subjected to GLC analysis without prior separation. Composition of commercial cinnamon bark oil is given in Fig. 3.2, Table 3.4.

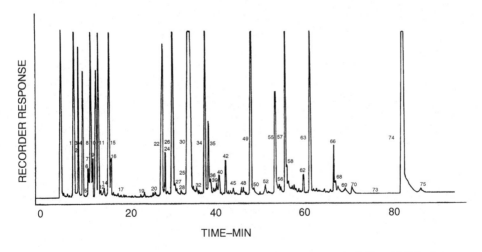

Figure 3.1 GLC chart of non-eugenol fraction of commercial cinnamon leaf oil on SCOT column.

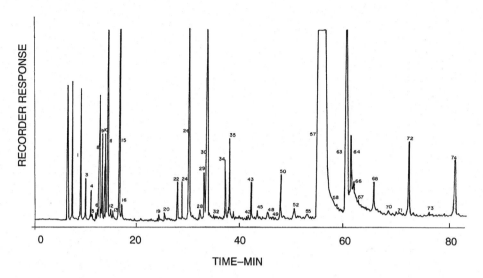

Figure 3.2 GLC chart of commercial cinnamon bark oil on SCOT column.

Following the development of solid injection techniques, Senanayake *et al.* (1975), Wijesekera *et al.* (1975) and Wijesekera (1978) employed this methodology for the study of volatile constituents of morphologically distinct parts of the cinnamon tree, employing as little as 15 mg of plant material. Use of this technique has revealed that during the normal steam distillation, cinnamon volatiles do not undergo any significant chemical changes, that leaf veins contain more cinnamic aldehyde (30%) than the rest of the leaf, and many other useful biosynthetic features.

Location of Cinnamon Volatiles in *C. verum*

C. verum is interesting in that it yields three types of oils from the leaf, stem bark and root bark. The major constituent in the leaf oil is eugenol, in the stem bark oil it is cinnamic aldehyde while camphor is the major constituent in the root bark oil. Solid injection techniques have revealed that the young and old leaves are deficient in volatiles; only mature leaves (about four leaves below the apical bud) have the general volatile pattern of the distilled oil. It is also found that the leaf petiole and veins contained more cinnamic aldehyde than in the leaf lamina. This is in agreement with the view (Neish, 1960) that cinnamic aldehyde will eventually be converted to lignin in the xylem vessels found in cinnamon stem, petiole and vein.

Cinnamic aldehyde and eugenol are components of the shikimic acid pathway leading to lignin formation. Cinnamic aldehyde is directly formed by the reduction of cinnamic acid, the latter being formed from phenylalanine (PA). It was shown that cinnamic aldehyde gets further reduced to cinnamyl alcohol, which is a precursor of lignin. Eugenol, due to the absence of a hydroxy group in the allyl side-chain, cannot readily contribute to the lignin formation. However, Siegel (1955) and Higuchi (1957) have shown that eugenol is capable of polymerising to lignin-like substances.

Since lignification occurs in the xylem tissues, it has to be established whether synthesis of cinnamic aldehyde and eugenol occurs in the xylem tissues or whether they are synthesised elsewhere in the plant and transported to the sites of lignification. Experimental evidence has shown that the stem (heartwood) has a low content of cinnamic aldehyde as well as eugenol. This is as expected since free eugenol and cinnamic acid would have been converted to non-volatile lignin.

Both cinnamic aldehyde and eugenol are relatively insoluble in water, hence they have to be converted to water-soluble glucosides for transportation within the plant (Senanayake, 1977; Senanayake *et al.*, 1977). There was a high content of these glucosides in the leaves in comparison to other parts of the plant. Glucoside formation was related to photosynthesis. This fact indicated that cinnamic aldehyde and eugenol should be in a site more accessible to the photosynthetic glucose present in the leaves, so that they could be converted to glucosides. The petiole-dip method clearly demonstrated that detached leaves were capable of forming these compounds; the bark had a low incorporation of precursors into these compounds when cuttings were used. This indicated that bark contributed very little to the biosynthesis of these two compounds.

No attempts were made by Senanayake and collaborators (1977) to isolate lignin and determine the labelling pattern (this would indicate the precursor of the lignin), however, the data in this study pointed to the possibility that both cinnamic aldehyde and eugenol were formed in the leaves. Weissenböck *et al.* (1971) and Loffelhardt and Kindle (1975), in similar biosynthetic studies with PA, have shown that chloroplasts and thylakoid membranes are capable of converting PA into cinnamic acid and related products. The latter authors have also indicated that the chloroplast could be the predominant, if not exclusive, site of benzoic acid formation in *Astilbe chinensis*.

Biosynthesis of phenylpropanoids

Both eugenol and cinnamic aldehyde belong to the group of compounds consisting of a benzene ring with a propane side branch, *viz.* C6-C3. The side branch may have some degree of unsaturation and the benzene ring may or may not contain some functional groups.

The origin of the aromatic ring of many natural phenylpropanoid compounds is now regarded to be the cyclohexane derivatives that arise by the cyclisation of sedoheptulose, a C7 sugar molecule (Neish, 1960; Geissman, 1963). The key compound in this biosynthetic scheme is the now well recognised shikimic acid.

The shikimic acid pathway was established largely from the experimental work of Davis and Sprinson, which has been exhaustively reviewed by Davis (1955, 1958), Sprinson (1960), Neish (1960, 1968), and Bohm (1965). The key to this scheme was the discovery by Davis (1951) that a requirement of a mutant strain of *Escherichia coli* for five aromatic compounds – phenylalanine, tyrosine, tryptophan, p-aminobenzoic acid and p-hydroxybenzoic acid – could be completely satisfied by the single compound, shikimic acid. Thus shikimic acid was established as an obligate intermediate for the biosynthesis of aromatic rings in *E. coli*.

Both cinnamic aldehyde and eugenol are formed through the Shikimic acid pathway leading to lignin. The exact point of cinnamic aldehyde and eugenol formation, however, is unclear. It would appear that cinnamic aldehyde should be formed by a single step reduction of cinnamic acid. A further step reduction would yield cinnamic alcohol, which could contribute towards the formation of lignin.

In species such as *Cinnamomum*, it may well be that a genetic block occurs at the conversion of cinnamic aldehyde to cinnamyl alcohol, and as a result there can arise an accumulation of cinnamic aldehyde. Birch (1963) suggested the loss of ability to

92   U.M. Senanayake and R.O.B. Wijesekera

introduce para-oxygen to the ring as a possible explanation for the cinnamic aldehyde remaining as such. Similarly a two-step reduction of the side-chain of ferulic acid will yield coniferyl-alcohol. Elimination of the terminal hydroxy group in the side-chain and re-arrangement of the double bond will yield eugenol. In *Cinnamomum* species there may be some enzyme responsible for such a transformation of coniferyl alcohol. Some further speculations have been made on the basis of mechanistic consideration (Wijesekera, 1978; Wijesekera *et al.*, 1975).

In view of the structural relationship with lignin, it is logical that cinnamic aldehyde should be widespread in the plant kingdom. It is found in Ascomycetes, Basidiomycetes and in many higher plants (Guenther, 1949; Birch, 1963; Zenk and Gross, 1972; Gottlieb, 1972). Among the commercially important essential oils, cinnamic aldehyde is found in cinnamon bark and leaf oils, cassia bark and leaf oils, patchouli leaf oil and myrrh oil (Guenther, 1949). Due, as postulated here, to a possible block in the pathway (as in the case of a mutation), both cinnamic aldehyde and eugenol appear to accumulate, especially in cinnamon and in cassia. Gottlieb (1972) pointed out that the family Lauraceae is at the bottom of the evolutionary ladder concerning phenylpropanoid biosynthesis and it is therefore reasonable to expect some anomaly in the shikimic acid pathway leading to lignin formation.

So far no reports have appeared on tracer studies of cinnamic aldehyde biosynthesis *in situ* in plants. However, the tracer studies of Kaneko (1960, 1961 and 1962), Canonica *et al.* (1971), Manitto *et al.* (1974, 1975a,b) and Senanayake (1977) have shown interesting developments in cinnamic aldehyde and eugenol biosynthesis. It is generally assumed that the allyl and propenyl groups attached to phenolic nuclei in many plant constituents, such as in anethole, chavicol, estrogole and eugenol, originate from the cinnamic acid side-chain through the reductive steps as shown in Fig. 3.3 (Birch, 1963; Bullock, 1965).

It now appears that formation of the side-chain may occur through a different mechanism (Canonica *et al.*, 1971). The difference seems to lie in the position of the double bond in the side-chain. In cinnamic aldehyde the double bond is between carbons 1 and 2 (Fig. 3.3). In such a situation the phenyl-propane skeleton of L-phenylalanine is incorporated into the molecule with the retention of all carbon atoms (Kaneko, 1962; Manito *et al.*, 1974). When DL-phenylalanine, labelled with (14C) at the side-chain in all three possible positions, was fed for example to *Pimpinella anisum* and *Foeniculum vulgare*, the anethole recovered was labelled in the same position as that of DL-phenylalanine originally fed to the plant.

In a separate study, Canonica *et al.* (1971) showed that when L-phenylalanine, labelled with (14C) in all three possible positions in the side-chain, was fed to *Ocimum basilicum* L., the eugenol recovered was not labelled when L-phenyl-(1–14 C) alanine was used. This indicated that the carboxyl group was lost during complete conversion of the ferulic acid to the allyl group. In their subsequent studies, Manitto *et al.* (1974, 1975a) showed that the decarboxylation of the side-chain took place at the ferulic acid stage and an extra carbon atom was introduced to the side-chain, probably donated by S-adenosyl-methionine or an equivalent compound. Decarboxylation at the ferulic acid stage was established as labelled ferulic acid and incorporated into eugenol in an appreciable quantity (Manitto *et al.*, 1975b). Thus it appears that the prophenyl and allyl side-chains have independent origins.

Furthermore, as in the case of eugenol, the allyl group only occurs when there is a p-oxygen attached to the ring. Thus Birch's hypothesis (Birch, 1963) has been rejected

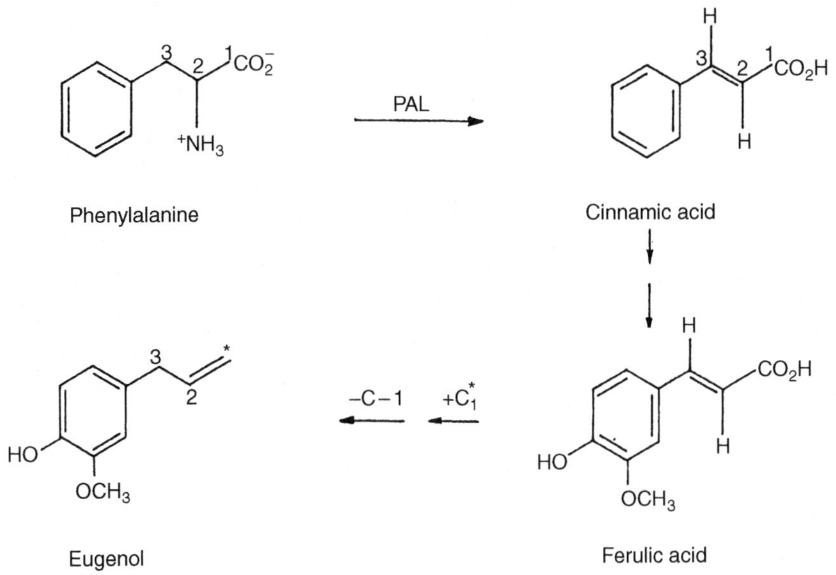

Figure 3.3 Formation of eugenol from phenylalanine.

with regard to the sequence: – cinnamic acid → cinnamyl alcohol → allyl derivative. The favoured pathway for eugenol biosynthesis is accepted to be: L-phenylalanine → cinnamic acid → P-coumaric acid → caffeic acid → ferulic acid → eugenol.

Manitto *et al.* also found that L-tyrosine was not incorporated into eugenol. Thus it is in agreement with the lack of L-tyrosine ammonia-lyase in dicolyledonous plants (Young *et al.*, 1966).

Cinnamic aldehyde biosynthesis

In cinnamon, cinnamic aldehyde was formed from phenylalanine without randomisation of the 14C label. The same degree of labelling was observed when PA1, PA2 and PA3 were separately used as a precursor (Senanayake *et al.*, 1977; Angmor *et al.*, 1979). Such a uniform labelling would not be observed if randomisation of the label occurred. The experimental evidence suggests that the C6-C3 skeleton of PA was retained as one unit in the biosynthesis of cinnamic aldehyde from PA. This is in agreement with the suggestion of Birch (1963) that cinnamic aldehyde can be formed from PA via cinnamic acid and cinnamic aldehyde, which can be further reduced to cinnamyl alcohol that then goes for lignin biosynthesis (Neish, 1968). The retention of the C6-C3 skeleton in the biosynthesis of propenylphenols, such as anethole (P-methoxy propenyl benzene), from PA has been observed by other workers. Kaneko (1960, 1961) and Manitto *et al.* (1974) observed that anethole was formed in *Foeniculum vulgare* and *Pimpinella anisum* (Umbelliferae) from PA without label randomisation, suggesting that the C6-C3 skeleton was retained as one unit. Thus the experimental evidence indicates that in the cinnamon plant cinnamic aldehyde is formed from pheylalanine via the shikimic acid pathway (Fig. 3. 4).

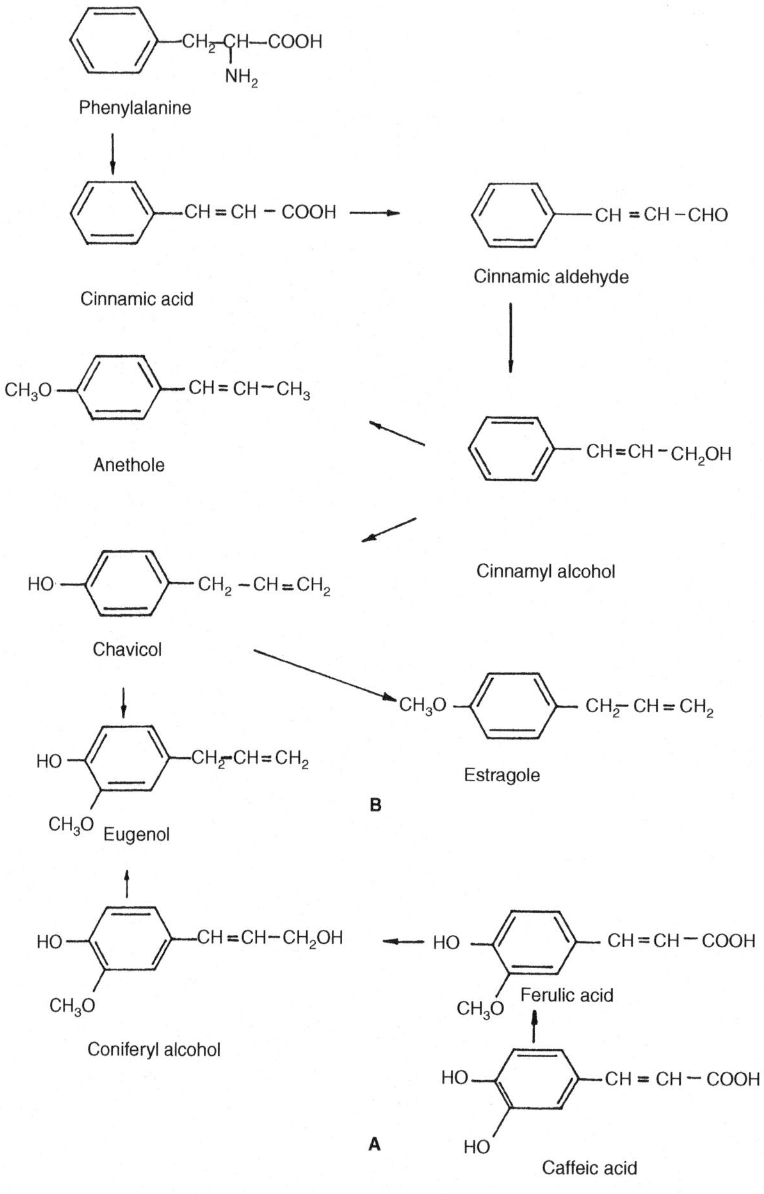

Figure 3.4 Pathways for the formation of (A) eugenol from ferullic acid and (B) eugenol and related compounds from cinnamic acid.

Eugenol biosynthesis

To investigate the mechanism of the transformation of the side-chain of L-phenylalanine into the allyl group of eugenol, DL-phenylalanines, each carrying a 14C label at one of the three positions in the side-chain, were administered to cinnamon cuttings. PA2 and PA3 but not PA1 were effective precursors of eugenol biosynthesis (Senanayake *et al.*, 1977).

The effectiveness of PA2 and PA3 can be explained if it is assumed that C-2 and C-3 of the PA side branch are retained in the allyl group of eugenol. The contribution of the label from L-(14C) methyl methionine and the subsequent loss of the label when this labelled eugenol was degraded to homoveratric acid, which was nonradioactive, indicated that C1 in the allyl group was not derived from PA but from a methyl donor such as L-methionine or an equivalent precursor. The homoveratric acid is produced with the loss of C1 in the allyl group in eugenol. When PA3 and 14C (me) methionine were fed together as precursors, the percentage of label in eugenol increased, as the label at C3 was derived from PA3 and the label at C1 was derived from 14C-(me) Methionine.

Biosynthesis of terpenoids in cinnamon

A major breakthrough in the study of terpene biogenesis was the discovery of mevalonic acid (MVA) by Wolf *et al.* (1956) and the recognition of it as an efficient precursor of cholesterol, (Tavormina *et al.*, 1956). Subsequent work by Park and Bonner (1958), Goodwin (1965), etc. indicated that "active isoprene" in plant and animal tissue was formed from mevalonic acid.

It is generally expected that MVA is converted to isopropenyl pyrophosphate (IPP) and dimethyl allyl pyrophosphate (DMAPP), which are regarded as "active isoprene". The first product formed from the union of IPP and DMAPP is geranyl pyrophosphate (GPP), which can isomerise to neryl pyrophosphate (NPP). Both GPP and NPP give rise to the acyclic monoterpenes. Cyclisation of NPP leads to the 1-p-menthene-8-carbonium ion of Ruzicka's scheme (Ruzicka *et al.*, 1953). From, this carbonium ion an array of mono-, bi- and tricyclic monoterpenes are formed. It is generally assumed that GPP is the immediate precursor of all monoterpenes (Goodwin, 1965). An isomer of GPP has been postulated as the more likely precursor of the cyclic monoterpenes (Valenzuela *et al.*, 1966). Experimental support for this hypothesis has been found in several biological systems (Cori, 1969; Beytia *et al.*, 1969). In citrus, linaloylpyrophosphate (LPP) has been found to be the precursor of both acyclic and cyclic monoterpenes (Attaway and Buslig, 1968; Potty and Bruemmer, 1970; George-Nascimento and Cori, 1971; Suga *et al.*, 1971) (Fig. 3.5).

Cinnamon oils contain a number of terpenes (Table 3.4) including both mono- and sesquiterpenes. In root bark oil over 90% of the total weight consists of terpenes while stem bark oil contains about 10% and that of leaf oil is about 8%. Camphor content of the root bark oil is about 65%.

Banthorpe *et al.* (1972) have studied the biosynthesis of (+)- and (−)-camphor in *Artemisia, Salvia* and *Chrysanthemum* species, and Battersby *et al.* (1972) have examined the biosynthesis of (−)-camphor in *Salvia officinalis*, in order to verify the validity of Ruzicka's (1953) scheme. Banthorpe and Baxendale fed labelled (2–14C) MVA to cut foliage of *Artemisia, Salvia* and *Chrysanthemum* by the infusion method. The time interval from feeding to harvesting ranged from between 4 and 50 days. The camphor was labelled by approximately 0.002 to 0.2% of the applied precursor. After degradation of the labelled camphor, it was concluded that in these species camphor biosynthesis occurred with the union of IPP and DMAPP via the 1-p-menthene-8-carbonium ion as postulated by Ruzicka.

Battersby *et al.* (1972) tested the validity of Ruzicka's scheme (Ruzicka *et al.*, 1953), particularly the point where geraniol isomerises to nerol on the route to camphor formation. They chose *Salvia officinalis* as a satisfactory plant for this study. When cuttings of this species were fed with 2-(14C) geraniol for 29 hours,

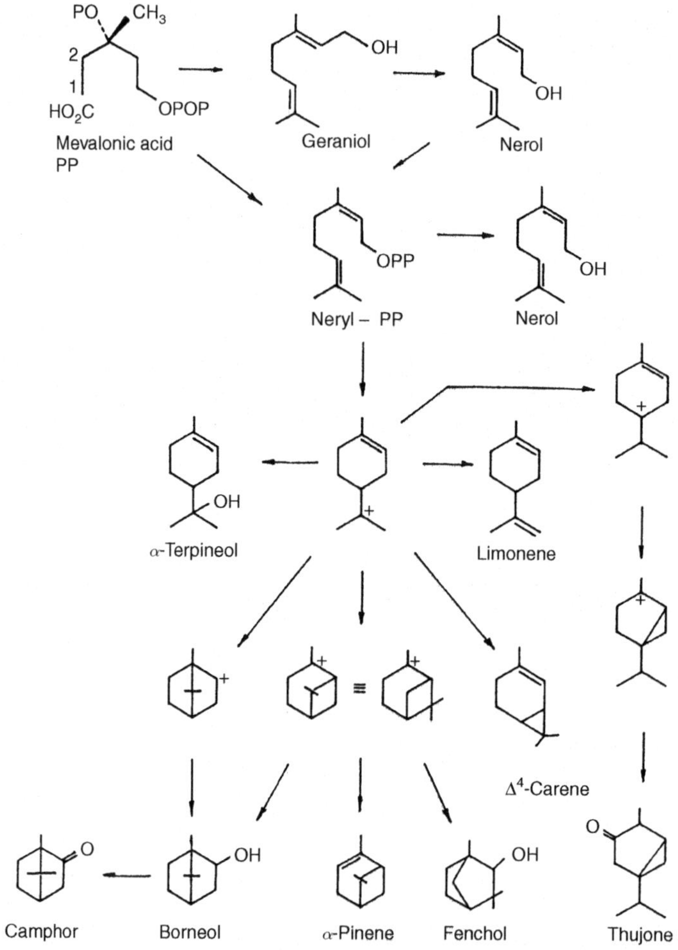

Figure 3.5 Ionic mechanism in the biosynthesis of monoterpenes. Ruzica (1953) as modified by Loomis (1967).

the camphor isolated contained 0.0025% of the applied label. Unambiguous degradation of the labelled camphor confirmed that labelled geraniol had been incorporated specifically into camphor by this plant, indicating that geraniol had isomerised to nerol.

Linalool, an acyclic monoterpene, is found in cinnamon leaf and stem bark oil to the extent of 3%. It has been convincingly shown that linalool is formed from MVA in several plant tissues. For example, Potty and Bruemmer (1970) and George-Nascimento and Cori (1971) demonstrated that linalool, the major monoterpene in orange peel, is biosynthesised from MVA, and Suga *et al.* (1971) obtained labelled linalool when (2–14C) MVA was fed to *Cinnamomum camphora*. They suggested that geraniol had been transformed to linalool, as linalool is a tertiary isomer of geraniol. Thus Fujita *et al.* (1975) obtained an increased quantity of geraniol, nerol, limonene and terpineol when (±) linalool (unlabelled) was administered to *Fatsia japonica*, *Iris pseudacorus* and *Quercus phillyracoides*.

There is no evidence, however, that these monoterpenes that occur in the cinnamon are synthesised in an analogous fashion.

Biosynthesis of the sesquiterpenoids

In cinnamon leaf, stem bark and root bark oils sesquiterpenes occur to an extent of about 4 to 5%. β-caryophyllene and α-humulene are the major sesquiterpenes, and together they contribute about 3 to 4%. Like monoterpenes, sesquiterpenes may occur in acyclic, monocyclic, bicyclic and tricyclic forms. Ruzicka's biogenetic isoprenoid rule (1953) also applies to sesquiterpenes. Most of these terpenoids can be divided into three isoprene units connected head-to-tail, and thus the molecular formula is $C_{15}H_{24}$. Parker *et al.* (1967) proposed that the trans-farnesyl pyrophosphate (FPP) is the fundamental starting unit for the formation of sesquiterpenes. Like monoterpenes, sesquiterpenes occur in plant as hydrocarbons with functional groups such as hydroxyl and aldehyde. Farnesol is widely distributed in the plant kingdom (Simonsen and Barton, 1952).

Since farnesol is formed by the union of three isoprene units, it follows that MVA should be a precursor of farnesol, in particular, and sesquiterpenes, in general. Many workers have reported the formation of farnesol from MVA. In particular Loomis (1967), Nicholas (1967), Rogers *et al.* (1968), Banthorpe and Wirz-Justice (1969), Staby *et al.* etc. (1973) have investigated both the formation and sites of sesquiterpene synthesis.

Even though the incorporation of MVA into monoterpenes is less than 1%, the observed incorporation into sesquiterpenes was in the range of 2–4% of the fed MVA. Rogers *et al.* (1968) pointed out that mono- and sesquiterpenes are synthesised at different sites, which could account for the later incorporation of label into sesquiterpenes. They have suggested that sesquiterpenes may be synthesised outside the chloroplast, where applied precursors are more accessible.

Mass Spectrometric Studies

A systematic study of the chemical composition of Sri Lankan produced spice oils and essential oils was carried out by Mubarak and Paranagama (Paranagama, M.Phil. thesis, 1991). In this study these researchers used GC-MS techniques with capillary columns to investigate *inter alia* the essential oils of cinnamon, (leaf oil, bark oil and root bark oil). The essential oils were laboratory distilled from material derived from commercial sources. The authors also analysed the volatile oil of the cinnamon fruits for the first time (Tables 3.5 and 3.6). The constituents identified by these workers are listed in Table 3.6. The GLC patterns of fruit and root bark oils on carbowax 20 M capillary columns are given in Fig. 3.6 and Fig. 3.7.

The δ and γ cadinenes (36%) are major constituents of the cinnamon fruits which have an array of compounds not found in the other sites explored. The fruits have relatively minor amounts of cinnamaldehyde and eugenol. Other compounds in the fruit volatile oil were cadinol T (7.7%), β-caryophyllene (5.6%), α-muurolene (4.4%), α-cadinene (2.8%), α-ylangene (2.7%) and α-gurjugene (2.3%). The oil contained 67% sesquiterpene hydrocarbons, 17% sesquiterpene alcohols and only 1% phenyl propanoids. The aroma of the fruit therefore defers considerably from that of the leaf or bark, due to the sesquiterpenoids.

Table 3.5 Constituents of cinnamon bark, leaf, root and fruit oils, SPB-1 column

Component	% C. bark	% C. leaf	% C. root	C. fruit
Unknown	–	–	0.03	–
α-Pinene	3.34	0.73	5.70	2.19
β-Thujene	1.10	0.08	0.57	–
Camphene	0.63	0.29	2.77	0.29
β-Pinene	0.61	0.26	3.45	1.61
Sabinene	0.26	–	1.51	–
α-Phellandrene	0.14	0.65	4.92	0.43
Myrcene	2.70	0.77	0.43	–
α-Terpinene	1.30	1.10	1.05	0.08
Limonene	1.21	0.32	6.16	1.00
β-Phellandrene	–	–	2.09	0.07
cis-Ocimene	0.14	–	0.28	0.03
γ-Terpinene	0.16	–	0.57	0.05
trans-Ocimene	0.13	–	0.94	0.02
p-Cymene	1.91	0.92	1.38	0.01
Terpinolene	0.21	0.61	0.47	0.30
Linalool	3.70	2.77	0.13	0.08
β-Terpineol	–	0.06	0.05	–
Terpinen-4-ol	0.40	0.11	1.90	0.27
α-Terpineol	0.70	0.28	3.94	0.64
Fenchyl alcohol	–	–	–	0.41
Isoborneol	0.08	–	0.68	0.70
Sabinol	–	–	0.20	–
1,8-Cineole	4.60	0.51	6.39	0.05
Methyl chavicol	–	–	0.19	–
Cumenol*	–	–	0.06	–
Cinnamyl alcohol	0.16	0.09	0.12	–
Phenyl ethyl alcohol	0.47	–	–	–
Coumarin	0.36	–	–	–
Benzaldehyde	0.61	0.14	–	0.50
Hydrocinnamaldehyde	0.80	0.12	0.09	–
Cinnamaldehyde	46.70	2.81	0.41	0.29
Camphor	–	–	47.42	–
Piperitone*	–	–	0.24	–
Phenyl ethyl acetate	0.18	–	0.05	–
Phenyl prophyl acetate	0.38	–	0.03	–
Methyl cinnamate	0.27	0.09	0.10	–
Cinnamyl acetate	8.78	1.00	0.12	0.10
Benzyl benzoate	1.10	6.01	0.16	–
Eugenyl acetate	0.40	0.64	0.10	1.00
Linalyl acetate	–	–	0.60	–
Eugenol	4.15	76.74	0.21	0.45
Methyl isoeugenol	–	–	–	0.22
Isoeugenol	0.08	0.07	0.04	–
Safrole	0.08	0.08	1.32	0.32
Methyl eugenol	0.15	–	0.12	1.79
Methoxy eugenol	–	–	0.17	–
α-Cubebene	–	–	0.68	0.20
α-Ylangene	0.70	0.14	0.03	2.71
β-Caryophyllene	8.00	3.47	0.62	5.63
Sesquiterpene hydrocarbons	–	–	–	0.50
Sesquiterpene hydrocarbon	–	–	0.08	1.41
Unknown	–	–	–	0.30

	1.30	0.57	0.12	1.40
α-Humulene	1.30	0.57	0.12	1.40
β-Cubebene	–	–	–	1.08
β-Farnesene	–	–	–	1.58
α-Gurjugene	–	–	–	2.32
β-Cadinene	–	–	–	2.78
β-Gurjugene	–	–	–	0.70
α-Narrolene	–	–	–	
δ-Cadinene & δ-Cadinene	–	–	–	36.00
Cadin-1,4-diene	–	–	–	1.59
α-Cadinene	–	–	–	2.84
Calamanene	–	–	–	0.17
α-Cadinene	–	–	–	2.8
Bisabolol	–	–	–	0.35
Unknown	–	–	–	0.39
β-Nerolidol	–	–	–	0.84
α-Cadinol	–	–	–	1.12
Unknown	–	–	–	0.61
Elemol	–	–	–	0.13
Sesquiphellandrol	–	–	–	0.30
Unknown	–	–	–	1.8
Cadinol-T	–	–	–	7.7
Muurolol-T	–	–	–	1.05
Unknown	–	–	–	4.34
Unknown	–	–	–	1.79

Source: Paranagama, 1999.

Note
* New compounds isolated.

Table 3.6 Constituents of cinnamon fruit oil

Peak no.	Component	Retention time Carbowax 20 M	Method of identification		% Relative abundance
			SPB – 1		
1	α-Pinene	3.6	5.5	GC	2.2
2	Unknown	4.1	–	–	0.2
3	Camphene	4.2	5.7	GC/MS	0.3
4	β-Pinene	5.1	6.7	GC/MS	1.6
5	Myrcene	6.6	7.4	GC/MS	0.4
6	α-Terpinene	7.0	–	GC	tr
7	Limonene	7.6	8.8	GC/MS	1.0
8	1,8-Cineole	7.7	–	GC	tr
9	β-Phellandrene	7.8	–	GC	tr
10	cis-Ocimene	9.0	–	GC	tr
11	γ-Terpinene	9.3	9.8	GC	tr
12	trans-Ocimene	9.7	–	GC	tr
13	p-Cymene	10.3	8.4	GC/MS	tr
14	Terpinolene	10.8	11.7	GC/MS	0.3
15	α-Cubebene	20.1	–	GC	0.2
16	α-Ylangene	21.6	29.7	GC/MS	2.7
17	Benzaldehyde	23.5	–	GC	0.5
18	Linalool	25.8	–	GC/MS	0.1
19	β-Caryophyllene	27.6	32.1	GC/MS	5.6
20	Fenchylalcohol	27.8	–	GC/MS	0.4

Table 3.6 (Continued)

Peak no.	Component	Retention time Carbowax 20 M	Method of identification SPB – 1		% Relative abundance
21	Sesquiterpene hydrocarbon	28.1	–	GC/MS	0.5
22	Terpinen-4-ol	28.4	–	GC	0.3
23	Sesquiterpene hydrocarbon	30.1	–	GC/MS	1.4
24	α-Humulene	31.4	34.1	GC/MS	1.4
25	β-Cubebene	31.6	–	GC/MS	1.4
26	Unknown	32.7	–	–	0.3
27	β-Farnesene	33.0	–	GC	1.6
28	α-Gurjugene	33.9	36.9	GC/MS	2.3
29	Isoborneol	39	–	GC/MS	0.3
30	β-Cadinene	34.6	35.6	GC/MS	2.8
31	β-Gurjugene	34.9	–	GC/MS	0.7
32	α-Muurolene	35.4	37.4	GC/MS	4.4
33	δ- and γ-Cadinene	37.8	38 & 38.9	GC/MS	36.0
34	Cadinene 1–4 diene	38.6	39.2	GC/MS	1.6
35	α-Cadinene	39.2	–	GC/MS	2.8
36	Unknown	41.0	–	–	0.2
37	Calamanene	41.2	–	GC/MS	0.3
38	Safrole	43.6	–	GC/MS	0.2
39	Unknown	45.5	–	–	0.2
40	Phenyl propyl acetate	46.9	–	GC	0.2
41	Unknown	50.7	–	–	0.2
42	Cinnamaldehyde	52.5	–	GC/MS	0.3
43	Bisabolol	52.6	–	GC/MS	0.4
44	Sesquiterpene alcohol	52.9	–	GC/MS	0.4
45	α-Nerolidol	53.2	42.7	GC	0.8
46	α-Cadinol	53.4	–	GC/MS	1.1
47	Unknown	53.9	–	–	0.1
48	Elemol	54.3	–	–	0.1
49	Sesquiphellandrol	56.0	–	GC	0.3
50	Unknown	56.4	–	–	0.2
51	Unknown	57.2	–	–	0.2
52	Cinnamyl acetate	58.0	32.6	GC/MS	0.1
53	Cadinol-T	59.2	45.4	GC/MS	7.7
54	Sesquiterpene alcohol	59.8	–	GC/MS	1.8
55	Muurolol-T	60.4	48.8	GC/MS	1.0
56	Sesquiterpene alcohol	61.9	–	GC/MS	1.0
57	Eugenyl acetate	63.4	–	GC	0.4

Syamasunder *et. al.* (2000) reported the constitution of essential oil from unripe and ripe fruits of cinnamon. The GC-MS study indicated more than 90 components. Variations exist in the composition of the two oils. The predominant components in the oil of unripe fruits are δ-cadinene, α-pinene and β-pinene, while in the oil of ripe fruits

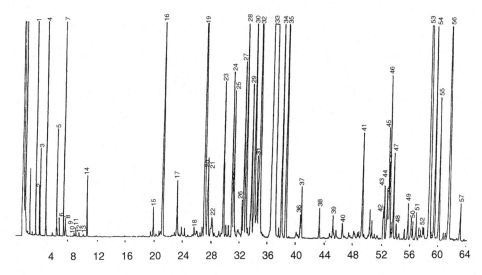

Figure 3.6 GLC chart of laboratory distilled fruit oil on Carbowax 20 M capillary column.

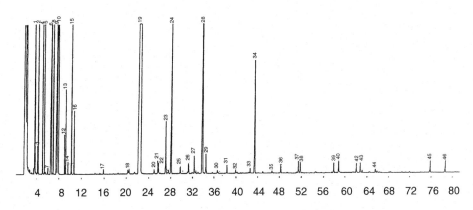

Figure 3.7 GLC chart of laboratory distilled cinnamon root bark oil on Carbowax 20 M capillary column.

γ-cadinene, α-pinene, (E)-cinnamyl acetate and α-muurolene are more prominent (Table 3.7). This study indicated the higher concentration of mono- and sesquiterpenoids and lower concentrations of phenolic and phenylpropanoids in the oil of cinnamon fruits.

Jayaprakash *et al.* (2000) analysed the chemical composition of oil distilled from the cinnamon flower. The cinnamon flower on distillation yielded about 0.5% oil (v/w); GC-MS analysis of this oil led to the identification of 26 compounds (constituting approximately 97% of the volatile oil) (Table 3.8). The major constituent was (E)-cinnamyl acetate (41.98%), the others being trans-α-berga-montene (7.97%), caryophyllene oxide (7.29%), α-cardinol (6.35%), globulol (3.80%), tetradecanol (5.05%), etc. Shikimic acid and mevalonic acid metabolites account for 52% and 37%, respectively. About, 9% was contributed from straight chain compounds. Esters are present to an extent of 45% ((E)-cinnamyl acetate and 2-phenyl-ethyl benzoate). Other compounds from the shikimic acid metabolites are benzaldehyde, hydrocinnamaldehyde, (E)-cinnamaldehyde and

Table 3.7 Chemical composition of the oil of cinnamon flowers

Sl. no.	Compound	%
1	(Z)-hex-3-en-1-ol	0.10
2	Benzaldehyde	0.35
3	Hydrocinnamaldehyde	0.18
4	Borneol	0.17
5	α-Terpineol	0.15
6	(E)-cinnamaldehyde	0.38
7	(E)-cinnamylalcohol	0.49
8	3-Phenylpropyl acetate	1.99
9	α-Copaene	2.40
10	Trans-α-bergamontene	7.97
11	(E)-cinnamylacetate	41.98
12	α-Humulene	2.40
13	Germacrene D	1.31
14	δ-Cadinene	2.97
15	Nerolidol	0.95
16	Caryophyllene oxide	7.29
17	Globulol	3.80
18	Tetradecanol	5.05
19	α-Cadinol	6.35
20	Cadalene	1.39
21	Epi-α-bisabolol	0.73
22	n-Heptadecane	2.14
23	Benzyl benzoate	3.19
24	Pentadecanol	0.71
25	2-Hexadecanone	0.71
26	2-Phenylethylbenzoate	0.44

Source: Jayaprakash *et al.*, 2000.

Table 3.8 Composition of *C. verum* fruit oils

Kovat's index	Name of the compound	Unripened fruit	Ripened fruit
838	(Z)-3-hexenol	0.06	0.06
924	Tricyclene	0.02	0.06
935	α-Pinene	11.47	11.52
947	Camphene	1.31	3.16
976	β-Pinene	10.51	5.64
983	Myrcene	1.31	1.72
994	α-Phellandrene	0.79	0.65
1003	δ-3-Carene	0.04	0.02
1008	p-Cymene	0.14	0.34
1011	α-Terpinene	0.25	0.06
1021	Limonene	2.96	3.87
1029	(Z)-β-Ocimene	0.08	0.14
1040	(E)-β-Ocimene	0.12	0.24
1049	γ-Terpinene	0.10	0.17
1076	Terpinene-4-ol	0.45	0.81
1082	Terpinelone	0.35	–
1085	Linalool	5.28	–
1101	Endo-fenchol	0.34	0.05
1130	β-Terpineol	2.91	0.08
1135	Citronellal	0.10	0.07

1147	Isoborneol	0.36	0.55
1158	Borneol	0.09	0.0
1165	Terpinen-4-ol	0.92	1.08
1206	Nerol	1.34	0.14
1241	(E)-cinnamyl aldehyde	1.48	0.12
1241	Geraniol	0.04	–
1258	Linalyl acetate	0.03	0.05
1325	Eugenol	0.11	0.04
1343	Thymol acetate	0.25	0.03
1369	(Z)-cinnamyl acetate	2.00	1.91
1386	β-Elemene	0.83	0.11
1413	(E)-cinnamyl acetate	7.11	8.62
1420	β-Caryophyllene	0.04	0.19
1432	(E)-methyl cinnamate	0.05	0.96
1449	(E)-β-farnesene	0.16	1.56
1451	α-Amorphene	1.40	1.31
1458	α-Humulene	1.84	2.38
1473	γ-Muuurolene	0.41	0.31
1486	Germacrene-D	0.55	1.95
1488	α-Patchoulene	0.55	0.27
1502	α-Muurolene	0.04	8.22
1513	γ-Cardinene	8.05	23.48
1521	δ-Cardinene	19.15	1.40
1527	Cis-calamene	1.07	2.33
1543	β-Elemol	1.81	–
1552	(E)-nerolidol	0.09	0.14
1555	Caryophyllene-alcohol	0.05	0.49
1567	Spathulenol	0.22	0.56
1576	Caryophyllene oxide	0.09	0.05
1581	Globulol	0.09	0.29
1599	Humulene epoxide I	0.36	0.42
1606	Humulene epoxide II	0.05	0.12
1612	Epi-cubenol	0.48	0.52
1624	Cubenol	4.23	5.52
1636	α-muurolol	1.29	1.71

Source: Syamasundar *et al.*, 2000.

(E)-cinnamyl alcohol. Among the mevalonic acid metabolites sesquiterpenes were the major compounds. The monoterpene portion was very low (21%); α-bergamontene, α-copaene, α-humulene, and δ-cadinenes were the major sesquiterpenes, α-cadenol and globulol were the major sesquiterpene alcohols. The only oxide present was caryophyllene oxide. Fifteen compounds present in the fruit volatiles were found in the flower oil constituting 74% of the oil. The chemical composition of flower oil is distinctly different from bark and leaf oils. (See Annex 3.1 for other secondary metabolites.)

Conclusion

The species *C. verum*, *C. cassia* and *C. camphora* contain volatiles both common and unique to each species. The three types of oils produced from cinnamon have the same array of aromatic and terpenoid compounds but in characteristically different proportions. Cinnamic aldehyde is the major compound in the stem bark oil containing

65–75% of the total volatiles, eugenol accounts for about 70–75% of leaf oil, while 70–90% of the root bark oil accounts for camphor.

Cassia produces only one type of oil, usually called bark oil, obtained by distilling leaves and bark together. Almost 95% of the oil consists of cinnamic aldehyde. *C. camphora* is not a spice, the oil of camphora usually contains only terpenes, camphor and cineole being the major compounds.

Tracer studies indicate that phenylpropanoids, eugenol and cinnamic aldehydes are synthesised via the Shikimic acid pathway. The phenylalanine skeleton is retained in the cinnamic aldehyde biosynthesis. The end carbon of the alanine chain of phenylalanine does not contribute to the eugenol biosynthesis. During the eugenol biosynthesis the end carbon atom of the allyl branch is not derived from phenylalanine, but from a methyl donor such as methionine.

The incorporation of labelled mevalonic acid into terpenes indicated that the normal mevalonic acid pathway was operating in the cinnamon plant during the biosynthesis of terpenoid compounds. Tracer studies also indicated that the leaf was the predominant site of biosynthesis of these compounds.

References

Angmor, J.E., Dicks, D.M., Evans, W.C. and Santra, D.K. (1972) Studies on *Cinnamomum zeylanicum* Part 1. The Essential Oil components of *C. zeylanicum* Nees grown in Ghana. *Planta Med.*, 21, 416–420.

Angmor, J.E., Dicks, D.M. and Evans, W.C. (1979) Chemical changes in Cinnamon oil during the preparation of the bark oil. Biosynthesis of cinnamaldehyde and related compounds. *Planta Med.*, 35, 342–347.

Asakawa, Y., Komatsu, T., Hayashi, S. and Matsuura, T. (1971) Chemical components of the benzene extracts of *Cinnamomum loureirii* Nees (Lauraceae). *The Flav. Ind.*, 2, 114–119.

Attaway, J. and Buslig, B.S. (1968) Conversion of linalool to α-terpineol in citrus. *Biochem. Biophys. Acta*, 164, 609–610.

Banthorpe, D.V., Charlwood, B.V. and Francis, M.J.O. (1972) The biosynthesis of monoterpenes. *Chem. Rev.*, 72, 115–126.

Banthorpe, D.V. and Wirz-Justice, A. (1969) Terpene biosynthesis. Part 1. Preliminary tracer studies on terpenoids and chlorophyll of *Tanacetum vulgare* L. *J. Chem. Soc.*, (C), 2694–2696.

Baruah, A.K.S., Bhagat, S.D., Hazarika, J.N. and Saikia, B.K. (1975) Volatile oil of *Cinnamomum camphora* grown in Jorhat, Assam. *Indian J. Pharm.*, 37, 39–41.

Battersby, A.R., Laing, D.G. and Ramage, R. (1972) *Biosynthesis*. Part XIX. Concerning the biosynthesis of (−) camphor and (−)-borneol in *Salvia officinalis*. *JSC Perkin*, 1, 2743–2747.

Beytia, E., Valenzuela, P. and Cori, O. (1969) Terpene biosynthesis: Formation of nerol, geraniol, and other prenols by an enzyme system from *Pinus radiata* seedling. *Arch. Biochem. Biophys.* 129, 346–356.

Bhramaramba, A. and Sidhu, G.S. (1963). Chromatographic Studies on Indian cinnamon leaf oil. *Perfumer & Essential oil Res.*, 54, 732–738.

Birch, A.J. (1963) Biosynthetic pathways. In: *Chemical Plant Taxonomy* (ed. T. Swain), pp. 141–166. Academic Press, London.

Bohm, B.A. (1965). Shikimic Acid. *Chem. Rev.*, 65, 435–466.

Brown, E.G. (1955) Cinnamon and cassia: Sources, production and trade, Part 1. Cinnamon. *Colonial Pl. Anim. Prod.*, 5, 257–280.

Brown, E.G. (1956) Cinnamon and cassia: Sources, production and trade, Part II. *Cassia. Colonial Pl. Anim. Prod.*, 6, 96–116.

Bullock, J.D. (1965) *The Biosynthesis of Natural Products*, pp. 77–93. McGraw Hill, London.

Canonica, L., Manitto, P., Monti., D. and Sanchez, M. (1971) Biosynthesis of allylphenols in *Ocimum basilicum* L. *Chem. Commun.*, 1108–1109.

Carrol, M.F. and Price, W.J. (1964) Routine analysis of essential oils by infrared spectroscopy. *Perfumer & Essential oil Res.*, **55**, 394–395.

Chalchat, J.C. and Valade, I. (2000) Chemical composition of leaf oils of *Cinnamomum* from Madagascar: *C. zeylanicum* Blume, *C. camphora* L., *C. fragrans* Baillon and *C. angustifolium J. Essential Oil Res.*, **12**, 537–540.

Cori, O. (1969) Terpene biosynthesis: Utilization of neryl pyrophosphate by an enzyme system from *Pinus radiata* seedlings. *Arch. Biochem. Biophys.*, **135**, 416–418.

Datta, P.R., Susi, H., Higman, H.C. and Filipic, V.J. (1962) Use of gas chromatography to identify geographical origin of some spices. *Food Technol.*, **16**, 116–119.

Davis, B.D. (1951) Aromatic biosynthesis 1. The Role of Shikimic Acid. *J. Biol. Chem.*, **191**, 315–325.

Davis, B.D. (1955) Intermediates in amino acid biosynthesis. *Adv. Enzymol.*, **16**, 247–312.

Davis, B.D. (1958) On the importance of being ionized. *Arch. Biochem. Biophys.*, **78**, 497–509.

Fujita, Y., Fujita, S. and Hasegawa, T. (1975) Biological transformation of linalool administered to the living plants. *J. Chem. Soc. (Japan)*, **4**, 711–713.

Geissman, T.A. (1963) The biogenesis of phenolic plant products. In: P. Bernfels (ed.), *Biogenesis of Natural Products*, p. 743, Pergamon Press, London.

George-Nascimento, C. and Cori, O. (1971) Terpene biosynthesis from geranyl and neryl pryophosphates by enzymes from orange flavour. *Phytochemistry*, **10**, 1803–1810.

Goodwin, T.W. (1965) Regulation of terpenoids synthesis in higher plants. In: J.B. Pridham and T. Swain (eds), *Biosynthetic Pathways in Higher Plants*, pp. 57–71, Academic Press, London.

Gottlieb, O.R. (1972) Chemosystematics of the Lauraceae. *Phytochemistry*, **11**, 1537–1570.

Guenther, E. (1949) *The Essential Oils*, Vol. 2, p. 516. van Nostrand Co., New York.

Guenther, E. (1950) *The Essential Oils.*, Vol. 4, pp. 213–329. van Nostrand Co., New York.

Higuchi, T. (1957) Biochemical studies of lignin formation II. *Physiol. Plantarum*, **10**, 621–632.

Jayaprakash, G.K., Rao, L.J.M. and Sakariah, K.K. (2000) Chemical composition of the flower oil of *Cinnamomum zeylanicum* Blume. *J. Agric. Food Chem.*, **48**, 4294–4295.

Jayewardene, A.L. (1987) Practical Mannual on the Essential Oil Industry, Wijesekera, R.O.B. (ed.) UNIDO, Vienna.

Kaneko, K. (1960) Biogenetic studies of natural products. IV. Biosynthesis of anethole by *Foeniculum vulgare*. *Chem. Pharm. Bull.* (Tokyo), **8**, 611–614.

Kaneko, K. (1961) Biogenetic studies of natural products. VI. Biosynthesis of anethole by *Foeniculum vulgare*. *Chem. Pharm. Bull.* (Tokyo), **9**, 108–109.

Kaneko, K. (1962) Biogenetic studies of natural products. VII. Biosynthesis of anethole by *Foeniculum vulgare*. *Chem. Pharm. Bull.* (Tokyo), **10**, 1085–1087.

Lawrence, B.M. (1967) A review of some commercial aspects of cinnamon. *Perf. Essent. Oil Rec.*, April, 236–241.

Lawrence, B.M. and Hogg, J.W. (1974) The Chemical composition of uncommon spices and condiments 1. Two *Cinnamomum* species from the Philippines. *Planta Med.*, **25**, 1–5.

Loomis, W.D. (1967) Biosynthesis and metabolism of monoterpenes. In: J.B Pridham (ed.), *Terpenoids in Plants*, pp. 59–82. Academic Press, London.

Loffelhardt, W. and Kindl, H. (1975) The conversion of L-phenylalanine into benzoic acid on the thylakoid membrane of higher plants. *Physiol. Chem.*, **356**, 487–493.

Manning, C.E.F. (1970) The market for cinnamon and cassia and their essential oils. *Tropical Products Institute, London, Report* G., **44**, p. 74.

Manitto, P., Monti, D. and Gramatica, P. (1974) Biosynthesis of phenylpropanoid compounds. Part 1. Biosynthesis of eugenol in *Ocimum basilicum. JCS Perkin*, **1**, 1727–1731.

Manitto, P., Gramatica, P. and Monti, D. (1975a) Biosynthesis of phenylpropanoid compounds. Part II. Incorporation of specifically labelled cinnamic acids into Eugenol. *JCS Perkin*, **1**, 1548–1551.

Manitto, P., Gramatica, P. and Ranzil, B.M. (1975b) Stereochemistry of the decarboxylation of phenolic cinnamic acids by *Saccharomyces cerevisiae*. *Chem. commun.*, 442–443.

Montes, A.L. (1963) Gas chromatography and adulteration of essential oils. *Soc. Arg. Invest. Prod. Arom. Bol.*, No. 5, p. 6.

Mubarak, A.M. and Paranagama, P.A. (1991) Analysis of Sri Lanka Essential Oils by Gas Chromatography and Mass spectroscopy, M.Phil. Thesis, Kelaniya University, Sri Lanka.

Nath, S.C., Modon, G. and Baruah (1996) Benzyl benzoate, the major component of the leaf and stem bark oil of *Cinnamomum zeylanicum* Blume. *J. Essent. Oil. Res.*, 8, 327–328.

Neish, A.C. (1960) Biosynthetic pathways of aromatic compounds. *Ann. Rev. Plant Physiol.*, II, 55–80.

Neish, A.C. (1968) Monomeric intermediates in the biosynthesis of lignin. In: A.C. Neish and K. Freudenberg (eds), *Constitution and Biosynthesis of Lignin*, pp. 1–43. Springer-Verlag, New York.

Nicholas, H.J. (1967) The biogenesis of terpenes in plants; In: P. Bernfeld (ed.), *Biogenesis of Natural Compounds*, p. 829. Pergamon Press, London.

Park, R.B. and Bonner, J. (1958) Enzymic synthesis of rubber from mevalonic acid. *J. Biol. Chem.*, 233, 340–343.

Parker, W., Roberts, J.S. and Ramage, R. (1967) Sesquiterpene biogenesis. *Quart. Rev.*, 21, 331–363.

Potty, V.H. and Bruemmer, J.H. (1970). Formation of isoprenoid pyrophosphates from Mevalonate by orange enzymes. *Phytochemistry*, 9, 1229–1237.

Raina, V.K., Srivastava, S.K., Aggarwal, K.K., Ramesh, S. and Kumar, S., (2001) Essential oil composition of *Cinnamomum zeylanicum* Blume leaves from Little Andaman, India. *Flavour fra. J.*, 16, 374–376.

Rogers, L.J., Shah, S.P.J. and Goodwin, T.W. (1968) Compartmentation of biosynthesis of terpenoids in green plants. *Photosynthetica*, 2, 184–207.

Ross, M.S.F. (1976) Analysis of cinnamon oils by high-pressure liquid chromatography. *J. Chromatog.*, 118, 273–275.

Ruzicka, L., Escheumoser, A. and Heuser, H. (1953) The isoprene rule and the biogenesis of terpenic compounds. *Experimentia*, 9, 357–396.

Senanayake, U.M., Edwards, R.A. and Lee, T.H. (1975) Simple solid injection method for qualitative and quantitative estimation of essential oils. *J. Chromatog.*, 116, 468–471.

Senanayake, U.M., Wills, R.B.H. and Lee, T.H. (1977) Biosynthesis of eugenol and cinnamic aldehyde in *Cinnamomum zeylanicum*. *Phytochemistry*, 16, 2032–2033.

Senanayake, U.M. (1977) *The Nature, description and biosynthesis of volatiles of Cinnamomum* spp. Ph.D. thesis. University of New South Wales, Kensington, Australia.

Senanayake, U.M. and Wijesekera, R.O.B. (1989) Volatiles of Cinnamomum Species In: Bhattacharyya, S.O., Sen, N. and Sethi, K.L. (eds) *Proceedings 11th International Congress of Essential oils, Fragrances and Flavours*. Vol. IV, pp. 103–120. Oxford & IHB Publishing Co. Pvt. Ltd. New Delhi, Bombay, Culcatta.

Shankaranarayana, M.L., Abraham, K.O., Raghavan, B. and Natarajan, C.P. (1975) Infrared identification of compounds separated by gas and thin layer chromatography: application to flavour analysis. *Critical Rev. in Food Sci. Nutrition*, 7, 271–315.

Siegel, S.M. (1955) The biochemistry of lignin formation. *Physiol. Plantarum*, 8, 20–32.

Simonsen, J.L. and Barton, D.H.R. (1952) *The Terpenes* Vol. III, *The sesquiterpenes, diterpenes and their derivatives*. 590 pp, Cambridge University Press, U.K.

Sprinson, D.B. (1960) The biosynthesis of aromatic compounds from D-glucose. *Adv. Carbohydrate Chem.*, 15, 235–270.

Staby, G.L., Hackett, W.P. and Hertogh, A.A. De (1973) Terpene biosynthesis in cell-free extracts and excised shoots from Wedgwood Iris. *Plant Physiol.*, 52, 416–421.

Suga, T., Shishibori, T. and Bukeo, M. (1971) Biosynthesis of linalool in higher plants. *Phytochemistry*, 10, 2725–2726.

Syamasunder, K.V., Ramesh, S. and Chandrasehara, R.S. (2000) Volatile constituents of *Cinnamomum zeylanicum* Blume fruit oil. In (Anon.) *Spices and Aromatic Plants: Challenges and opportunities lies in the new century*, Indian Soc. For Spices, IISR, Calicut, India, pp. 284–286.

Tavormina, P.A., Gibbs, M.H. and Huff, J.W. (1956) The utilization of β-Hydroxy-β-Methyl-δ-Valerolactone in cholesterol biosynthesis. *JACS*, **78**, 4498–4499.

Ter Heide, R. (1972) Qualitative analysis of the essential oil of cassia (*Cinnamomum cassia* Blume). *J. Agric. Food Chem.*, **20**, 747–51.

Valenzuela, P., Beytia, E., Cori, O. and Yudelevich, A. (1966) Phosphorylated intermediates of terpene biosynthesis in *Pinus radiata*. *Arch. Biochem. Biophys.*, **113**, 536–539.

Weissenbôck, G., Tevini, M. and Reznik, H. (1971) Unber das-Vorkommen von Flavonviden in Chloroplasten von *Impatiens balsamina* L. *Z. pflanzenphysiol.*, **64**, 274–277.

Wijesekera, R.O.B. and Jayewardena, A.L. (1972) Recent development in the production of spices and their essential oils in Ceylon. In: *Proc. Conf. Spices*, pp. 159–167, Tropical Products Institute, London.

Wijesekera, R.O.B. and Fonseka, K.H. (1974) Essential Oils II. Infrared spectroscopy in the analysis of the volatile oils in cinnamon. *J. Natl. Sci. Coun. Sri Lanka*, **2**, 35–49.

Wijesekera, R.O.B. and Jayewardene, A.L. (1974) Essential Oils II. Chemical constituents of the volatile oil from the bark of a rare variety of cinnamon. *J. Natl. Sci. Coun. Sri Lanka*, **2**, 141–146.

Wijesekera, R.O.B., Jayewardene, A.L. and Rajapakse, L.S. (1974) Volatile constituents of leaf, stem and root oils of cinnamon (*C. zeylanicum*). *J. Sci. Food Agric.*, **25**, 1211–1220.

Wijesekera, R.O.B., Ponnuchamy, S. and Jayewardena, A.L. (1975) *Cinnamon*, CISIR, Colombo M.D. Gunasena Printers 1975.

Wijesekera, R.O.B. (1978) The Chemistry and Technology of Cinnamon. *CRC Critical Review in Food Science and Nutrition*, **10**, 1–30.

Wolf, D.E., Hoffman, C.H., Aldrich, P.E., Skeggs, H.R., Wright, L.D. and Folders, K. (1956) β-Hydroxy-β-Methyl-δ-Valerolactone, A new biological factor. *JACS*, **78**, 4499.

Young, M.R., Towers, G.H.N. and Neish, A.C. (1966) Taxonomic distribution of ammonia-lyases for L-phenylalanine and L-tyrosine in relation to lignification. *Can. J. Botany*, **44**, 341–349.

Zenk, M.H. and Gross, G.G. (1972) The enzymic reduction of cinnamic acids. *Recent Adv. in Phytochemistry*, **44**, 87–106.

Zurcher, K., Hadorn, H. and Strach, Ch. (1974) Versuche zur Beurteilung der Zimtqualitat durch gas chromatographische Trennung des atherischen Oles. *Mitt. Gebiete Lebensm Hyg.*, **65**, 440–452.

Annex 3.1 Additional information provided by the editors

Secondary metabolites from cinnamon and cassia

Many interesting chemical substances, other than those present in the essential oils, have been isolated from the bark of cinnamon and cassia. Some of these compounds are responsible for the pharmacological–toxicological properties of cinnamon and cassia.

Isogai *et al.* (1976, 1977, 1979) isolated cinnzeylanine and cinnzeylanol from cinnamon bark. Cinnzeylanine[1] has a molecular formula of $C_{22}H_{34}O_8$, is a diterpene, having a pentacyclic skeleton (M_P 265 ~ 267°C; $[\alpha]^{27}_D$ + 45 ($C = 20$, CH_3OH). Cinnzeylanine[2] is derived from the above, and has a molecular formula $C_{20}H_{32}O_7$ (MP 125 ~ 127°C, $[\alpha]^{15}_D$ + 18). Alkali hydrolysis converts (1) to (2) (Fig. 3A.1). Other compounds derived from (1) and (2) are anhydrocinnzeylanine, anhydrocinnzeylanol, dioxocinnzeylanine, dioxocinnzeylanol, anhydrodioxocinnzeylanol and trioxocinnzeylanol. Cinnzeylanine and cinnzeylanol show strong insecticidal properties, as a dose of 16 ppm in the diet leads to complete mortality of silkworm larvae.

108   *U.M. Senanayake and R.O.B. Wijesekera*

A group of compounds known as cinncassiols have been isolated from cinnamon and cassia barks (Nohara *et al.*, 1980a, 1980b, 1980c, 1981, 1982, 1985; Yagi *et al.*, 1980). They are diterpenoids, designated as: Cinncassiol A, Cinncassiol B, Cinncassiol C1, Cinncassiol C2, Cinncassiol C3, Cinncassiol D1, Cinncassiol D2, Cinncassiol D3, Cinncassiol D4, Cinncassiol E and their glucosides (Fig. 3A.2). These compounds exhibit various biological activities. Cinncassiol C1 and D4 and their glucosides exhibit antiallergic activity. Cinncassiol B and its glucoside show anti-complement activity (Fig. 2–9).

Gellert and Summons (1970) identified the important alkaloids present in cinnamon bark as benzylisoqinolines: (1) (+)-1-(4′-hydroxybenzyl)-6,7-methylenedioxy-1,2,3,4-tetrahydroisoquinoline (norcinnamolaurine); (2) (−)-cinnamolaurine; (3) (+)-reticuline; (4) aporphine, and, (5) (+)-corydine. Norcinnamolaurine is unique to the cinnamon bark (Fig. 10–12). The alkaloid's reticuline and laurolitsine were isolated by Tomita and Kozuka (1964) from *C. camphora*. (+)-corydine was isolated by Craig and Roy (1965).

Cinnamolaurine and norcinnamolaurine have optical rotations of opposite sign at the sodium D-line. However, they both possess the same absolute configuration and that N-methylation of (+)-norcinnamolaurine gives (–)-cinnamolaurine. Both belong to the D-series of benzyl iso-quinoline alkaloids.

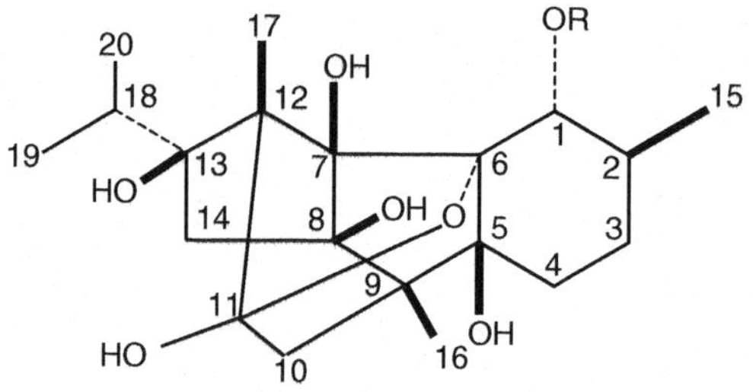

1   Cinnzeylanine 1 R = Ac
    Cinnzeylanine 2 R = H

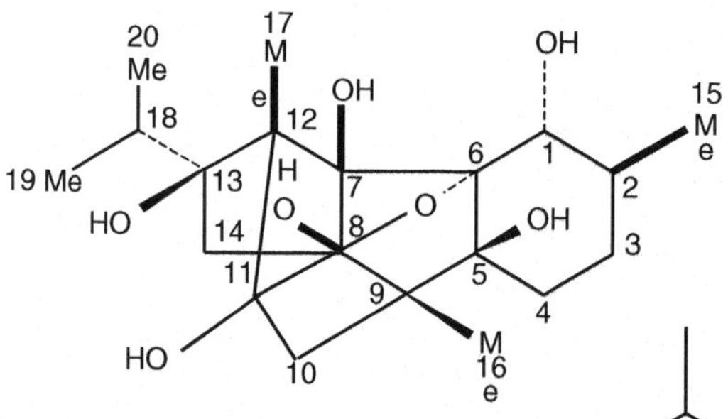

2   Cinncassiol A

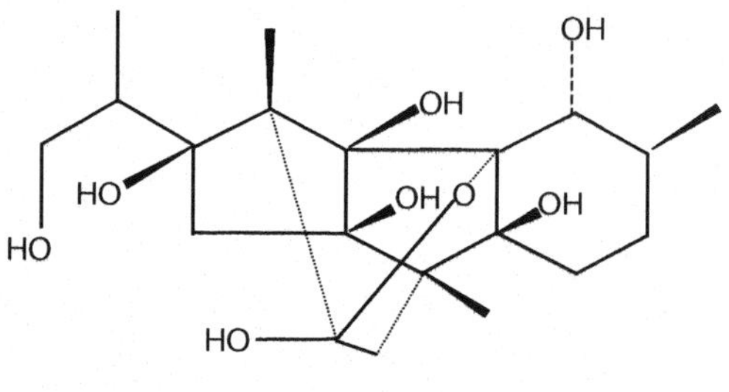

3   Cinncassiol B

Figure 3A.1 Secondary metabolites from cinnamon and cassia barks (see text for explanation).

Chemistry of Cinnamon and Cassia 109

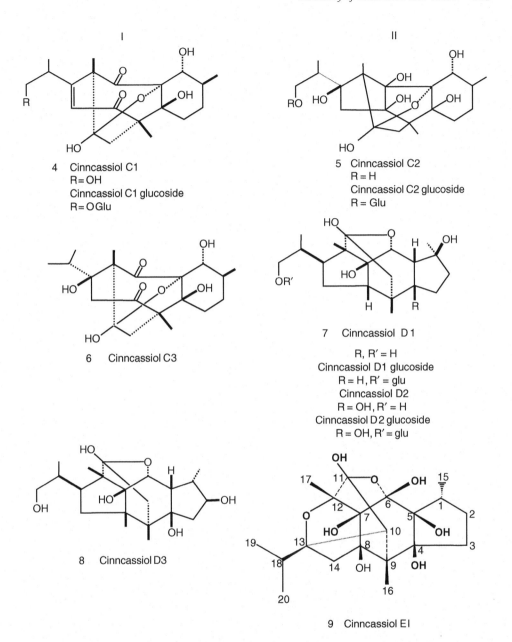

Figure 3A.2 Secondary metabolites from barks of cinnamon and cassia (see text for explanation).

Cinnamon bark also contains large amounts of condensed tannins. Morimoto *et al.* (1986a,b) suggested that condensed tannin is the main component responsible for the medicinal effect of cinnamon bark. Tannins form a large group of poly-phenolic compounds, and are generally classified into two groups – (1) hydrolysable tannins which are esters composed of polyphenolic acids (eg: gallic acid and hexahydroxydiphenic acid) and glucose (or some other sugar or polyalcohol); and (2) condensed tannins, which are condensates of flavan units, such as (+)-catechin and (−)-epicatechin. Cinnamon and cassia barks contain condensed tannins, which are also known as proanthocyanidins, since they release anthocyanidin when heated in presence of acid. In condensed tannins, the condensation occurs mostly between C_4 of a flavan unit and C_8 and/or C_6 of another unit. A large variety of condensed tannins exists in the plant kingdom, and in cinnamon and cassia bark the tannins are the condensates of (+)-catechin and (−)-epicatechin (Fig. 3A.3).

Nonaka and Nishioka (1980, 1982), Nonaka et al. (1985) and Morimoto et al. (1985a,b, 1986a,b) conducted detailed analysis of tannins and related compounds from the barks of cinnamon, cassia and related species. They have isolated many tannin-related compounds such as flavone-3-ol glucosides, oligomeric procyanidins (named as cinnamtannins), several known proanthocyanidins, epicetechin etc. The cinnamtannins (designated as A_2, A_3 and A_4) are tetrameric, pentameric and hexameric procyanidins respectively, consisting exclusively of (−)-epicatechin units linked linearly through

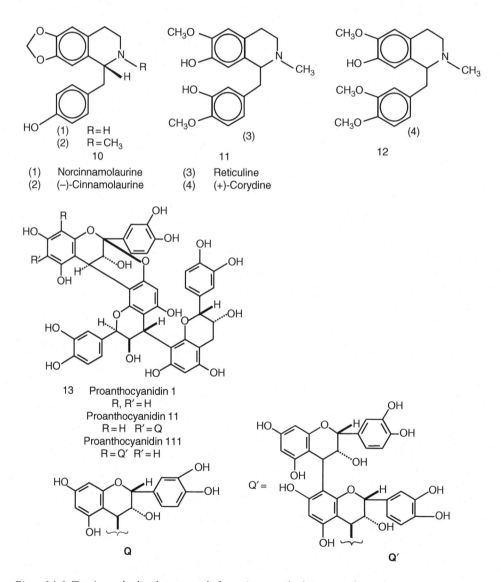

Figure 3A.3 Tannins and related compounds from cinnamon bark (see text for explanation).

C(4β)–C(8) bonds. Cinnamtannin A_2 is a pale amorphous powder, with anisaldehyde-sulphuric acid reagent it gives an orange-red colouration, characteristic of procyanidins. On acid catalysed thiolytic degradation A_2 gives ($-$)-epicatechin moiety. Similar degradation products are obtained in the case of A_3 and A_4 also. From analytical studies the interflavanoid linkage involved was shown as C (4β)-C8 (Morimoto *et al.* 1986a,b). The chemical nature of cinnamtannin A_2, A_3 and A_4 are thus shown to be:

A_2-epicatechin – ($4\beta \rightarrow 8$)-epicatechin – ($4\beta \rightarrow 8$)-epicatechin – ($4\beta \rightarrow 8$)-epicatechin (tetrameric); A_3-epicatechin ($4\beta \rightarrow 8$)-epicatechin ($4\beta \rightarrow 8$)-epicatechin ($4\beta \rightarrow 8$)-epicatechin ($4\beta \rightarrow 8$)-epicatechin (pentameric).

Morimoto *et al.* (1986b) further characterised procyanidins B-1, B-2, B-5 and B-7 from *C. cassia* bark. They also for the first time identified two new procyanidins, procyanidin B-2–8-C and 6-C-β-D-glucopyranosides (Fig. 3A.4).

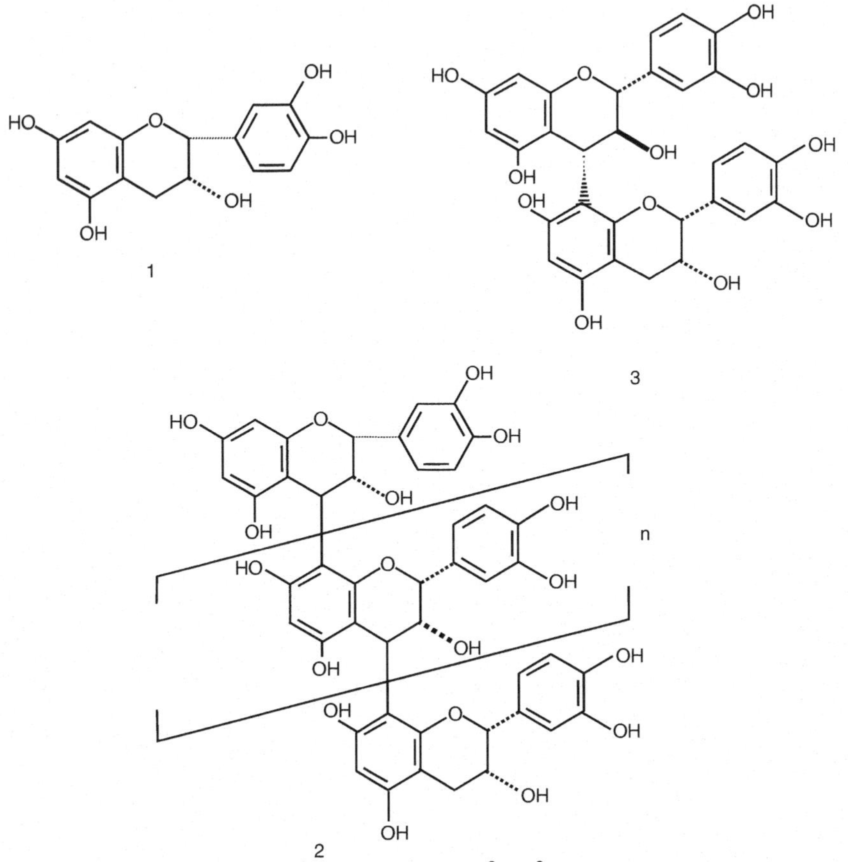

2: n = 0
4: n = 1

1 : (–)-epicatechin
3 : procyanidin B4

2 : procyanidin B2
4 : procyanidin C1

Figure 3A.4 Procyanidins from cassia bark (see text for explanation).

Yazaki and Okude (1990) reported that callus and suspensions cultures of *C. cassia* produced large amount of condenced tannin. They have isolated (–)-epicatechin and procyanidins B2, B4 and C1, which are precursors of condensed tannins, from callus cultures.

Cinnamon bark contains a significant amount of mucilaginous substances, which consists mainly of a water-extractable L-arabino-D-xylan and an alkali extractable D-glucan. Gowda and Sarathy (1987) studied the nature of the arbinoxylan from cinnamon bark. The arabinoxylan is composed of L-arabinose and D-xylose in the molar ratio 1.6:1.0. It contained a $(1 \rightarrow 4)$-linked β-D-xylan backbone in which each D-xylopyranosyl residue is substituted at both 0–2 and 0–3 with L-arabinofuranosyl, 3–0-α-D-xylopyranosyl-L-arabinofuranosyl and 3–0-L-arabinofuranosyl-arabinofuranosyl-L-arabinofuranosyl groups.

References

Craig, J.C. and Roy, S.K. (1965) Tetrahedron, **21**, 401 (cited from Gellert and Summons 1970).

Gellert, E. and Summons, R.E. (1970) Alkaloids of the genus *Cinnamomum*. II. The alkaloids of the bark of *Cinnamomum* sp. *Aust. J. Chem.*, **23**, 2095–2099.

Gowda, C. and Sarathy, C. (1987) Structure of an L-arabino-D-xylan from the bark of *Cinnamomum zeylanicum. Carbohydrate Res.*, **166**, 263–269.

Isogai, A., Suzuki, A., Tamura, S., Murakoshi, S., Ohashi, Y. and Sasada, Y. (1976) Structure of cinnzeylanine and cinnzeylanol, polyhydroxylated pentacyclic diterpenes from *Cinnamomum zeylanicum* Nees. *Agr. Biol. Chem.*, **40**, 2305–2306.

Isogai, A., Suzuki, A. and Tamura, S. (1977) Cinnzeylanine, a new pentacyclic diterpene acetate from *Cinnamomum zeylanicum. Acta Cryst.*, **B33**, 623–626.

Isogai, A., Murakoshi, S., Suzuki, A. and Tamura, S. (1979) Chemistry and biological activities of cinnzeylanine and cinnzeylanol, new insecticidal substances from *Cinnamomum zeylanicum* Nees. *Agric. Biol. Chem.*, **41**, 1779–1784.

Morimoto, S., Nonaka, G.-I., Nishioka, I., Ezaki, N. and Takazawa, N. (1985a) *Chem. Pharm. Bull.*, **33**, 2281 (cited from Morimoto, 1986a).

Morimoto, S., Nonaka, G.-I., Nishioka, I., Ezaki, N. and Takazawa, N. (1985b). Tannins and related compounds xxix. Seven new methyl derivatives of flavan-3-ols and 1,3-diaryl propan-2-ol from *Cinnamomum cassia, C. obtusifolium* and *Lindera umbellata* var. *membranacea. Chem. Pharm. Bull.*, **33**, 2281–2286.

Morimoto, S., Nonaka, G.-I. and Nishioka, I. (1986a) Tannins and related compounds XXXVII. Isolation and characterization of flavan-3-ol glucosides and procyanidin oligomers from cassia bark (*Cinnamomum cassia* Blume). *Chem. Pharm. Bull.*, **34**, 633–642.

Morimoto, S., Nonaka, G.-I. and Nishioka, I. (1986b) Tannins and related compounds. XXXIX. Procyanidin c-glucosides and an acylated flavan-3-ol glucoside from barks of *Cinnamomum cassia* Blume and *C. obtusifolium* Nees. *Chem. Pharam. Bull.*, **34**, 643–649.

Nohara, T., Tokubuchi, N., Kuroiwa, M. and Nishioka, I. (1980a). The constituents of Cinnamomi cortex. III. Structures of cinncassiol B and its glucoside. *Chem. Pharm. Bull.*, **28**, 2682–2686.

Nohara, T., Nishioka, I., Tokubuchi, N., Miyahara, K. and Kawasaki, T. (1980b) The constituents of cinnamomi cortex. II Cinncassiol C. a novel type of diterpene from Cinnamomi cortex. *Chem. Pharma. Bull.*, **28**, 1969–1970.

Nohara, T., Kashiwada, Y., Yomimatsu, T., Kido, C., Tokobuchi, M. and Nishioka, N. (1980c) Cinncassiol D_1 and its glucoside, novel pentacyclic diterpenes from cinnamomi cortex. *Tetrehedron Lett.*, **21**, 2647–2648.

Nohara, T., Kashiwada, Y., Murukami, T., Tomimatsu, T., Kido, M. and Yagi, A. (1981) Constitution of cinnamomi cortex. V. Structure of five novel diterpenes, cinncassiols D_1, D_1 glucoside, D_2, D_2 glucoside and D_3. *Chem. Pharma. Bull.*, **29**, 2451–2459.

Nohara, T., Kashiwada, Y., Tomimatsu, T. and Nishioka, I. (1982) Studies on the constituents of Cinnamomi cortex. VII. Two novel diterpenes from bark of *Cinnamomum cassia*. *Phytochemistry*, 21, 2130–2132.

Nohara, T., Kashiwada, Y. and Nishioka, I. (1985) Cinncassiol E, a diterpene from the bark of *Cinnamomum cassia*. *Phytochemistry*, 24, 1849–1850.

Nonaka, G. and Nishioka, I. (1980, 1982) (Cited from Morimoto *et al.*, 1986).

Nonaka, G., Nishimura, I. and Nishioka, I. (1985) (Cited from Morimoto *et al.*, 1986).

Tomita, M. and Kozuka, M. (1964) *J. Pharm. Soc.* Japan, 84, 365 (cited from Gellert and Summons, 1970).

Yagi, A., Tokubuchi, N., Nohara, T., Monaka, G., Nishioka, I. and Koda, A. (1980) The constituents of cinnamomi cortex. 1 Structures of cinncassiol A and its glucoside. *Chem. Pharm. Bull.*, 28, 1432–1436.

Yazaki, K. and Okuda, T. (1990) Condensed tannin production in callus and suspension cultures of *Cinnamomum cassia*. *Phytochemistry*, 29, 1559–1562.

Annex 3.2 Additional information provided by the editors on the chemical composition of cinnamon

Vernin *et al.* (1994) carried out detailed studies on cinnamon and cassia essential oils. They conducted GC-MS analyses and also headspace analyses. These workers separated a total of 142 compounds, and they could identify 90 compounds definitely and 11 compounds tentatively, while 41 compounds remained unidentified. Among the 90 identified compounds, 25 were reported for the first time in cinnamon oil. The more important compounds present in commercial samples of cinnamon and cassia oils are given in Table 3A.1 and in Annex 3.3.

The criteria used to distinguish between cinnamon and cassia oils are the small amounts of monoterpenes and sesquiterpenes in cassia oil as compared to cinnamon oil, and an increased content of benzaldehyde, *o*-methoxycinnamaldehyde, and coumarin in cassia oil.

Vernin *et al.* (1994) also carried out headspace analyses of the above essential oils, as well as the quills and powdered cinnamon and cassia. The results of the analyses provided valuable information about the content of volatile components responsible for the desirable characteristics of these spices and consequently for their quality (Table 3A.2).

Vernin *et al.* (1994) also carried out a detailed analysis of cinnamon and cassia obtained from different sources. The headspace composition for cinnamon and cassia quills of different origin is given in Table 3A.3.

In 1997 Koketsu *et al.* reported the leaf oil composition of *C. verum* grown in Parana, Brazil. Using GC and GC-MS, these authors resolved two chemical races among the trees, one rich in eugenol (94–95%), the other rich in eugenol and safrole (Table 3A.4).

Jirovetz *et al.* (1998) reported the comparative chemical composition of cinnamon leaf oils produced from leaves obtained from Cameroon compared with that of commercial sample of oil from Seychelles. The results of this GC and GC-MS study are summarised in Table 3A.5.

Jirovetz *et al.* (2001) recently reported the oil composition of a sample of oil produced from cinnamon leaves from Calicut, India. This GC and GC-MS study gave the results shown in Table 3A.6. The major components of this oil were found to be linalool (85.7%). It is the first time that cinnamon oil has been found in which linalool and not eugenol was the major component.

Mallavarapu and Ramesh (2000) analysed an oil produced from fruits of cinnamon using GC and GC-MS (Table 3A.4). The composition of fruit oil is found to be different from that of leaf oil.

Table 3A.1 Composition of cinnamon essential oils (bark oil A and B; leaf oil C; and, cassia essential oil (leaf and stem oil, D) (in %)[a]

Compound	Cinnamon			Cassia D
	A	B	C	
α-Pinene	0.07	0.035	0.3	traces
Camphene	0.04	0.049	0.14	traces
β-Pinene	0.03	0.02	0.12	traces
δ-3-Carene	0.01	0.015	0.025	traces
A-p-menthene	0.06	0.06	–	traces
Myrcene	0.08	0.07	traces	traces
Limonene	0.09	0.095	0.18	traces
β-Phellandrene (+1,8-cineole)	0.4	0.64	0.28	traces
p-Cymene	0.35	0.25	0.79	traces
α-Copaene	0.05	0.04	0.68	traces
Benzaldehyde	0.05	traces	0.17	0.94
Linalool	0.7	1.06	2.4	traces
β-Caryophyllene	1	2.0	3.33	traces
α-Humulene	0.2	0.62	0.61	–
Salicylaldehyde	–	–	–	–
α-Terpineol (+ borneol)	0.35	0.62	0.35 (0.065)	0.02 (borneol)
Hydrocinnamaldehyde	0.1	0.7	–	0.06
δ-Cadinene	–	–	0.2	–
Phenethyl acetate	–	–	traces	0.24
cis-Cinnamaldehyde	0.12	0.98	–	0.14
Safrole	–	–	1.05	–
o-Methoxybenzaldehyde	0.15	0.09	–	0.7
Caryophyllene oxide	traces	0.12	0.64	0.02
trans-Cinnamaldehyde	72	82.15	5.7	65.45
trans-Cinnamyl acetate	3.65	3.24	0.2	3.55
Eugenol	13.3	1.07	68.5	traces
trans-Cinnamyl alcohol	0.6	0.50	–	0.24
Acetyleugenol	–	–	–	0.15
o-Methoxycinnamaldehyde	0.8	0.3	traces	2.65
Isoeugenol	–	–	1.1	–
Coumarin	–	–	–	8.73
Benzyl benzoate	1.0	0.4	4.06	0.63
Phenethyl benzoate	–	–	traces	0.07
Total	95.18	95.12	90.89	83.79

Source: Vernin *et al.*, 1994.

Note

a Essential oils A, C, and D were commercial products. Oil B was obtained by steam distillation from commercially available quills (ESPIG). The composition was determined by gas chromatographically.

Table 3A.2 Head space analysis of cinnamon and cassia essential oils[a]

Compound	Cinnamon (%)		Cassia (%) Bark
	Bark	Leaves	
α-Pinene	0.18	–	0.1
Limonene	–	–	0.28

p-Cymene	13	21.35	0.82
α-Copaene	0.40	3.46	0.67
Benzaldehyde	<u>5.9</u>	<u>3.07</u>	<u>23.8</u>
Linalool	0.40	10.50	0.05
Bornyl acetate	0.90	–	0.62
Linalyl acetate	0.48	–	0.44
Salicylaldehyde	0.58	0.55	1.27
α-Terpineol + borneol	0.68	1.36	3.8
Hydrocinnamaldehyde	–	–	0.69
Phenethyl acetate	0.35	–	0.96
cis-Cinnamaldehyde	–	–	0.21
Phenethyl alcohol	–	–	0.24
o-Methoxybenzaldehyde	–	–	0.82
trans-Cinnamaldehyde	<u>41.5</u>	<u>1.44</u>	<u>39.5</u>
trans-Cinnamyl acetate	0.24	–	0.24
Eugenol	<u>traces</u>	<u>16.7</u>	–
	64.6	58.53	74.5
Miscellaneous	35.4	41.47	25.5

Source: Vernin *et al.*, 1994.

Note
a The most important components are underlined.

Table 3A.3 Headspace composition of cinnamon and cassia quills of different origin (in %)

Origin	Form[a]	Benzaldehyde	*trans*-Cinnamaldehyde	R[b]
Madagascar (1998)	Q	0.38	83.5	<u>220</u>
Ducros (Marseilles, 1988)	Q	0.66	86.2	<u>130</u>
Espig (Marseilles, 1988)	Q	0.13	75	<u>577</u>
	P	0.05	77	<u>1540</u>
Epi-Exotis	P	0.28	73	<u>261</u>
China (Rabelais)	P	1.2	73	61
Sri Lanka (Marseilles)	Q	0.27	72	<u>267</u>
Indonesia	Q	0.05	68.3	<u>1366</u>
Guadelupe	Q	1.38	62.4	45
Epi-Exotis	Q	0.2	62	<u>310</u>
Sri Lanka (Vence, Cave B)	Q	2.5	60	24
Sri Lanka (Vence, Market)	Q	5.8	50	8.6
Egypt	P	10.3	51	5
Ducros (prepared in laboratory)	P	12.6	48.2	3.8
Reunion	Q	0.5	36	<u>72</u>
	P	0.4	34	<u>85</u>
Ducros (commercial)	P	6.7	28.4	4
Thailand	B	36	12.5	0.35
	P	33	5	0.15
Vietnam	B	40.5	2.3	0.06
	P	38	4.6	0.2

Source: Vernin *et al.*, 1994.

Notes
a Q: quills; P: powder obtained from quills or bark; B: bark (crude).
b Ratio *trans*-cinnamaldehyde/benzaldehyde.

Table 3A.4 Percentage composition of leaf oil components from two chemovars of cinnamon from Brazil

Compound	Type 1	Type 2
α-Pinene	0.12–0.15	t–0.2
Camphene	t	t
β-Pinene	t	t
α-Phellandrine	t	t–0.69
α-Terpinene	t	t
Liminene	t–0.20	t–0.11
1,8-Cineole + b-phellandrine	t–0.38	0.22–0.46
p-Cymene	t–0.26	t–0.26
Terpinolene	t	t
α-Ylangene	t–0.21	t
Benzaldehyde	0.36–0.50	t
Linalool	0.51–1.33	1.44–1.67
β-Caryophyllene	t	t
Terpinene-4-ol	1.03–1.25	1.68–4.28
α-Humulene	t	0.26–0.70
α-Tepineol	t	t
Safrole	t–0.56	29.57–39.52
Caryophyllene oxide	0.28–0.42	t
(E)-cinnamaldehyde	0.59–0.98	0.65–1.03
Eugenol	94.14–95.09	55.08–58.66
Benzyl benzoate	0.57–0.92	t–0.98

Source: Koketsu *et al.*, 1997.

Note
t = trace (<0.01%).

Table 3A.5 Comparative percentage composition of cinnamon leaf oil (Cameroon type) and two commercial oils of Seychelles

Compound	Cameroon leaf oil	Seychelles leaf oils
Eugenol	85.2	64.3–72.4
(E)-cinnamaldehyde	4.9	0.5–11.1
Linalool	2.8	0.3–2.7
β-Caryophyllene	1.8	2.4–3.8
α-Phellandrine	0.9	0.1–1.8
Caryophyllene oxide	0.5	0.2–0.5
α-Pinene	0.5	0.1–0.6
p-Cymene	0.4	0.1–2.3
1,8-Cineole	0.3	0.2–1.1
Limonene	0.3	0.1–0.5
Camphene	0.2	0.5–0.6
β-Pinene	0.2	0.1–0.7
β-Phellandrine	0.1	0.1–0.2
α-Thujene	0.1	0–0.3
γ-Terpinene	0.1	0.2–0.4
(E)-β-ocimene	0.1	0.1–0.6
Terpinolene	0.1	0.1–0.3
Myrcene	0.1	0.1–0.6
Terpinen-4-ol	0.1	0–1.2
Borneol	0.1	0.1–0.3
cis-Linalool oxide*	0.1	0–0.4

(E)-cinnamyl acetate	0.1	0.2–2.2
δ-3-Carene	0.1	0–0.2
α-Tepinene	0.1	0–0.7
Eugenyl acetate	0.1	0.1–1.4
α-Tepineol	0.1	0.1–0.9
Safrole	t	0.1–0.2
Benzyl benzoate	t	0.1–0.3
(E)-cinnamyl alcohol	t	0–0.2
α-Humulene	t	0.1–0.5
γ-Cadinene	t	0–0.3
(Z)-β-ocomene	t	0–t

Source: Jirovetz *et al.*, 1998.

Notes
* correct isomer not identified; t = trace (<0.1%).

Table 3A.6 Composition of oil from cinnamon leaves from Calicut, India

(E)-2-hexenol (0.1%)	Borneol (0.1%)
(Z)-3-hexenol (0.1%)	Terpinen-4-ol (0.3%)
1-Hexen-3-ol (0.1%)	α-Terpineol (1.1%)
Hexanol (0.1%)	Dihydrocarveol (t)
α-Pinene (t)	Linalyl acetate (0.1%)
(Z)-3-hexenyl acetate (0.1%)	(E)-cinnamaldehyde (1.7%)
(E)-2-hexenyl acetate (0.1%)	Safrole (t)
p-cymene (t)	(E)-cinnamyl alcohol (0.1%)
β-phellandrene (t)	Eugenol (3.1%)
(E)-β-ocimene (t)	(E)-cinnamyl acetate (0.9%)
1,8-cineole (0.1%)	β-Caryophyllene (2.4%)
Limonene (0.2%)	α-Humulene (0.2%)
cis-Linalool oxide* (0.1%)	Eugynyl acetate (0.1%)
Terpinolene (0.1%)	Caryophyllene oxide (0.1%)
trans-Linalool oxide* (0.1%)	Spathulenol (0.2%)
Linalool (85.7%)	
Nonanol (0.3%)	

Source: Jirovetz *et al.*, 2001.

Notes
* furanoid form; t = trace (<0.01%).

Table 3A.7 Composition of cinnamon fruit oil

(E)-2-hexenal (t)	(E)-cinnamyl acetate (0.4%)
Tricyclene	β-Caryophyllene (11.0%)
α-Pinene (11.2%)	(E)-β-farnesene (0.8%)
Camphene (0.6%)	α-Humulene (2.2%)
β-Pinene (9.2%)	γ-Muurolene (0.2%)
Myrcene (1.6%)	Germacrene D (0.2%)
α-Phellandrene (0.7%)	α-Muurolene (6.1%)
α-Terpinene (0.2%)	δ-Cadinene (7.1%)
p-Cymene (0.1%)	δ-Cadinene (13.1%)
Limonene (2.8%)	*cis*-Calaminnene (2.2)
1,8-Cineole (0.1%)	α-Cadinene (1.2%)
(Z)-β-ocimene (0.1%)	Elemol (1.9%)
(E)-β-ocimene (0.2%)	(E)-nerolidol (0.1%)
γ-Terpinene (0.1%)	Isocaryophyllene oxide (0.2%)

118 U.M. Senanayake and R.O.B. Wijesekera

Table 3A.7 (Continued)

Tepinolene (0.5%)	Spathulenol (0.8%)
Linalool (0.2%)	Caryophyllene oxide (0.4%)
α-Fenehyl alcohol (0.5%)	Globulol (0.4%)
Isoborneol (t)	Humulene epoxide 1(0.5%)
Borneol (0.5%)	Humulene epoxide 11 (0.6%)
Terpinen-4-ol (0.1%)	1-Epi-cubenol (0.1%)
α-Terpineol (0.5%)	T-cadinol (0.2%)
Nerol (t)	Cubenol (0.9%)
Geraniol (t)	α-Muurolol (9.8%)
Isobornyl acetate (0.1%)	Selin-11-en-4a-ol (0.1%)
(Z)-cinnamyl acetate (0.1%)	α-Cadinol (3.1%)
α-Copaene (2.1%)	4-Hydroxy-3,4-dihydrocalacorene* (0.2%)
β-Elemene (0.4%)	4-Hydroxy-3,4-dihydrocalacorene* (0.1%)

Source: Mallavarapu and Ramesh, 2000.

Notes
* correct isomer not identified; t = trace (<0.1%).

References

Jirovetz, L., Buchbaner, G., Ruzika, J., Shaft, M.P. and Rosamma, M.K. (2001). Analysis of *Cinnamomum zeylanicum* Blume leaf oil from South India. *J. Essent. Oil res.*, **13**, 442–443.

Jirovetz, L., Buchbaner, G., Ngassoum, M.B. and Eberhardt, R. (1998) Analysis and quality control of the essential oil of the leaves of *Cinnamomum zeylanicum* from Cameroon. *Ernahrung*, **22**, 443–445.

Koketsu, M., Goncalves, S.L., de Oliveira godoy, R.l., Lopes, D. and Morsbach, N. (1997) Oleos essenciais de cascas e folhas de Canela (*Cinnamomum verum* Presl.) cultivada no Paranai *Ciene. Technol. Aliment.*, **17**, 281–285 (cited from Lawrence, 2002).

Lawrence, B.M. (2002) Progress in essential iols. *Perfumer and Flav.*, **27**, 48–69.

Mallavarapu, G.R. and Ramesh, S. (2000) Essential oil of the fruits of *Cinnamomum zeylanicum* Blume. *J. Essent. Oil Res.*, **12**, 628–630.

Vernin, G., Vernin, C., Metzger, J., Pujol, L. and Parkanyi, C. (1994) GC/MS analysis of cinnamon and cassia essential oils: A comparative study. In G. Charalambous (ed.) *Spices, Herbs and Edible Fungi*. Elseiver Science, B.V. Amsterdam, pp. 411–425.

Annex 3.3 The chemical structure of important constituents of cinnamon oil

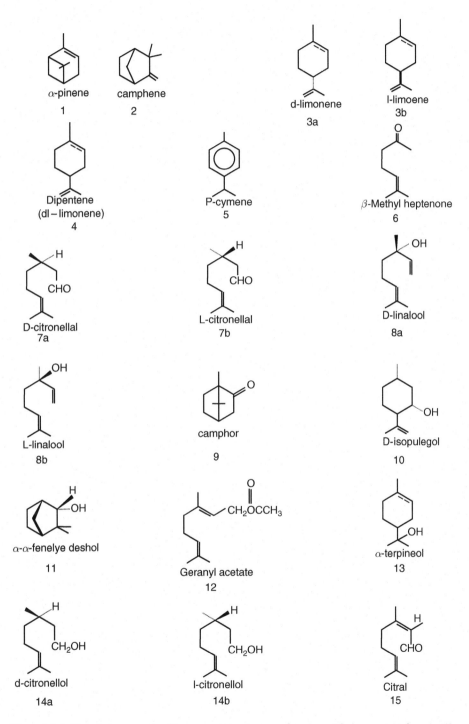

Figure 3A.5 Chemical structures of important constituents of commercial cinnamon and cassia oils.

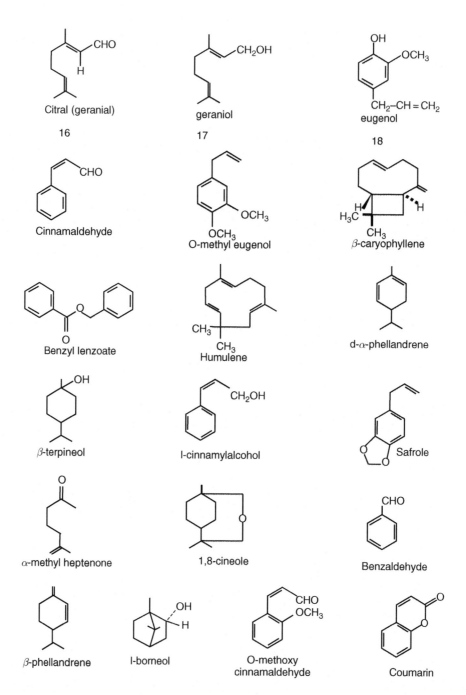

Figure 3A.5 (Continued)

4 Cultivation and Management of Cinnamon

J. Ranatunga, U.M. Senanayake and R.O.B. Wijesekera

Introduction

In Sri Lanka, cinnamon has originated in the central hills where several species of related taxa occur, especially in places such as Kandy, Matale, Belihull Oya, Haputale and the Sinharaja forest range. Currently, cinnamon cultivation is concentrated along the coastal belt stretching from Negom to Matara. Of late, cultivation has spread to the inland areas of Kalutara, Ambalangoda, Matara and Ratnapura (Fig. 4.1). The area under cinnamon cultivation in Sri Lanka is around 15,500 ha (Anon, 1996a). The bulk of the cinnamon plantation is about 70–80 years old, most of which belongs to small holders. Only about 10% of the plantations exceed 8–10 ha. Sri Lanka commands about 60% of the world export market and exports about 7,000 t of quills and chips per year, apart from cinnamon leaf oil and bark oil.

In the systematic cultivation methods prevalent in plantations in Sri Lanka, cinnamon is maintained as a bush with four to five slender shoots growing up two to three metres. (Fig. 4.2). The economic life span of a cinnamon plant is around 30–40 years (Senanayake, 1977; Wijesekera *et al.*, 1975). In two to three years after planting, depending on the climatic factors, the plants reach a height of 1.5–2 m, with three to four shoots, and are then ready for harvesting. Generally, cinnamon can be harvested two to three times per year depending on the rainfall and soil fertility. After harvesting, the leaves are separated and the stems are peeled to remove the bark. The processing of different products then begins.

Sri Lanka dominates in the supply of cinnamon. Other important sources are the Seychelles, Madagascar and India. Cassia is found in China, Indonesia, India, Taiwan and other countries in South-East Asia, and the Pacific Ocean Islands. However, most cassia oil in international trade is of Chinese origin.

Soil, climate and varieties

Cinnamon is a hardy plant, which can grow well in almost all types of soils under a wide variety of tropical conditions. In Sri Lanka, it is cultivated under varying conditions ranging from semi-dry to wet zone conditions and soils varying from the silver sands of Kadirana, Ekala and Ja-ela to the loamy, lateritie, and gravelly soils of the Kalutara, Galle and Matara districts (Fig. 4.1). The ideal temperature for growing cinnamon is between 20 °C and 30 °C. Rainfall should be in the range of 1250–2500 mm. Generally, cinnamon does not thrive well in the drier parts of the low-country. It thrives well as a forest tree at 300–350 m above sea level.

0-415-31755-X/04/$0.00 + $1.50
© 2004 by CRC Press LLC

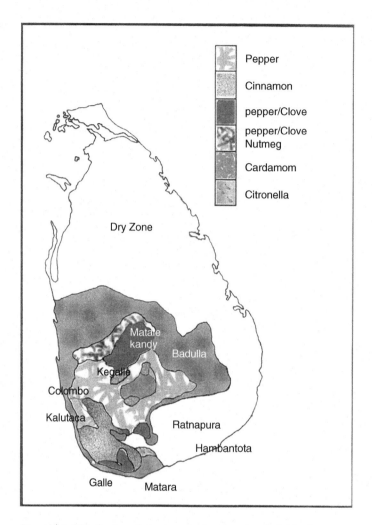

Figure 4.1 Distribution of cinnamon and other spices in Sri Lanka.

The quality of bark is greatly influenced by soil and ecological factors. The best quality cinnamon is produced on white silicatious sandy soils like the 'silver sands' of the Negombo district. Yield is higher in other soils but the quality is coarser than in sandy soils.

Varieties

There are several species of *Cinnamomum* found in South and South-East Asia. In addition to the true cinnamon (*C. verum*), there are other species which are endemic to Sri Lanka (Wijesekera *et al.*, 1975). It has been reported that eight different types (cultivars) of cinnamon are grown in Sri Lanka. These are distinguished by pungency of bark and petiole, texture of bark and the structure of leaves (Anon, 1996b). However, no botanical description of such varieties are available. They are known by the following local (Sinhalese) names:

Figure 4.2 Field plantation of cinnamon.

Panni Kurundu, *Pat Kurundu* or *Mapat Kurundu*
Naga Kurundu
Panni Miris Kurundu
Veli Kurundu
Sewala Kurundu (or *Sevel Kurundu*)
Kahata Kurundu
Penirasa Kurundu
Peiris Kurundu

In addition, there are 19 high quality, high yielding selections, identified through screening of 210 different accessions by the Department of Export Agriculture of Sri Lanka.

Propagation and Field Planting

Cinnamon can be propagated from seeds and cuttings of young three-leaved shoots. However, propagation by seeds is easier and is the most common practice. Ripe seeds are collected from selected mother plants with desired characteristics such as:

1. erect stem with smooth bark;
2. vigorous growth;
3. ease of peeling the stem bark;

124   *J. Ranatunga, U.M. Senanayake and R.O.B. Wijesekera*

4.   resistance towards pests and diseases;
5.   chemical composition of the oil (*viz.* high oil content of the bark and leaves and desired chemical characteristics of oil).

Propagation by seeds

Ripe seeds collected from mother plants having desired characteristics are heaped in the shade for two to four days until the pulp turns black and disintegrates. The seeds are then separated mechanically, washed and dried in the shade. Over-fermented and light seeds are discarded and the rest are used for planting. According to the Department of Export Agriculture, the seed viability diminishes rapidly with time, and the viability is very low after five weeks. As such, it is important to use fresh seeds for germination.

There are three methods for establishing seedlings for cultivation:

1.   Direct sowing of seeds in groups of four to five in the field;
2.   Transplanting of nursery raised plants (root balling method);
3.   Seedlings raised in polythene bags.

Direct sowing of seeds

When seeds are sown directly in the field, $0.3 \times 0.3 \times 0.3$ m holes are dug at a spacing of 1.2 m $\times$ 1 m and filled with top soil. Seeds are sown in groups of 15–20 in each hole. The main disadvantages of this method are:

1.   unnecessary competition among seedlings;
2.   a longer period of germination;
3.   ill effects of adverse environmental and climatic conditions.

Nursery or root ball method

Nursery beds 1 m wide and of a convenient length are prepared. Seeds are planted 20 cm apart in rows and 10 cm apart within a row. Seeds are planted in lots of seven to ten in a 4 cm diametre hole at a depth of 4–8 cm and covered with a thin layer of soil. Beds are shaded with polythene or coconut leaf and watered daily in dry weather. After the seedlings have reached a height of 12 cm, the shade is removed gradually. When seedlings are three to four months old it is necessary to remove weak seedlings leaving only three well grown seedlings in a hole. Plants are ready for planting out in the field in eight to ten months times. Before removing seedlings, nursery beds should be watered thoroughly to enable the removal of the seedlings with soil without damaging the root system. Then the seedlings are removed from the bed in such a way that the root system is well covered with a ball of surrounding soil, keeping it in the centre of the soil ball. The soil ball with three seedlings can be transported to the field for planting. This is the most common method adopted by farmers.

Raising seedlings in polythene bags

Cinnamon seedlings are also raised in polythene bags (10 cm $\times$ 20 cm) filled with a mixture of top soil/cow dung (1:1). About five to seven seeds are sown in a bag and at the age of three to four months excess seedlings are removed leaving the best three seedlings in the bag. The plants are ready for field planting in about six months time.

Vegetative propagation

Propagation by shoot cuttings

Cinnamon can also be propagated by cuttings of young three-leaved shoots or by layering. Partially matured shoots with a node are removed from selected mother plants with the desired characteristics. Single node shoots are prepared by making a sloping cut at the node. Cuttings should be put into water immediately and maintained in water until planted in polythene bags. Polythene bags 10 cm diameter by 20 cm long should be filled and pressed to be firm. The filled polythene bags should be put together within frames made of bamboo or suitable supports to give beds not more than 1 m wide. The soil under the pots should be forked over to ensure good drainage. One cutting should be placed in each bag. The bed of polythene bags and cuttings must be kept cool and moist. In order to prevent water losses through evapo-transpiration the bags must be covered with polythene. It is also important to provide shade as protection from direct sunlight. After three to four months the shade has to be removed gradually for the hardening of the plantlets. The rooted plantlets are ready for planting after 10–12 months. Vegetative propagation techniques for *C. verum* and *C. cassia* have been reported from the Indian Institute of Spices Research, Calicut, Kerala, India (Rema *et al.*, 1997). For further details see Chapter 2).

Propagation in vitro

In vitro propagation has the potential for rapid multiplication of selected plant types with desired characteristics. The possibility of using plant tissue culture techniques for the rapid multiplication of cinnamon has been established. Rai and Jagadishchandra (1987) induced multiple shoots from hypocotyl segments of seedlings on Murashige and Skoog's basal medium, supplemented with α-naphthalene acetic acid (NAA) and 6-Benzylamino purine (BAP) at 0.5 mg/l. However, there is no information about the adoption of tissue culture technology for the commercial micropropagation of cinnamon so far, and this is an area needing research and development. Nirmal Babu *et al.* (1997) also reported micropropagation of *C. verum*, *C. cassia* and *C. camphora* from mature trees. Mini *et al.* (1997) have reported the induction of somatic embryogenesis in seedling cultures (see Chapter 2 for details).

Air layering

The possibility of propagation of cassia through air layering has been recorded. According to Krishnamoorthy and Rema (1994) propagation of cassia has been achieved through air layering with 50% to 87.5% success, depending on the time of the year (see Chapter 2 for more details).

Field Planting

On flat land cinnamon can be planted in straight lines and on sloping land planting on contour lines is recommended. Holes (30 cm $\times$ 30 cm $\times$ 30 cm) for planting are dug at 1.2 m × 0.6 m spacing and filled with top soil. At this spacing 14,000 holes can be dug in one hectare. Planting must be done during the onset of rains at a rate of three plants per hole. A single hectare plantation requires 42,000 plantlets. Planting holes could also be sown direct with prepared seeds during the rainy season. If the inter plant spacing is too close then a situation results where the plants tend to grow elongated as they seek sunlight. This impairs the eventual flavour characteristics of the bark.

Maintenance

Soil management

When the land is sloppy the following soil conservation measures should be adopted:

- Digging of contour drains at suitable distances depending on the slope and the rainfall. Drains should be deep enough to allow for the settlement of eroded soil.
- Mulching with pruned branches and weeds.

Burying weeds and pruned leaves around rootstocks is helpful for better stooling. Gathering earth up to the rootstock without mounting should be done as cinnamon is a surface feeder. Rootstocks should be exposed to sunlight to allow new shoots to develop and to prevent termite attack.

Fertilizer Application

Fertilizer application is important for the commercial cultivation of cinnamon for higher productivity, as cinnamon is normally grown as a long-term monoculture. In the absence of fertilizer application the supply of nutrients in the soil will become exhausted, leading to mineral deficiencies and a drop in yield, which includes dry matter yield as well as the oil content of bark and leaves and also poor stooling. The fertilizer requirement may be based on the results of field experiments, soil and plant tissue analysis or the symptoms of mineral deficiencies or toxicities. As such, the optimum applications will vary from one region, plantation or field to another in accordance with local conditions. The Department of Export Agriculture of Sri Lanka recommends the following fertilizer mixture and quantities.

Composition	Ratio by Weight	Mineral content in the Mixture
Urea	2	$N - 23\%$
Rock phosphate (28% P_2O_5)	1	$P_2O_5 - 07\%$
Muriate of Potash (60% K_2O)	1	$K_2O - 15\%$

The fertilizer requirement will also vary according to the age of the plantation. Following are the recommended rates of fertilizer for young plantations:

First year – 200 kg/ha/annum
Second year – 400 kg/ha/annum
Third year – 600 kg/ha/annum

The above quantities have to be applied as two splits at six month intervals. It is important to apply fertilizer when the soil is moist or at the commencement of rains. Fertilizer should be applied at a 50 cm radius around the plant or between rows. After application of the fertilizer it is important to fork it into soil. In addition, when there are symptoms of magnesium deficiency the application of Dolomite at the rate of 500 kg/ha two to three months before applying the recommended fertilizer mixture would be advantageous.

After three years when the cinnamon is mature, the dose of fertilizer should be doubled for every successive application thereafter. However, the fertilizer requirement may be determined according to the yield potential.

Training of Plants and Pruning

The objective of training cinnamon plants is to establish a strong base which is capable of producing a greater number of healthy stems. When seedlings attain the age of about two years, and the basal diameter is about 4–6 cm, the main stem is coppiced or cut back to a height of about 4–6 cm from ground level. Cutting is done with a sharp knife having a long handle (*keththa*) at a 30° angle in such a way that the cut faces inwards. This will promote the tillering from the base towards the outside. Only three strong and straight tillers are retained while all others should be removed so as to promote the growth of these shoots as the main stems. After one to two years, the main stems are harvested (pruned) alternately. It is necessary to allow each base of the pruned stem to initiate three healthy new stems. This practice generates five to eight stems from a single bush for harvesting per year after eight to ten years.

It is also important to remove side branches of the main stem to promote strong growth of stems. In addition, the pruning of side branches will expose the base of the plant to sunlight, which is believed to initiate more tillers from the base. It is also very important to make sure that all the harvesting cuts are made at an angle of 30° inward as this encourages the clump to spread outwardly.

Replanting

When cinnamon plants are about 40–50 years old, their regenerating ability gets reduced considerably, resulting in decreased yields. Hence replacement of old or low yielding plantations becomes necessary. If large-scale replanting of a plantation is contemplated, then it becomes worthwhile to consider the possibility of distilling the root bark to produce the camphor-rich root bark oil used in pharmaceutical preparations.

Pest and Diseases

There are no major pests affecting the production of cinnamon. However, in certain areas of Sri Lanka minute insects attack leaves causing gall formation, which results in a decrease in the leaf oil yield by about 20%. The gall-forming insects could be brought under control by spraying a systemic insecticide. According to the Department of Export Agriculture, Sri Lanka, two types of galls are formed (Anon, 1996b).

128 *J. Ranatunga, U.M. Senanayake and R.O.B. Wijesekera*

Gall-forming mites – Erioplytes boisi *Gerber*

Galls can be found on the upper surface of leaves. The galls are pinkish in colour at the beginning which, as they mature, turn green. Mites lay eggs in the leaf tissue and gall formation starts immediately.

Jumping plant louse – Trioza cinnamini *Boelli*

Galls can be found on the lower surface of leaves. The eggs are laid on the leaf surface and the gall formation appears to be due to the feeding effects of emerging nymphs.

In addition, two other minor pests have been found in Sri Lanka. The first, the clear wing moth causes damage to the base of the old cinnamon plants when its larvae makes holes through feeding. The other pest, a Shoot Borer, feeds on the tender shoots which results in the death of the upper part of the shoot. Singh *et al.* (1978) reported that there are several major pests, including cinnamon butterfly (*Chilasa clytie*), shoot and leaf webber (*Sorolopha archimedias*), leaf miner (*Acrocercops* sp.) and chafer beetle (*Popillia*) in the cinnamon growing tracts of India.

Apart from these pests, several diseases of cinnamon have also been reported.

Leaf spot or leaf blight – Colletotrichum gloeosporiodes

Symptoms are seen as brownish leaf spots and these may enlarge to make large lesions. Spraying 1% Bordeaux mixture or any other copper fungicide may be necessary to control the disease.

Black sooty mould – Stenella *sp.*

The blackish growth on the leaf surface is the characteristic symptom of the disease. The fungal growth is confined only to the surface and no penetration into the leaf tissue occurs. As this disease does not affect the yield severely, application of fungicides is not necessary.

Harvesting

Cinnamon is ready for harvesting after two to three years when the plant reaches a height of 1.5–2 m with three to four shoots and the bark turns brown in colour. The main shoot is coppiced or cut back to a height of about 6 cm from ground level. Two to three crops are taken annually depending upon the rainfall.

Normally the harvesting of mature sticks is done following the two rainy seasons when the new flush of leaves have hardened. At this time the bark peels off easily. Under good management conditions, harvesting could be done more than twice per year (Anon, 1996b). Such a practice, coupled with the split application of fertilizer, can help to increase the yield.

References

Anonymous (1973) Cinnamon cultivation and processing. Department of Minor Export Crops, Sri Lanka.
Anonymous (1996a) Cinnamon. Department of Export Agriculture, Sri Lanka.

Anonymous (1996b) *Cinnamon Cultivation and Processing*, Technical Bulletin No. 5. Department of Export Agriculture, Sri Lanka.

Coppen, J.J.W. (1995) Cinnamon oils (including cinnamon and cassia). In *Non-wood Forest Products 1/Flavours and Fragrances of Plant Origin*, FAO, Rome, pp. 7–15.

Ibrahim, J. and Goh, S.H. (1992) Essential oils of cinnamon species from Peninsular Malaysia. *J. Essential oil Research*, 4, 161–171.

Krishnamoorthy, B. and Rema, J. (1994) Air layering of cassia (*Cinnamomum aromaticum* Nees). *J. Spices and Aromatic Plants*, 3, 48–49.

Mini, P.M., John, C.Z., Samsudeen, K., Rema, J., Nirmal Babu, K. and Ravindran, P.N. (1997) Micropropagation of *Cinnamomum verum* (Brecht and Presl.). In S. Edison, K.V. Ramana, B. Sasikumar, K. Nirmal Babu and J. Eapen Santhosh (eds) *Biotechnology of Spices, Medicinal and Aromatic Plants*, Indian Society for Spices, Calicut, India, pp. 35–38.

Nirmal Babu, K., Ravindran, P.N. and Peter, K.V. (1997) *Protocols for Micropropagation of Spices and Aromatic crops*, Indian Institute of Spices Research, Calicut, Kerala, India, p. 35.

Rai, R.V. and Jagadishchandra, K.S.J. (1987) Clonal propagation of *Cinnamomum zeylanicum* by tissue culture. *Plant Cell Tissue and Organ. Culture*, 9, 81–88.

Rema, J., Krishnamoorthy, B. and Mathew, P.A. (1997) Vegetative propagation of major spices- a review. *J. Spices and Aromatic Crops*, 6, 87–105.

Senanayake, U.M. (1977) Ph.D. Thesis. *The Nature, Description and Biosynthesis of Volatiles of Cinnamomum* spp. University of New South Wales, Australia.

Singh, V., Dubey, O.P., Nair, C.P. and Pillai, G.B. (1978) Biology and bionomics of insect pests of cinnamon. *J. Plantation Crops*, 6, 24–27.

Wijesekera, R.O.B., Ponnuchamy, S. and Jayewardene, A.L. (1975) *Cinnamon* – Ceylon Institute of Scientific and Industrial Research (CISIR), Colombo, Sri Lanka, p. 48.

5 Harvesting, Processing, and Quality Assessment of Cinnamon Products

K.R. Dayananda, U.M. Senanayake and R.O.B. Wijesekera

Introduction

Cinnamon plants are grown as bushes. When plants are two years of age they typically measure about 2 m in height and about 8–12 cm at the base. At this stage they are ready for harvesting. The commercial products of cinnamon are quills, quillings, featherings, chips, cinnamon bark oil and cinnamon leaf oil (Fig. 5.1). The most commonly produced product is cinnamon quills.

The term quills is defined as scrapped peel of the inner bark of mature cinnamon shoots, joined together by overlapping tubes, the hollow of which has been filled with smaller pieces of cinnamon peels which is thereafter dried first in the sun and thereafter in shade for a certain length of time. Quillings are broken pieces and splits of all grades of cinnamon quills. The feather like pieces of inner bark consisting of shavings and small pieces of bark left over from the quill-making process are called featherings. Cinnamon chips are obtained from rough unpeelable bark scraped off from thicker stems. Cinnamon leaf and bark oils are obtained by distilling the leaf and bark separately.

Harvesting and Preparation of Shoots for Peeling

The harvesting of cinnamon shoots is undertaken by skilled workers. Mature shoots are coppiced or cut back to a height of about 5–8 cm from the ground. Two to three crops are taken annually depending on the rainfall. The bark is relatively easy to remove immediately after the rainy season. The most valuable products are obtained from the bark of the cinnamon tree. Removal of the bark is a traditional process, requiring considerable skill and is normally done by trained peelers.

Shoots that are sufficiently mature are selected for harvesting and the side branches are pruned off about three months before harvesting. These shoots are cut at the base by an inward cut, which encourages sprouting from the outside portion of the stump. The freshly harvested sticks are carried to the peeling shed. Several tools are used to remove bark from the stick with minimum damage (Fig. 5.2). These tools include:

(a) a curved knife for scrapping the outer dead bark surface;
(b) a knife having a point on one side for ripping side branches;
(c) brass or wooden rod to loosen bark;
(d) a specially designed pointed knife to remove the loosened bark.

0-415-31755-X/04/$0.00 + $1.50
© 2004 by CRC Press LLC

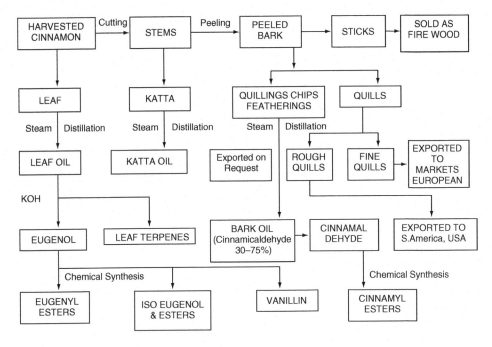

Figure 5.1 Flow diagram of cinnamon processing.

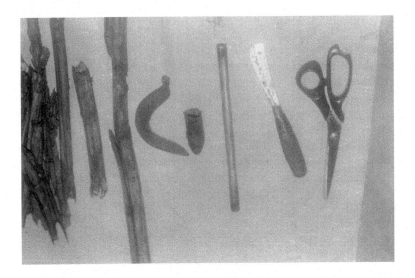

Figure 5.2 Tools used in cinnamon peeling.

Extraction and processing of the bark

The process of bark extraction involves scraping the bark with a steel scrapper, peeling, filling up and drying. The outer cork from the bark is scraped prior to peeling. Then the bark is thoroughly rubbed with a wooden or brass rod to loosen it from the woody stem. A longitudinal line is then drawn on the bark from end to end with a sharp pointed knife. A similar line is drawn on the opposite side of the bark. By skillfully maneuvering the knife, two equal halves of long bark are peeled off from either side of the stick (Fig. 5.3a).

The bark halves are packed one inside the other until cigar like quills are obtained. The worker sits with a board and a measuring stick 107 cm long, to which is attached a wooden lifter (known in the local language as 'Pethi Kotuwa'). The compound quill is assembled on the board (Fig. 5.3b). When it has reached the required length the end is trimmed with scissors and it is gently lifted and kept on the mat for drying. The hollow inside of the quill is then packed with pieces of thin bark, which are unsuitable for forming a single quill. The prepared quills are arranged in rows suspended by ropes and dried in the shade for a few days. They are never dried in direct sunlight. The dried bark turns a pale yellow colour. Sometimes the bark is fumigated with sulphur dioxide to obtain a uniform yellow colour and to kill any insects and microorganisms in the quills. The dried bark is quite hard and can be handled without much breakage. The moisture content at this stage will be about 10%. They are then packed in polythene lined jute bags.

Cinnamon quills prepared by experienced peelers are of uniform thickness from end to end. They have a characteristicly smooth skin and uniform yellow colour. The dried quills are tightened into small bundles, each bundle containing about 30–40 quills. The edges are then trimmed, making them ready for marketing (Fig. 5.4).

Cinnamon quills produced in Sri Lanka are specially designated as *Ceylon Cinnamon*, as there are differences in the flavour quality of the product. Products produced in Seychelles and Madagascar are also found in the market and are named accordingly. Dark brown patches found on the surface of quills are called foxing. Depending on the amount of foxings, size and colour they are graded. Quills free from foxing are considered to be the best.

After systematic grading the cinnamon quills are exported. The characteristics considered for the grading of Sri Lankan cinnamon quills are given in Annex 5.1. It is significant that grading is carried out by the exporter and not by the producers. The finest and smoothest quality quills are graded as (00000) five zeros and the coarsest as zero (0). Over-matured barks are generally coarser and producers obtain an advantage in terms of weight. However, since high quality five zero quills fetch the highest price, producers should be induced to supply the better quality quills without consideration of weight. Another grade called *Alba*, fine thin bark rolled in to pencil like quills, are also produced for special export orders.

Cinnamon Products

As discussed earlier, cinnamon quills represent the most valuable product originating from the cinnamon bush. However, there are several other by-products generated during the processing of quills (Fig. 5.5). They are classified into three major commercial groups.

Quillings

Quillings are made from broken pieces and splits of all grades of cinnamon quills. The main characteristics of quillings are their shapes and sizes. The aroma and taste of

Figure 5.3 (a) Cinnamon peelers at work (tools used to peel the bark are in the foreground and harvested shoots ready for peeling are in the background); (b) preparation of compound quills.

Figure 5.4 A bundle of cinnamon quills ready for export.

quillings are the same as the quills, even though they are marketed as medium quality cinnamon. They contain featherings and chips but their quantities should not exceed 3% by mass. If proper precautions are not taken during the processing, extraneous matter including pieces of wood, stems or twigs may get mixed with the quillings. Quillings are separated from the quills in the preparation of quills and they are separately dried in shade followed by sun drying.

Featherings

Featherings are feather-like pieces of inner bark consisting of shavings and small pieces of bark left over from the process of making quills. Scrapings from the bark or small twigs and stalks of cinnamon shoots, including a minimum quantity of chips, are also considered as featherings. The product is marketed as medium quality cinnamon.

Figure 5.5 Cinnamon products – unscrapped bark, quills, quillings and powder.

Chips

Chips are not peeled out from the stem. Instead they are scraped off from the greenish brown, mature, thick pieces of bark, which are inferior quality cinnamon. The outer bark, which has been obtained by beating or scraping the shoots is also considered to be chips.

Chips are graded into two categories,

Grade 1 – Those containing small featherings obtained by scraping very small twigs. They contain a small amount of other bark material.

Grade 2 – Those containing inner and outer bark and pieces of wood.

Depending on the extent to which chips are free from extraneous matter such as refuse and dust, the chips are cleaned by washing or bleaching and are further divided into four types (type 1, 3, 0 and 00).

136   K.R. Dayananda, U.M. Senanayake and R.O.B. Wijesekera

Table 5.1 Chemical requirements for cinnamon bark

	Requirement		
	Sri Lanka	*Seychelles type and Madagascar type cinnamon*	*Reference to ISO Recommendations*
Moisture % w/w	12.0	12.0	ISO – R 939
Total ash % w/w on dry basis max.	5.0	7.0	ISO – R 928
Acid insoluble ash % (w/w) on dry basis max.	1.0	2.0	ISO – R 930
Volatile oil ml/100 g on dry basis min.	Whole 1.0 ground 0.7	0.4 0.3	

Source: Anon, 1997.

The cleanliness and levels of dust, mould and other pathogenic organisms are important aspects of the production of cinnamon quills, quillings, featherings and chips. Consumers directly use them in most of the food formulations. It has been reported that compounds such as cinnamic aldehyde and eugenol found in them inhibit the growth of micro-organisms (Bullernan *et al.*, 1977). The general requirements for cinnamon bark are given in Table 5.1.

Products Based on Cinnamon Bark and Leaf

Cinnamon quills and remnants of bark are prepared into various other forms, such as small pieces of quills 7–15 cm long or as a powder. For most baked products, cinnamon is used in powdered form. Finely ground cinnamon quills are the best, however, flavour and odour characteristics become less than those of quills. The essential oil content of the powder is less compared to the bark due to losses during the process of grinding. Adulteration of cinnamon powder with lower grade bark is commonly practiced (Llewellyn *et al.*, 1981).

Airtight containers made of glass or wooden boxes with an outer wrapping of polythene are generally acceptable as packaging for cinnamon products. However, a shelf-life of over two years is generally not recommended for cinnamon bark or their products since flavour and odour characteristics tend to change with time.

Cinnamon oils

In the world trade, the oils from the cinnamon species are classified as cinnamon oil, cassia oil and camphora oil, according to the place of origin. The true cinnamon oils come from Sri Lanka and they are accordingly classified as "Cinnamon Leaf Oil" (ISO 3524). Cassia oils come from China, Taiwan and Burma and are called Chinese, Taiwanese or Burmese cassia oils, respectively. Unlike the Sri Lankan cinnamon oils where a category for both bark and leaf oil exists, only one category exists in cassia oils. This is because during distillation cassia bark, leaves and twigs are distilled together to obtain one category of oil. Camphor oil, although chemically different from either cinnamon or cassia oils, is obtained from distilling the bark and wood of the *Cinnamomum camphora* tree. China and Japan are the main producers of this oil.

Cinnamon bark oil

Cinnamon bark oil is one of the expensive essential oils in the world market. The price or value of bark oils, largely depend on the material used to distill the oil. Even though quills are the best to obtain quality bark oil, quills are not always used for the distillation. The quillings when fresh provide oil of equal quality and are usually employed in distillation as the preferred raw material on grounds of economy as well as quality.

Like most of the other essential oils, cinnamon oils are also produced by hydrodistillation (Wijesekera, 1989). Hydro-distillation implies water-cum-steam distillation. In Sri Lanka traditional distillation units are built on the hydro distillation principle (Fig. 5.6). To obtain commercial cinnamon bark oil, broken pieces of quills, quillings, pieces of inner bark from twigs and twisted shoots are distilled.

Figure 5.6 Traditional cinnamon bark still.

Figure 5.7 Modern cinnamon bark still.

Hydro-distillation units are heated by direct firing. Bark or other parts of the inner bark are placed in the still and direct fire is introduced. The distillate produced from the still body is first passed through a pre-cooler and then through a large static water condenser. In Sri Lanka a large number of existing cinnamon oil distillation units are still of traditional design. Although present day distillers introduced several modifications to modernise the design (Fig. 5.7), they have not made any significant differences to the system. These old stills are still adequate to distill the oil from the bark. They are made out of copper or tin lined copper or steel and hold around 50 kg of bark. The still body is built over a brick or mortar hearth and direct fire is introduced from the bottom. Cinnamon bark is placed in the still, which is partly filled with water. The condenser is a static copper tubing immersed in a large water tank. Recently a precooler has been introduced in between the still body and static condenser. The precooler cools the distillate to about 70 °C when it is sent into the main condenser. All the time the precooler is kept cold by cold water. Certain distillers have a battery of still bodies with a common static water condenser. Five to six copper, aluminium or stainless steel Florentine vessels in a cascade arrangement are placed to collect the distillate. As cinnamon oils are heavier than water, condensed oils collect at the bottom of vessels. Loading and unloading of the bark is done manually. The time taken to distill a batch of 40–50 kg of bark is around four to five hours and a yield of 0.5–0.7% of oil on dry weight basis is obtained (Anon, 1993).

Cinnamon bark oil is graded according to its cinnamic aldehyde content (CS 185, 1972).

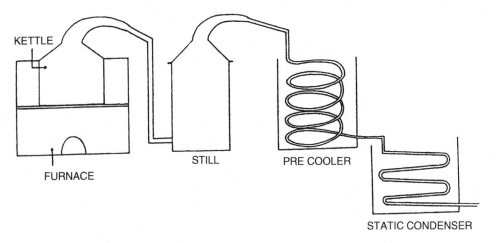

Figure 5.8 Diagram of a traditional cinnamon leaf oil distillation unit.

Cinnamon leaf oil

Leaf oil distillations are normally carried out in large distillation vessels (200–500 kg) (Fig. 5.8). Cinnamon leaves are obtained as a by-product of the cinnamon industry. When the shoots are harvested for bark, the leaves and twigs are trimmed in the field. Before distilling the leaf oil, the trimmed leaves are allowed to remain in the field for three to four days. Traditional leaf oil distillation units consist of a wooden still body with a copper goose neck lid. The condensing system is similar to the bark oil-distilling unit. The steam is generated in a separate kettle like external boiler. The condensed distillate is collected in a series of Florentine vessels (Fig. 5.8). Due to the low solubility of leaf oil in water it turns milky white in colour. Oil separated water is kept stored for days before complete separation of oil from the water. Since leaf oil is heavier than water, oil collects at the bottom of the Florentine vessel. Due to oxidation and to reactions with copper or steel vessels, the oil turns usually dark brown in colour. It takes about six to seven hrs to distill one batch of leaf, normally a yield of 1.0% oil on dry weight basis is obtained. After months of operation, a dark resinous material is found in condenser tubes. This is removed by burning condenser tubes in open fire.

Cinnamon leaf oil is graded according to the eugenol content (CS 184, 1972).

Supercritical fluid extraction (SCFE) of cinnamon

SCFE is a versatile tool for addressing the disadvantages of the conventional extraction technologies of essential oils and oleoresins. In this process the ground spice is brought into contact with a supercritical solvent at a relatively high pressure ranging from 60–300 bar at a temperature 35–70 °C. The supercritical solvent, after getting enriched with the solute, is passed through a micrometer valve or pressure reducing valve, where the pressure on the separator side is much less and the temperature much lower due to the adiabatic expansion of the supercritical solvent. This results in lower solubility of the solute leading to the separation of the solute from the solvent. Once the material is separated the gas is compressed back to extract the material

(Tiwari, 1989). CO_2 is the most commonly employed SCF. CO_2 is available freely, it is cheap, non-flammable, non-toxic and non-corrosive. It behaves either as a polar or non-polar solvent depending on pressure and temperature employed. CO_2 is liquid below its critical point (31.2 °C, 7.38 milli pascal pressure) and above that it exists as a super critical fluid. The technical advantages of using SCFE for spices were studied by Udayasankar (1989).

Coarsely ground cinnamon bark is extracted with supercritical CO_2 using pressures of 300, 400, 500 and 600 bar. The yield is around 1.4% as compared to steam distilled (0.5–0.8%). No significant differences were reported sensorily between the SCF extract and steam distilled oil and extracts (Pruthi, 2001). L CO_2 extract of powdered bark is a mobile oil without the dark coloured waxes and resins normally encountered with the bark oleoresin prepared with chemical organic solvent. The CO_2 selective extract is more soluble in foods, while the resinous compounds have a slight fixing effect on the flavour components of cinnamon and cassia bark extracts. However, the SCFE process is not practiced for commercial cinnamon or cassia oleoresin production because the process is very costly and the product does not differ much in quality from the solvent extracted product.

Cinnamon oleoresin

Oleoresins are solvent extracts of spices that contain the volatile oil, non-volatile resinous material and the active ingredient that characterises the spice as hot or pungent when such an ingredient is present. The solvents commonly used for the preparation of the oleoresins are acetone, ether, ethanol, propanol or methylene chloride. Ethanol was at one time the preferred solvent but is no longer used due to its high cost. Acetone is now the most commonly used solvent for oleoresin production. In solvent extraction the selected solvent is allowed to percolate down through coarsely ground spice which is loaded in a percolator. The bottom drain of the percolator is kept open for the escape of air. When all the material is soaked the bottom drain is closed and sufficient contact time is given to reach the solutes into the solvent. After the contact time, the extract, called "micella", is drained and collected.

For oleoresin production either the Soxhlet extraction method or the batch counter current extraction method (CCE) is industrially practiced. In the Soxhlet method the solvent is allowed to percolate through the bed of material several times, which involves continuous heating of the extract. In the CCE method, which is very widely used, the powdered spice is packed in to a series of extractors and extracted with a suitable selected solvent. The solvent moves from extractor to extractor and the dilute extract is allowed to percolate through the spice. The concentrated extract obtained first from each extractor is withdrawn for solvent removal in the production of oleoresin.

The micelle obtained from the extractor is distilled to get the finished product. Most of the solvent (90–95%) present in the micella is recovered by normal atmospheric pressure distillation, while the remaining solvent is taken off by distillation under reduced pressure. The trace amounts of solvent are finally removed by either the azeotropic or extractive distillation method using an innocuous solvent like ethyl alcohol. Alternatively, bubbling nitrogen into the thick viscous material is carried out to drive away the residual solvent. The maximum permitted residual limits for some of the solvents are 30 ppm for acetone and chlorinated solvents, 50 ppm for methyl alcohol and isopropyl alcohol and 25 ppm for hexane (CFR, 1995). After distillation the oleoresin is stored in suitable containers such as S.S. (stainless steel) drums or epoxy coated HDPE (high density polyethylene) containers.

Pharmaceutical preparations

The following preparations utilising cinnamon are listed in pharmacopoeias such as the British Pharmacopoeia (BPC) the Indian Pharmacopoeia (IPC) and the Argentinean Pharmacopoeia (AP).

Compound cinnamon powder (BPC)
Distilled cinnamon powder (BPC)
Concentrated cinnamon water (BPC)
Tincture of cinnamon (BPC)
Compound tincture of cinnamon (TPC)
Aromatic chalk mixture (IPC)
Ammoniated quinine and cinnamon elixir (BPC)
Cinnamon spirit (BPC)
Cinnamon syrup (AP)

Cinnamon Standards and Specifications

Details of Indian and international standards laid down for Sri Lankan cinnamon products are given in Annex 5.1. The Indian standards cover cinnamon (whole) and the ISO specifications cover cinnamon whole and cinnamon powder (Sri Lankan cinnamon, Seychelles cinnamon), in addition to their storage, sampling and testing. Annex 5.2 gives the safety regulations for cassia leaf oil and cinnamon bark oil. Annex 5.3, provides the cleanliness specifications prescribed by ASTA, ESA and the honest trading practices of Germany.

References

Anonymous (1977) Sri Lanka Standards, SLS 81 Part 1 and Part 2. Specification for Cinnamon. Sri Lanka standards-Specifications for cinnamon oil:

- CS 184 – Specification for cinnamon leaf oil
- CS 185 – Specification for cinnamon bark oil

Anonymous (1993) Proceedings of a Workshop on the Technological Aspects of the Production and Processing of Essential Oils in Sri Lanka, Institute of Chemistry, Ceylon.
Bullernan, L.B., Lieu, F.Y. and Seier, S.A. (1977) Inhibition of growth and aflatoxin production by cinnamon and clove oils, cinnamic aldehyde and eugenol. *J. Food Sci.*, 42, 1107–1108.
CFR (1995) Code of Federal Regulations, 21, CFR 173.2, Washington, USA.

International Standard Organization Standards (ISO):

ISO 3524 – Cinnamon Leaf oil
ISO 6538 – Cassia Leaf Oil
ISO 6539 – Cinnamon specification
ISO/R 928 – Spices determination of water insoluble ash

142 K.R. Dayananda, U.M. Senanayake and R.O.B. Wijesekera

ISO/R 930 – Spices – determination of acid insoluble ash
ISO/R 939 – Spices – determination of moisture content

Llewellyn, G.C., Burkett, M.L. and Eardie, T. (1981) Potential mould growth, aflatoxin production and antimycotic activity of selected Natural spices and Herbs.
Pruthi, J.S. (2001) Advances in Super-Critical Fluid Extraction (SCFE) Technology of Spices – A global overview and future R&D needs. Beverage & Food World, Jan. 2001, pp. 44–55.
Tiwari, K.K. (1989) Technological aspects of supercritical fluid extraction (SCFE). In Trends in Food Science and Technology, AFSTI/CFTRI, Mysore, pp. 59–70.
Udayasankar, K. (1989) Supercritical carbon dioxide extraction of spices: The Technical advantages. Proc. Recent Trends and Development of Post Harvest Technologies for Spices, CFTRI, Mysore, pp. 56–78.
Wijesekera, R.O.B. (1989) Practical Manual on the Essential oils Industry – UNIDO, Vienna.

Annex 5.1

Indian Standard (IS 4811: 1992)

Cinnamon Whole-Specification

(First Revision)

Scope

This standard prescribes the requirements for whole cinnamon (*Cinnamomum verum* Bercht. & Presl.), for use as a spice and in condiments (syn. *Cinnamomum zeylanicum* Blume).
    This standard does not cover the requirements for cinnamon powder.

References

The Indian Standards listed below are necessary adjuncts to this standard:

IS No.	Title
1070: 1992	Reagent grade water – Specification (third revision)
1797: 1985	Methods of test for spices and condiments (second revision)
13145: 1991	Spices and condiments – Methods of sampling

Terminology

Quills

The long compound rolls of cinnamon bark measuring up to 1 m in length.

Quillings

The breakages during grading and transportation and small pieces of bark left after the preparation of quills.

Chips

The bark obtained from thick branches and stems, trimmings of the cut shoots, shavings of the outer and inner barks and odd pieces of outer bark.

Harvesting, Processing, and Quality Assessment of Cinnamon Products 143

Grades

The Cinnamon bark shall have four grades. The designations of the grades and their requirements are given in Table 5A.1.

Requirements

Description

The cinnamon shall consist of layers of dried pieces of the inner bark of branches and of young shoots from *Cinnamomum verum* Bercht. & Presl. syn *C. zeylanicum*, which are obtained on removal of the cork and the cortical parenchyma from whole bark. The thickness of the bark shall range from 0.2 to 1.0 mm. It shall be free from insect damage.

Flavour or taste and aroma

Cinnamon, whole, shall have a fresh aroma and the delicate and sweet flavours characteristic of the spice. The material shall be free from foreign odour including mustiness.

Freedom from moulds, insects, etc

The cinnamon, whole, shall be free from living insects and moulds and practically free from dead insects, insect fragments and rodent contamination visible to the naked eye (corrected, if necessary, for abnormal vision), with the aid of magnification (not exceeding 10 ✕).

Extraneous matter

The proportion of extraneous matter like dust, dirt, stones, earth, chaff, stem, straw and outer bark of the shoots of the cinnamon plant shall not exceed the limits prescribed in Table 5A.2 for the relevant grades, when determined in accordance with the method given in 4 of IS 1797: 1985.

Table 5A.1 Grade designation of cinnamon bark and their requirements

Grade designation (1)	Length (2)	Diameter (3)	Colour (4)	Texture (5)
Quills fine	Up to 1 m	Not more than 10 mm	Pale, Brownish, Yellow	Brittle and smooth
Quills average	Up to 1 m	Not more than 20 mm	Slightly reddish	Brittle and smooth
Quillings	Large pieces of quills less than 1 m	Not more than 30 mm	Pale brown to reddish	Brittle and smooth
Chips	10 to 30 mm	–	Pale brown to reddish	Brittle

144 *K.R. Dayananda, U.M. Senanayake and R.O.B. Wijesekera*

Table 5A.2 Chemical requirements for whole cinnamon

Sl no. (1)	Characteristics (2)	Requirement		Method of test, ref to clause no. of IS 1797: 1985 (5)
		Quills fine/ quills average (3)	Other grades (4)	
(i)	Moisture content, percentage by mass, *max*	11.0	12.0	9
(ii)	Volatile oil content, percentage by mass, *min*	1.0	0.8	15
(iii)	Extraneous matter content, percentage by mass, *max*	1.0	2.0	4

Chemical requirements

Cinnamon, whole, shall also comply with the requirements given in Table 5A.2.

Packing

Cinnamon, whole, shall be packed in sound, clean and dry containers made up of jute/ cotton/paper/polythene. The container used for packing shall be free from any odour that might affect the characteristics of cinnamon.

Marking

The following particulars shall be marked or labelled on the container:

(a) name of the material and grade designation;
(b) variety of the trade name;
(c) batch or code number;
(d) net mass;
(e) date of packing;
(f) country of origin; and
(g) any other details required by the purchaser.

Sampling

Representative samples of cinnamon shall be drawn by the method specified in IS 13145: 1991.

Tests

Tests shall be carried out in accordance with the methods prescribed under "*Extraneous matter*" above and in column 5 of Table 5A.2.

Quality of Reagents

Unless specified otherwise, pure chemicals and distilled water (see IS 1070: 1992) shall be employed in tests.

Note – "Pure chemicals" shall mean chemicals that do not contain impurities which affect the results of analysis.

International Standard ISO6539–1983 (E)

Cinnamon [type Sri Lanka (Ceylon), type Seychelles and type Madagascar], whole or ground (powdered) – Specification

Scope and field of application

This International Standard specifies requirements for whole or ground (powdered) cinnamon [type Sri Lanka (Ceylon), type Seychelles and type Madagascar] constituted by the bark of *Cinnamomum verum* Bercht. & Presl.

Recommendations relating to storage and transport conditions are given in Annex 5.1b.

Note – Requirements for cassia [type China, type Indonesia and type Vietnam (Saigon)] are given in ISO 6538.

References

ISO 927, Spices and condiments – Determination of extraneous matter content.

ISO 928, Spices and condiments – Determination of total ash.

ISO 930, Spices and condiments – Determination of acid insoluble ash.

ISO 939, Spices and condiments – Determination of moisture content – Entrainment method.

ISO 948, Spices and condiments – Sampling.

ISO 1208, Ground spices – Determination of filth (Reference method).

ISO 2825, Spices and condiments – Preparation of a ground sample for analysis.

ISO 6571, Spices and condiments – Determination of volatile oil content.[1]

Definitions

Cinnamon quills (full tubes): Scraped peel of the inner bark of mature cinnamon plant on shoots joined together by overlaps, the hollow of which has been filled with small pieces of the same peel and thereafter dried in sun after air curing.

Cinnamon quillings (broken tubes): Broken pieces and splits of varying sizes of all grades of cinnamon quills.

Cinnamon featherings: Pieces of inner bark obtained by peeling and/or scraping the bark of small twigs and stalks of plantation cinnamon shoots which may include a quantity of chips as specified.

Cinnamon chips: Dried unpeelable bark of plantation cinnamon inclusive of the outer bark which has been obtained by beating or scraping the shoots.

Cinnamon powder: Ground cinnamon.

Whole cinnamon: All commercial forms of cinnamon except cinnamon powder.

146 *K.R. Dayananda, U.M. Senanayake and R.O.B. Wijesekera*

Types and classification

Types

Cinnamon, type Sri Lanka (Ceylon): The dried bark of cultivated varieties of the species *Cinnamomum verum* Bercht. & Presl., Synonym *Cinnamomum zeylanicum* Blume. (family Lauraceae).

Cinnamon, type Sri Lanka (Ceylon), is produced in four forms:

(a) quills;
(b) quillings;
(c) featherings;
(d) chips.

Cinnamon, type Seychells: The bark of trunks or branches of *Cinnamomum verum* Bercht. & Presl. (syn. *C. zeylanicum* Blume.) cultivated in the Seychelles.

Cinnamon, type Seychelles, is produced in three forms:

(a) rough cinnamon bark, which consists of slightly curved, elongated, irregular, medium or small pieces of the whole unscraped bark;
(b) scraped cinnamon bark, which is obtained from the younger shoots of bushes of the same species. The shoots are scraped with a curved knife before the bark is detached from the wood; and
(c) quills and quillings, which are prepared from the young shoots of bushes, in a way similar to that used for cinnamon, type Sri Lanka (Ceylon).

Cinnamon, type Madagascar: The bark of trunks or branches of *C. verum* Bercht. & Presl. (syn. *C. zeylanicum* Blume.), growing wild in Madagascar. It is produced either in the form of simple, hollow tubes about 30 cm long of unscraped or scraped bark of a rather coarse appearance, cut from smaller branches with a knife, or more usually in the form of unscraped or scraped pieces of bark from the larger branches and trunks, broken off with the flat side of a hatchet.

Commercial grades

CINNAMON, TYPE SRI LANKA (CEYLON)

Quills For classification, see Table 5A.3.

Quillings Quillings may contain up to 3% (m/m) of featherings and chips.

Featherings Featherings may contain up to 5% (m/m) of chips.

Chips Chips shall consist of well dried, hand-picked and clean unpeelable cinnamon bark.

CINNAMON, TYPE SEYCHELLES AND TYPE MADAGASCAR

For classification, see Table 5A.4.

Table 5A.3 Classification of cinnamon quills, type Sri Lanka (Ceylon)

Commercial designation of the grades and qualities	Diameter of quills mm, max.	Number of whole quills (1050 mm) per kg min.	Extent of foxing[1] (% max[2])	Minimum length of quills in a bale[3] mm	Pieces of tubes and broken pieces of the same quality per bale % (m/m) max.
Alba	6	45	nil	200	1
Continental					
C 00000 Special	6	35	10		
C 00000	10	31	10		
C 0000	13	24	10	200	1
C 000	16	22	15		
C 00	17	20	20		
C 0	19	18	25		
Mexican					
M 00000 Special	16	22	50		
M 00000	16	22	60	200	2
M 0000	16	18	60		
Hamburg					
H 1	23	11	25		
H 2	25	9	40	150	3
H 3	38	7	65		

Notes

1 Foxing: the occurrence of reddish-brown patches on the surface of the quills, which may become dark brown with time. Foxing can be (a) superficial ("malkorahedi") or (b) heavy ("korahedi"). This sub-division is based on the depth of the patches.

2 Extent determined by visual examination.

3 Bale: a package of any one particular grade of quills wrapped with suitable material for export.

Cinnamon powder

The powder shall be constituted solely by the types of the cinnamon listed above under 'Types and classification'.

Note – If there is a designation of origin, the powder shall be prepared exclusively from the barks concerned.

Table 5A.4 Classification of cinnamon, type Seychelles and type Madagascar

Commercial designation of the grade	Physical characteristics of the bark
Whole tubes (full tubes)	Tubes of length about 15 cm and bark thickness up to 1 mm
Pieces of scraped bark	Broken pieces, rough and grooved
Pieces of unscraped bark	Broken pieces, rough and grooved, of width up to about 3 cm and length up to 20 cm. The bark can be up to 5 mm thick
Chips, flakes of unscraped bark	Small pieces of unscraped bark of cinnamon stems

148   *K.R. Dayananda, U.M. Senanayake and R.O.B. Wijesekera*

Requirements

Flavour

The flavour of cinnamon shall be fresh and characteristic of the spice of the origin concerned. It shall be free from foreign flavours, including mustiness.

Colour

Cinnamon powder shall be of a yellowish to reddish brown colour.

Freedom from moulds, insects, etc.

Whole cinnamon shall be free from living insects, moulds, mites and insect remains, for example cocoons, and shall be practically free from dead insects, insect fragments and rodent contamination visible to the naked eye (corrected, if necessary, for abnormal vision), with such magnification as may be necessary in any particular case. If the magnification exceeds $\times 10$, this fact shall be stated in the test report.

In case of dispute, contamination in cinnamon powder shall be determined by the method described in ISO 1208.

Extraneous matter

The proportion of extraneous matter in whole cinnamon shall not exceed 1% (m/m) when determined by the method described in ISO 927.

Extraneous matter comprises leaves, stems, chaff and other vegetable matter. The only mineral matter permitted is sand, earth and dust.

In the case of cinnamon quills, type Sri Lanka, take about 110 g of quills per bale of Continental grade and 230 g of quills per bale of Mexican or Hamburg grades, break them up and inspect the filling. Unscraped inner bark, scrapings, foreign matter, bark of wild cinnamon and other genera shall not be present.

Chemical requirements

Whole cinnamon and cinnamon powder shall comply with the requirements given in Table 5A.5.

Sampling

Sample cinnamon quills by the method described in Annex 5.1A.

Sample all other forms of cinnamon in accordance with ISO 948.

Methods of test

Samples of whole cinnamon and cinnamon powder shall be tested for conformity with the requirements of the International Standard by the methods referred to in Tables 5A.4 and 5A.5.

Table 5A.5 Chemical requirements

Characteristic	Requirements		Method of test
	Cinnamon, type Sri Lanka	Cinnamon, type Seychelles and type Madagascar	
Moisture content, % (m/m), max.	12.0	12.0	ISO 939
Total ash, % (m/m) on dry basis, max.	5.0	7.0	ISO 928
Acid insoluble ash, % (m/m) on dry basis, max.	1.0	2.0	ISO 930
Volatile oils (ml/100 g) on dry basis,			
– Whole cinnamon	1.0*	0.4*	
Ground (powdered) cinnamon	0.7*	0.3*	ISO 6571

Note
* Tentative values.

Ground cinnamon shall be examined by microscope. It shall not contain any morphological extraneous matter.

For the preparation of a ground sample for analysis, coarsely crush the product until particles of 5 mm or less are obtained, before applying the general method described in ISO 2825.

Packing and marking

Packing

WHOLE CINNAMON

Whole cinnamon shall be packed in clean, sound and dry containers made of a material which does not affect the product or its flavour.

The different commercial classes are usually packed as follows:
– Cinnamon, type Sri Lanka: in cylindrical bales of about 45 kg;
– Cinnamon, type Seychelles quills and quillings: in wooden boxes of about 100 to 150 kg.

CINNAMON POWDER

Cinnamon powder shall be packed in the same type of containers as specified for whole cinnamon. In addition, the containers shall protect the cinnamon powder against moisture and loss of volatile matter.

150 *K.R. Dayananda, U.M. Senanayake and R.O.B. Wijesekera*

Marking

WHOLE CINNAMON

The following particulars shall be marked or labelled on each container:

(a) name of the material, and the trade name or brand name, if any;
(b) name and address of the manufacturer or packer;
(c) batch or code number;
(d) net mass;
(e) grade of the material;
(f) producing country; and
(g) any other marking required by the purchaser, such as year of harvest and date of packing (if known).

CINNAMON POWDER

The following particulars shall be marked or labelled on each container:

(a) name of the material, and the trade name or brand name, if any;
(b) name and address of the manufacturer or packer;
(c) batch or code number;
(d) net mass; and
(e) any other marking required by the purchaser, such as date of packing (if known).

Annex 5.1A

Sampling of cinnamon quills

Definitions

Consignment: The quantity of packages of cinnamon quills submitted at one time and covered by a particular contract or shipping document. It may be composed of one or more lots.

Lot: All the packages in a single consignment of cinnamon quills pertaining to the same grade.

Defective package: Any package of cinnamon quills not conforming to any one or more of the requirements of this International Standard.

Number of packages to be taken for inspection

The number (n) of packages to be taken from a lot depends on the size of the lot and shall be in accordance with Table 5A.6.

Testing of samples and criterion for conformity

All the packages taken for inspection shall be used individually to test for conformity with all the requirements given for this International Standard.

The lot shall be considered as conforming to the requirements of this International Standard if the number of defective packages in the sample tested is less than or equal to the corresponding acceptance number given in Table 5A.7.

Harvesting, Processing, and Quality Assessment of Cinnamon Products 151

Table 5A.6 Number of packages to be taken for inspection

Number of packages in the lot (N)	Number of packages to be taken (n)
1 to 5	All
6 to 49	5
50 to 100	10% of the number of packages
Over 100	The square root of the number of packages rounded to the nearest whole number.

Table 5A.7 Acceptance number

Number of packages tested	Acceptance number
Up to 12	0
13 to 20	1
21 to 35	2
36 to 50	3
51 to 75	4
Over 75	5

Annex 5.1B

Recommendations Relating to Storage and Transport Conditions

(This annex does not form part of the standard.)

Containers of cinnamon should be stored in covered premises, well protected from the sun, rain and excessive heat.

The store room should be dry, free from objectionable odours and proofed against entry of insects and vermin. The ventilation should be controlled so as to give good ventilation under dry conditions and to be fully closed under damp conditions. In a storage warehouse, suitable facilities should be available for fumigation.

The containers should be handled and transported such that they are protected from rain, sun or other sources of excessive heat, objectionable odours and from cross-infestation, especially in the hold of ships.

Annex 5.2

Material Safety Data for Cassia Leaf Oil (USA)

Product name	: Cassia leaf oil (Chinese). *Cinnamomum cassia.*
US DOT Hazard class	: N/A, Keep away from children.

Physical data:

Appearance/odour	: Yellowish to brownish liquid/woody, sweet, cinnamon spicy.
Solubility in water	: Insoluble.
Specific gravity	: 1.04500–1.06300 at 25 °C.

Flash point (°F)   : 194.00 (90.00 °C) TCC.
Extinguishing media   : Carbon dioxide, dry chemical, universal-type foam.

Health hazard data

Primary routes of exposure   : Skin contact, eye contact, ingestion.
Carcinogenecity   : No.
Effects of overexposure   : Irritating to skin and eyes.
Eye contact effects   : Irritating to eyes.
Skin contact effects   : Contact dermatitis may occur.
Inhalation effect   : Irritating if inhaled – mucous membrane reaction.
Ingestion effects   : Harmful if swallowed.
(Note: The above effects are based on evaluation of individual components and the relevancy to the mixture as a whole or to humans is unknown).

Emergency first aid and procedures

Eye contact   : Flush immediately with clean water for at least 15 minutes. Contact a physician immediately.
Skin contact   : Remove any contaminated clothing or shoes, wash effected areas thoroughly with soap and water. Contact a physician if necessary.
Inhalation   : Remove from the exposure to fresh air. If breathing has stopped administer artificial respiration and oxygen if available. Contact a physician.
Ingestion   : Wash out mouth with water and give water to dilute provided person is conscious. Contact a physician or local poison control room immediately.

Reactivity data

Chemically stable   : Yes.
Conditions to avoid   : This product presents no significant reactivity hazard. It is stable and will not react violently with water. Hazardous polymerisation will not occur.
Incompatibility with other materials   : Avoid contact or contamination with strong acids, alkalies or oxidising agents.
Hazardous combustion or decomposition products   : Carbon monoxide and unidentified organic compounds may be formed during combustion.

Spill or leak procedures

Spill, leak and waste disposal procedures:

Eliminate all ignition sources. Ventillate area, contain spill and recover free product. Absorb remainder on vermiculite or other suitable adsorbent material. Use of self contained breathing apparatus is recommended for any major chemical spills. Report spills to appropriate authorities if required.

Special protection information

Respiratory protection	: None generally required.
Ventillation protection	: Local exhaust.
Protective clothing	: Chemical resistant clothing is recommended.
Protective gloves	: Use of chemical resistant gloves is recommended.
Eye protection	: Use of goggles or face shield is recommended.

Other protective equipment:

Avoid inhalation and contact with skin and eyes. Good personal hygiene practices should be used. Wash after any contact, before breaks and meals, and at the end of the work period. Contaminated clothing and shoes should be cleaned before re-use.

Special precautions:

Precautions to be taken in handling and storage: Place material and absorbent into sealed containers and dispose of in accordance with current applicable laws and regulations.

Other precautions:

Good manufacturing practices dictate that an eyewash fountain and/or safety shower should be available in the work area.

(Source: TGSC Occupational Health and Safety Department, The Good Scents Company, Wisconsin, USA.)

Annex 5.3

Cleanliness Specifications

Cassia and cinnamon

ASTA cleanliness specifications.

	Whole insects dead by count	Excreta mammalian mg/LB	Excreta other mg/LB	Mold % by wt.	Insect defiled/ infested % by wt.	Extraneous foreign % by wt.
Cinnamon	2	1	2.0	1.0	1.0	0.50
Cassia	2	1	1.0	5.0	2.50	0.50

Defect action levels prescribed by USFDA.

	Defect (method)	Defect action level
Cassia or cinnamon bark, whole	mold (MPM-V32)	Average of 5% or more moldy. pieces by weight are moldy.

	Defect (method)	Defect action level
Cinnamon ground	insect filth (MPM-V32)	Average of 5% or more pieces by weight are insect infested.
	Mammalian Excreta (MPM-V32)	Average 1 mg or more mammalian excreta per pound.
	Insect filth (AOAC 968.38 B)	Average of 400 or more insect fragments per 50 g.
	Podel filth (AOAC 968.38 B)	Average of 11 or more rodent hairs per 50 g.

Requirement according to honest trading practices in Germany.

	Maximum moisture	Minimum oil content	Ash	Maximum sand content (Part of ash not soluble in HCl)
Cinnamon (whole)	13%	2%	5–7%	1–8%

Dutch regulations

	Ash	Sand	Oil	Mean
Cinnamon	8.0	2.0	1.0	–
Ceylon Cinnamon	5.0	2.0	1–5%	–

Source: European Spice Trade Association.

ESA product specification for quality minima.

	Ash % w/w *max.*	AIA* % w/w *max.*	H_2O % w/w *max.*	V/o** % v/w *min.*
Cassia	7	2	14	1.0
Cinamon	7	2	14	0.4

Notes
* Acid insoluble ash.
** Volatile oil content.

European Spice Trade Association specification of quality minima for herbs and spices.

Extraneous matter	Herbs 2%, Spices 1%
Sampling	(For routine sampling) Square root of units/lots to a maximum of 10 samples. (For arbitration purposes) Square root of all containers, e.g. 1 lot of pepper may = 400 bags, therefore square root = 20 samples.
Foreign matter	Maximum 2%

Ash	Refer to Annex 5.3.
Acid insoluble ash	Refer to Annex 5.3.
H_2O	Refer to Annex 5.3.
Packaging	Should be agreed between buyer and seller. If made of jute and sisal, they should conform to the standards set by CAOBISCO Ref C502–51– sj of 20-02-95 (see Annex 5.2). However, these materials are not favoured by the industry, as they are a source of product contamination, with loose fibres from the sacking entering the product.
Heavy metals	Shall comply with national/EU legislation.
Pesticides	Shall be utilised in accordance with manufacturers recommendations and good agricultural practice and comply with existing and/or EU legislation.
Treatments	Use of any EC approved fumigants in accordance with manufacturersinstructions, to be indicated on accompanying documents. (Irradiation should not be used unless agreed between buyer and seller).
Microbiology	*Salmonella* absent in (at least) 25 g. Yeast & Moulds 10^5/g absolute maximum. Other requirements to be agreed between buyer and seller.
Off odours	Shall be free from off odour or taste.
Infestation	Should be free in practical terms from live and/or dead insects, insect fragments and rodent contamination visible to the naked eye (corrected if necessary for abnormal vision).
Aflatoxins	Should be grown, harvested, handled and stored in such a manner as to prevent the occurrence of aflatoxins or minimise the risk of occurrence. If found, levels should comply with existing national and/or EU legislation.
Volatile oil	Refer to Annex 5.1.
Adulteration	Shall be free from adulteration.
Bulk density	To be agreed between buyer and seller.
Species	To be agreed between buyer and seller.
Documents	Should provide: details of any treatments the product has undergone; name of product; weight; country of origin; lot identification/batch number; year of harvest.

6 Chinese Cassia

Nguyen Kim Dao

Introduction

Cinnamon and Cassia* are among the oldest spices reaching ancient Egypt, reportedly by the seventeenth century B.C. But it seems that the bark that was entering the cinnamon trade in ancient times was of different botanical origin. Probably the Greek and Romans had used both cinnamon and cassia, but the Arab traders who dominated and controlled the spices trade shrouded the sources in mystery. Cinnamon and cassia were among the spices sought after by most fifteenth- and sixteenth-century European explorers (Dao *et al.*, 1999).

In Vietnam, about 2000 years ago, Giao Chi Ngoc Que paid valuables in tribute to the Chinese Emperor, and among these was cassia bark. Cassia is found both wild and cultivated in South-East Asia, south China (Kwangxi and Kwangtong provinces), Burma (Myanmar), Laos and Vietnam. It was introduced into Indonesia, Sri Lanka, South America and Hawaii. In Vietnam, it is found in many provinces from the North to the South, but is concentrated in the provinces of Quangninh, Yenbai, Nghialo, Tuyenquang, Ninhbinh, Thanhhoa, Nghean Hue, Quangnam and Quangngai and in Taynguyen plateau.

Taxonomy

Chinese cassia/Vietnam cassia (*Cinnamomum cassia*)

Berch. & Presl, 1825, *Prir. Rostlin* 2: 36, 44–45, t.6. Synonyms: *Laurus cassia* L. (1753); *Cinnamomum aromaticum* C. Nees (1831) (for nomenclature citations see Chapter 2).

Vernacular name: Chinense cassia, Chinense cinnamon, Cassia lignea (En); Cannellier de chine; Canellier casse (Fr); Kayu manis cina (Indonesian); S'a: chwang (Laos); Kaeng (Thailand); Qu[ees] thanh, Qu[ees] don, qu[ees] qu[ar]ng (Vietnam).

An evergreen, medium sized tree, 18–20 m high and 40–60 cm in diameter, with a straight and cylindrical trunk and grey brown bark that is 13–15 mm thick when mature. Branchlets of the previous year are dark brown, longitudinally striated, slightly pubescent, and those of the current year are more or less tetragonal, yellow-brown, longitudinally striated, densely tomentose, with grey-yellow hairs. Terminal buds are small, about 3 mm long, with broadly oval scales. The terminal buds are acuminate and are densely tomentose with grey hairs. Leaves are simple, alternate

---

* The name cassia is used to denote cassia cinnamon and should not be confused with the genus *Cassia* belonging to the family Cesalpiniaceae.

0-415-31755-X/04/$0.00 + $1.50
© 2004 by CRC Press LLC

Figure 6.1 Cinnamomum cassia. 1. Flowering branch; 2. Fruit and tepal; 3. Fruit; 4. Dissected flower; 5. Outer stamen and inner stamen with basal glands; 6. Staminode.

or subopposite, oblong-oval or narrowly elliptic to sublanceolate, 8–16 × 4–7 cm, coriaceous, thick, shining green, glabrous upper, greenish opaque and sparsely hairy under, triplinerved from the base, the apex slightly acute. Leaf base is acute, the margins are cartilaginous and involute, the basal lateral nerves subopposite, arising from 5–10 mm above the leaf base, arcuate-ascendent and evanescent below the leaf apex. Petioles are 1.5–2.5 cm long. Panicle's axillary or subterminal, 8–16 cm long, triplicate-branched, the end of the branch being 3-flowered, peduncle as long as half of inflorescence. Flowers are white, 4.5 mm long; pedicels 3–6 mm long, yellow-brown tomentellate. Perianth is densely yellow-brown-tomentellate outside and inside, the tube is obconical, about 2 mm long, the lobes are ovate-oblong, subequal. There are nine fertile stamens, about 2–3 mm long; (of 1st and 2nd whorls) lateral extrorse; the 3rd whorl anther ovate oblong, four-celled introse. Staminodes 3, of the innermost whorl (Fig. 6.1). Ovary ovoid is about 1.7 mm long, glabrous, style slender, as large as ovary, stigma small and inconspicuous. Fruit ellipsoid, is about

10 × 7 mm, pink-violet when mature, glabrous, perianth-cup in fruit shallowly cupuliform, 4 mm long, truncate or slightly dentate and up to 7 mm broad at the apex (Li and Li, 1998).

Vietnam cassia: a case of misidentification

Cinnamomum loureirii Nees, 1836, Lecomte, 1913, deser. emend. – *Laurus cinnamomum* Lour. 1790. non L. (1753). Nhuc que, que Thanh, que quan.

The earlier publications indicated that Vietnam cassia is *Cinnamomum loureirii*. This is also considered to be a variety of *C. obtusifolium* (Roxb.) Nees, occurring from the central and northern Himalayas, which is also confused with *C. japonicum* (Syn. *C. pedunculatum* Pres.) native to China, Korea and Japan, and with *C. sieboldii*, that is widely seen in Japan (Weiss, 1997). But from over 20 years of study on collection and classification of specimens of *Cinnamomum* from the north to the south of Vietnam, we reached the conclusion that the Vietnam cassia is nothing but *C.cassia*. This is a case of misrepresentation that has continued for some time. When Loureiro proclaimed *Laurus cinnamomum*, his opinion was *"Habitat agrestris in altis montibus cochinchine ad occidenterm versus Laosios"*, but the specimens on which Loureiro based his study could be mislayed or lost. Merrill (1935) was of the opinion that Lecomte (1913) based his study on specimens of Chinese origin, not specimens of Indochina origin. Merrill (Allen 1939) assigned the name of species that Loureiro studied as *Cinnamomum cassia* Blume. But the opinion of Li (1984) was that the name *Cinnamomum loureirii* sensu Lecomte (1913) non Nees (1836) was a synonym of *C. contractum* H.W. Li (1978) (see also Chapter 2). The Vietnam government introduced the name Vietnamese cassia as a standard without reference to regional origin, correct botanical source or official grades (Weiss, 1997). However, what is mostly traded are *C. cassia*. It was also pointed out that *C. loureirii* is very poorly represented in Vietnamese or Chinese herbaria, indicating that this is not a very common species. It is rather difficult to believe that a commonly cultivated/or a popular tree like Vietnam cassia is represented so rarely in all herbaria. The inevitable conclusion reached by taxonomists is that *C. loureirii* is a very rare species and it cannot be the botanical source for the commercially traded Vietnam cassia. What is currently grown in Vietnam is *C. cassia*. But the harvesting–processing practiced in Vietnam is very much different and this leads to a product that looks distinctly different from the product from China and, which possibly came to be considered as derived from another species.

Ecology

The main production areas of cassia cinnamon in China are characterised by mean daily temperature of about 22 °C and an annual rainfall of 1250 mm in about 135 wet days. The absolute maximum temperature is about 38 °C and the absolute minimum is 0 °C. It is grown in southern China at altitudes up to 300 m. North Vietnam has the same ecological condition as south China. However, South Vietnam is characterised by a mean daily temperature of about 23.1 °C. The absolute maximum temperature is about 35.5 °C and the absolute minimum is 17.6 °C. The cassia tree is naturally distributed in primary dense forests below 800 m altitude. It is a light demanding tree, slightly shade tolerant when young, preferring cool and wet condition with a mean

Figure 6.2 A 15-year old plantation of *C. cassia*.

annual rainfall of 1500 mm (2500–3000 mm in South Vietnam). The crop needs acidic soil (pH: 4.5–5.5) and prefers undulating hills. The root system is deep and strong (Fig. 6.2).

Planting and Husbandry

Cassia cinnamon is usually grown from seeds, but can be grown from cuttings also. However, cassia plants grown from cuttings are not good, as the bark of such trees is not thick and essential oil contents in leaves and barks are less. Ripe fruits from mother trees producing thick bark of good aroma should be selected for propagation. The harvesting cycle of cassia cinnamon is 20 years in Vietnam and that is divided into two main periods (Khoi, 1991):

Basic planting period: ten years from planting.
Harvest period: ten years later.

Basic planting periods of ten years are divided into:

Young tree planting period (growing seedlings).
Forest growing period – establishing the 'cassia forest'.
Forest care period.

Harvesting period is divided into:
Fist prune: after 10 years, adjustment of tree density at 1000–1250 trees/ha.
Second prune: after 15 years, stable density after pruning to harvest is over 600–625 trees/ha.

The details and norms of the planting technique are shown in Table 6.1.

160 *Nguyen Kim Dao*

Table 6.1 Main norms of planting cassia cinnamon in Vietnam

No.	Norms	Unit	Per ha of forest
	A. Technique		
1	Sowing area	m^2	40–50
2	Transplanting area	m^2	450–500
3	Ratio of standard trees	% trees	80%
4	Ratio of reserve trees for inter-crop	%	20%
5	Time to nurse seedlings	month	12–24
6	Number of standard trees/1kg seeds	trees	1500
7	Method to treat the vegetation		cutting
8	Method to prepare soil		by hoeing
9	Dimension of holes for growing	cm	$40 \times 40 \times 40$
10	Dimension of trees for growing	cm	H: 20–30
			D: 0,8–1
11	Seedling density	trees	3300
	B. Material		
12	Seed	kg	2.5
13	Seedling for growing	trees	4950
14	Fertilizer		
	Phosphorus	kg	300
	Manure	kg	700
	Nitrogenous	kg	20
	Potassium	kg	50
15	Insectisides		
	Basudin, Wofatox	kg	50
	Powdered lime + 666	kg	100
16	Material for making greenhouse	kg	200

In China, cassia cinnamon is grown more like the Sri Lankan cinnamon. Here the plants are coppiced four to five years after planting, and regularly thereafter. In such cases plants are maintained as thick bushes and are retained in the field for a considerable period of time. This difference in harvesting makes the final products from China and Vietnam very much different in quality and appearance.

Sowing and nursing

Trees for raising seedlings must be strong, with thick, sweet smelling bark. Seeds are collected in July–August, are put in water, and only seeds that sink are used for sowing. Before sowing, seeds are to be kept in warm water (30–35 °C) with 1% permanganate for six to eight hours. They are then taken out and kept in shade in humid sand in multi-layers: one layer of 8–10 cm thick sand is intercalated with one layer of seeds, and three such sand layers are usually made. Seeds are left for seven to ten days and receive occasional sprinklings of water. When seeds germinate the sprouts are transplanted to nursery beds or polybags. Alternatively direct seeding in polybags can also be employed.

Preparation of soil: Land is ploughed 30–40 days before sowing seeds, adding powdered lime at the rate of 0.2 kg/m^2. The soil is plouged for a second time followed by careful harrowing. Beds, 10 m long and 1–1.2 m wide are made. When seeds germinate, the sprouted seeds are transplanted to beds at the spacing of 10 × 10 cm. Alternatively sprouts can be transplanted in polybags filled with soil.

Covering frame: Cinnamon seedlings prefer shady places, so seedlings in the nursery must be covered. The covering frames should be 1.5–2 m high for easy watering and tending. About 80% shade is provided before transplanting.

Watering: The watering regime differs based on the humidity in each area. In Tra Mi, in Vietnam, it is rainy with high humidity, so watering is necessary only on dry days. In other areas watering should be adjusted based on rainfall.

Tending of seedlings in the first one to three months: Weeding needs to be done at least every month. In case of pests/disease, spraying the appropriate insecticide (insecticides Basudin or Wofatox 666: 1–2%, 4 l/m^2) or fungicide chemical to prevent soil borne diseases (Bordeaux solution: 1–1.5%, 4 l/m^2) is essential. Spraying should be done every 15 days, once or twice per month.

Fertilizers: The following are given – basal fertilization: Muck + 30% phosphorum 2kg/m^2. Additional fertilizers: potassium 0.1kg/m^2; nitrogenous or muck 0.3% through irrigation water or spraying.

Tree training: Seedlings before transplantation need to be trained by stopping all aspects of the cultivating regime, such as the application of additional fertilizers, spraying chemicals, taking off the coverings, stopping weeding, etc.

Forest growing: Being a long-term crop, choosing land for the extensive growing of cinnamon is very important. Cassia cinnamon seedlings prefer shady places, so a vegetation cover is essential for the healthy development of plants in the initial stage. Such vegetation cover is found in forests and in the fallow land under growing trees.

The vegetation cover in the forest must be cut down into stripes of 1 m wide, and the treeless stripes are intercalated with tree bands parallel to each other. The remaining trees in the forest must be exploited before growing cinnamon. The covering canopy of the forest should be around 60%.

Digging holes: The hole dimension should be 40 × 40 × 40 cm with a row interval of 2 m and a tree interval of 1.5 m. The holes should be dug at least 30 days prior to planting.

Planting: Seedlings must be fully trained before planting. A seedling is removed from the nursery bed together with a ball of earth and planted in the pit. Care is needed to prevent bud and root breaking. After planting in the pit, the piling up of soil at the base is important to prevent the stagnation of water during heavy rains. The best time for planting is October to November.

Tending a growing cassia forest

First year: Tend only once in September and October. The work required includes cutting the vegetation cover, weeding, lightly turning over the soil at the base, piling up soil 5–10 cm high and 0.6–0.8 cm wide, adjusting the shade at 50%, and replacing dried up seedlings.

Second to fourth years: Tend twice every year, first in January, February, March as done in the first year. The shade intensity required for a two-year old forest is 40%, that of three-year old forest is 30% and that of four-year-old forest is 20%. The second time in August, September and October. Trimming the lower and diseased branches, piling up soil at the foot of the plant and replacing dead trees are the important activities.

Fifth to tenth years: Tend once every year in May, June and July. Clear all the vegetation cover, shading for a five-year old forest is 20%. From the sixth year shade is completely removed, followed by the clearing and cutting of diseased and stunted branches. In a ten-year old plantation, thinning is done for adjusting the density of trees to 1000–1250/ha.

Harvesting period: Ten years later (20 years after planting) in Vietnam. In China, cassia harvesting is done as in the case of cinnamon (extracting bark from copieced shoots) and the first harvesting starts about four to five years after planting and subsequent harvests are conducted in three- to four-year cycles. However, in Vietnam, the harvesting is done 15–20 years after planting.

Protection and management

Protect the plantation from forest fire, cattle trampling and monitor for disease occurrence. Prune to regulate tree structure: the first pruning should be done in May–June of the tenth year, and the density is maintained at 1000–1250 trees/ha. The tree interval should be 4 m and the row interval should be 2 m following the quincunx. The aims of pruning are proper development of trees that are free from disease, and which have symmetrical leaf canopies and trunks over 8 cm in diameter. All branches must be pruned at a distance of 2 m from the base to produce straight trunks and better bark. The second pruning is in February/March or July/August in the fifteenth year when the density is reduced to 625 trees/ha; that means the tree as well as the row interval is 4 m. Clean the forest periodically.

Harvesting

Twenty-year old cinnamon trees are harvested in Vietnam. There are two crops: the first crop is harvested from February to March (in the south) and from April to May (in the north); the second crop is harvested from July to September (in the south), from September to October (in the north). The bark extracted from the first crop has more scales than that of the second one, and the second crop bark is of better quality. In many growing areas in China, harvesting starts four to five years after planting and is continued with every three to four year cycle.

The harvesting technique

In Vietnam

The harvesting of cassia bark depends on the accumulated experience of growers in each area. In the north of Vietnam scaffoldings of 4–5 m high are erected around the trees to make it easier to climb and to avoid damaging the bark. For peeling the bark, people tie a string around the stem and big branches at a distance of 40–50 cm for marking. Sharp knives and chisels are used to cut the bark around stem and branch. Two such round cuts are made 40 cm apart. The bark is then cut in longitudinal strips 40 to 50 cm apart with a sharp knife. These strips are then peeled off using knives or bamboo splits. In the south, prior to harvesting, a circle of bark of 5–10 cm wide from the base of the trunk is taken for facilitating the subsequent peeling. Afterwards people use rattan to tie around the trunk for making a ladder for climbing. The peeling is done from the top downwards in strips 40–50 cm long and a convenient width (40–50 cm). People avoid cutting down the trees because then the peeling becomes difficult. After peeling the trunk and big branches, the tree is cut down for collecting the bark from smaller branches and branchlets. Leaves are also collected for distillation (Khoi, 1991).

In China

In China, cassia is grown as a coppiced bush and harvested like Sri Lankan cinnamon, by cutting above the ground level initially four to five years after planting and every three to four years thereafter. The strongest shoots are sometimes retained for growing and later used for seed and cassia bud production. Bark is extracted from cut stems as in the case of cinnamon. The main shoots are ringed at intervals of 30–60 cm or else cut into pieces. Longitudinal slits are then made and the bark is then separated by using a special type of curved knife. The bitter tasting outer bark is then scrapped of and the bark is dried in sun. On drying the bark turns brown and curls into a hollow tube or quill. Quills are tied into bundles. In cases where peeling is difficult, the bark is chipped off from the stem (often together with a piece of wood) and dried in the sun. Such chips give inferior quality cassia.

The two types of harvesting techniques prevalent in China and Vietnam in effect produce two different products that differ in appearance, texture and quality. This difference in appearance and quality might be one reason for regarding Vietnam cassia as the product of a different species. The Chinese product is more similar to Ceylon (Sri Lankan) cinnamon of commerce, as this product is collected from four- to five-year old shoots. On the other hand the Vietnam product is collected from trees almost 15 years older, and the bark is much thicker which has a more pungent taste and stronger flavour.

Cassia buds (*Kuitsz*) are the dried immature fruits, which includes calyx and often pedicel. Cassia buds are harvested from trees left uncut in plantations or from wild trees during October–November, and are then dried in the sun or shade. Leaves for distillation are collected from cut shoots and from prunings of excess foliage from standing trees. The main harvest season is June–July, when the oil yield is highest. Large mature leaves are said to yield the best quality oil (Weiss, 1997). Often small twigs and branchlets are also mixed with leaves for distillation. Harvested leaves are allowed to partially dry before distilling.

Handling After Harvest

After peeling, the bark must be dried immediately by spreading it on bamboo screens 0.7–0.8 m high in well aerated areas during the day and at night they must be kept covered to avoid rainwater entering the bark and adversely affecting its quality. For uniform drying of both surfaces the bark is occasionally turned over until the drying is complete.

A better method for the preparation of bark, but which is more complex, is prevalent in certain areas in Vietnam. Here peeled cassia bark is put in fresh water right after peeling for about 24 hours. The bark is then cleaned, washed and dried in a shady place until it is apparently dry. Such bark is then kept in layers in a bamboo basket lined and separated with dry banana leaves. The top layer is covered with dried banana leaves and compressed by stones for 24 hours. The basket is then turned upside down and kept for 20–24 hours more. The bark is then washed and put on bamboo screen and dried for about 15 days in summer and in spring and about 30 days in autumn. In this way, the quality of the cinnamon bark is good with a better fragrance and taste. It also acquires an attractive red color. Such processing is practiced only in Vietnam (Khoi, 1991). Among cinnamon bark categories, the most valuable is the one having the shape of number 3. It is uniquely produced in Quang Nam (Tra Mi) and Quang Ngai (Tra Bong) areas of South Vietnam. To get this kind of cinnamon, the bark must be fresh, its thickness must be over 1.2 cm and its width over 40 cm. Only then the bark may take the form of number 3 on drying. People in Hong Kong and Taiwan are used to this kind of cinnamon at the price of $50 (US)/kg. Besides drying, there are some complicated procedures involved in the processing of this brand, such as waxing the cut ends of the extremities of the bark, covering the bark with soft cloth and then storing in zinc tubes. People with large quantities of cassia bark, usually make a double compartmented wooden box. A honey bowl is kept at the bottom to keep suitable level of humidity, and the upper compartment has a bottom made of a metal net for keeping the bark. In so doing, fragrant flavour and taste are ensured. Such elaborate post-harvest handling is practiced by some Vietnam growers.

In China the peeled bark (from three- to five-year old shoots) is dried in the sun and bundled. Usually no further post-harvest operations are done by Chinese producers. When a cleaner product is required the dried bark is washed in fresh clean water, and dried thoroughly and packed in bundles. In cases where peelings do not come off easily, the bark is either scraped or chipped (often together with a piece of wood) and dried in the open and then packed in bags.

Production and International Trade

Cassia bark and leaf oil are economically important. China (Kwangsi and Kwangtung Provinces) is the main producer and exporter of Chinese cassia. The harvested area in 1998 was estimated by FAO at 35,000 ha with a production of 28,000 t (Coppen, 1995). Exports of dry bark from China during the period 1966–1976 amounted to 1250–2500 t annually. In 1987–1993 the United States imported annually about 200 t of cassia bark and 340 t of cassia leaf oil. Given the considerable domestic consumption in China, the production of leaf oil must be in excess of 500 t. In 1991–1994 cassia leaf oil fetched a price of about $30–35 (US)/kg (Dao et al., 1999).

In Vietnam, from the early times to the present day, only the bark has been economically important, while essential oil extracted from leaves and bark has been used only locally. The harvested area in 1998 was estimated by FAO at 6100 ha with total production of 3400 t. Exports of dry bark from Vietnam up to 1966–1976 varied from 300 to 500 t. The export, has gone up subsequently and currently it is around 3000 t (Dao et al., 1999).

Adulterations and substitutes

In early times, cassia was usually described as a somewhat inferior substitute for cinnamon but now it has its own market. Cassia (Chinese) bark is interchangeable in many applications with cinamon bark and oil. Cassia bark oil and leaf oil contains cinnamaldehyde, while cinnamon leaf oil contains eugenol and bark oil cinnamaldehyde. Detection methods have much improved and therefore adulteration has become less common. Methods of distinguishing cinnamon bark oil from cassia oil are based on the presence/absence of minor components such as ortho-methoxy cinnamaldehyde, eugenol and coumarin (Dao *et al.*, 1999).

Yield

Yields are generally higher during the dry season than during the wet season. Yield varies so much from tree to tree, and from year to year, that it is practically impossible to give the normal values. It is, however, clear that yields depend on the climate of an area. Bark yield is higher in South Vietnam than in North Vietnam and south China. The Tra Mi and Tra Bong cassias of South Vietnam are of the highest quality (Khoi, 1991). The average yield of cassia bark depends on the age of the plant (Table 6.2). The data on yield from thin prunning and main crops are shown in Table 6.3.

Cinnamon bark yield (t/ha) in South Vietnam (for the 20-year old cycle) is around 6.5 t; in south China and North Vietnam it is about 3–5 t.

Table 6.2 Chinese cassia bark yield in Vietnam

Year	Diameter (cm)	Height (m)	Bark weight (kg)		Essential oil yield
			Fresh	Dry	
10	7	6	6	1.4	0.06
15	10	8	11.5	2.38	0.15
20	16	12	20.5	5.63	0.24
25	20	14	26.5	6.87	0.31
30–40	25	15	40.5	11.67	0.33

Table 6.3 Cassia average yield/ha during thin prunning and main crops in a cycle of 20 years

Standard	The first thin prunning	The second thin prunning	Main crops	Yield during the whole cycle
Plant number	1000–1250	375–625	625	
Fresh bark (kg)	600–6750	4257–7026	14062	24229–28624
Dry bark (kg)	1646	1378	3576	6540
– "Kep" Cinnamon (kgs)		276	1758	2034
(The bark made into 3 shape)	–	–	–	–
– Pipe Cinnamon (kgs)	988	689	1055	2732
– Cinnamon in loose bits (kg)	658	413	703	1174
Whole essential oil (kg)	70	59	151	280

Cassia Oil Production

The cassia oil of commerce is distilled from the leaves and twigs of *C. cassia*. Cassia oil is produced mainly in China, where hydrodistillation is employed for oil production. Cassia oil was earlier produced as a primitive cottage industry, but now modern factories owned by communes have been established in many cassia growing regions.

The distillation material consists of partially dried leaves and branchlets from the shoots used for bark production. Depending upon the proportion of leaf and stem (twigs) the quality of the oil produced may vary. Usually 3/4 leaves and 1/4 twigs are used. The best quality oil is obtained from material harvested in the summer and autumn from trees that are five to seven years old (Purseglove *et al.*, 1981).

The local stills used are the directly fired type, consisting of a shallow boiling pan made of cast iron. The rest of the body is usually made of tin or iron sheets, on which sits the still head surmounted by a condenser in the form of a shallow pan into which water flows. The still can accommodate about 60 kg leaf and twigs and 150 l of water. The still is heated, often by using debarked cassia wood. The distillation process goes on for about three to four hours. Steam and oil vapours passing through holes in the still head are liquified at the bottom of the cooling vessel (over which there is a shallow pan through which water flows). The condensed water and oil drip from the bottom of the still head into the lower part of the still head, from where the oil and water mixture flows through a pipe into an oil tank. In the oil tank, the oil and water separate. The oil is heavier than water and collects at the bottom of the oil tank from where it is periodically removed. The supernatent water which is milky because of suspended oil is sent back to the still. The yield of oil by this procedure is about 0.31 to 0.33%, but could be much higher by using more efficient equipments. The factories owned by the communes use larger coal-fired stills which are more efficient and which yield more oil.

The leaves yield more oil than twigs (0.54% versus 0.2%). The highest oil yield is from vigorous, five to seven year old trees. The quality of the oil depends upon the leaf material used and varies according to season. In winter and early spring the oil quality is said to be inferior to that from leaves harvested in mid-summer and autumn (Guenther, 1950).

Economic Value

Cassia has great economic value both in Vietnam and China. Long ago Hai Thuong Lan Ong appreciated it as one of the four tonic medicinal sources, (Ginzeng, Budding antler, Cassia cinnamon, Aconite). The cassia tree is commonly grown in mountainous forest regions of Vietnam, however, the trees grown in Tra Mi district (Quang Nam province) and Tra Bong district (Quang Ngai province) have been valued as the best quality cassia. The Chinese call Tra Mi cassia cinnamon "*Cao Son Ngoc Que*" (gem cinnamon of high mountain). With long-standing experience, people in the Tra Mi and Tra Bong districts have made cassia growing their main profession. Nowadays income from cassia growing accounts for 60% of the total income of the people in these districts. Economically, income from the cassia cinnamon business has long-term and short-term advantages, and it takes the main role in socio-economic development of ethnic people in high mountainous regions of Vietnam. Cassia growing is important as a means of reforestation, for transforming poor, desolate and waste forest into a growers' forest ensures the ecological and environmental balance of the whole region. The depleting forestry resources in the various regions make the establishment of forest resources extremely urgent. Reforestation by special trees such as cassia offers a convenient solution. The economical value of cassia trees helps to provide permanent agriculture and permanent settlement for ethnic minority people in high mountainous regions of Vietnam.

Diseases and Pests

The major destructive disease of cassia cinnamon nurseries in Vietnam is foot rot, caused by the soil borne fungus *Fusarium oxysporum*. This fungus thrives under warm and humid conditions. It fatally attacks the vascular system of mainly the roots. The disease usually arises after rain, when leaf infection is caused by *Treptomyces*. Application of *Trichoderma* is an effective control measure. The other major disease is caused by a Mycoplasma-like organism (MLO/Phytoplasma), called "witches broom" disease, which is found in nurseries, seedlings, young and old trees, and especially in the cassia gardens in Tra Giac hamlet (Tra Mi district, Quang Nam province) (Fig. 6.3). Studies were undertaken

Figure 6.3 Witches broom disease in *C. cassia*.

through the Witches Broom Disease Project of Tra Mi cinnamon by a study group from the Institute of Ecological and Biological Resources (IEBR, 1996). While the general cause of witches broom disease is Mycoplasma (Phytoplasma), infections by viruses, fungii, *Azotobacter* and insects are pre-disposing factors (Luc, 1999).

Preventative measures for this disease include soaking the seeds in warm water (70 °C) containing an antibacterial medicine. To exterminate microorganisms, the chemicals suggested are: Daconil – 0.14%; Viben – 0.07%; Anvil – 0.15%; Tilt – 0.006%; Sconee – 0.02%; Formaldehyde – 0.37%; SEAL – 0.05%; Chloramphenicol – 0.25%; or pulverized petroleum and soap at a ratio of 1:1 at the concentration of 5% (Luc, 1999). The solution can exterminate thrips and does not influence the plant growth. The diseases of cassia are more common in the south than in the north of Vietnam and the south of China, because of the different weather conditions. The weather in the south is hot, wet, and the rainfall and light regime is good for insect and disease development.

The incidence of diseases in private farms is lower than in state-owned farms because the farmers have much experience in growing and tending cassia as it is an asset and source of income for each family from one generation to another. Betrothal ceremonies have often included the cassia gardens, so the gardens are cared for greatly. Effective control measures suitable for small holders are not yet available. Witches broom disease has spread from nursery to gardens. Here the most important protective measure is phyto-sanitation. The seeds are selected from healthy mother plants and the land should be cleaned. Supplying sufficient nutrients and minerals and additional lime can be useful. Most diseases and pests occuring in the region can be effectively controlled by simple treatments with suitable fungicides and insecticides.

Insect pests

There are 14 insect species that are harmful to cassia gardens and forests (Table 6.4), two of which are serious pests: *Liothrips* sp. (thrips) and *Anatkima insessa*. Results from experiments proved that thrips cause yellow leaves, curly leaves, stunted excrescence in buds and treetops, and they transmit diseases from the exterior to the interior through stinging and juice sucking. Insecticide application is seldom practiced by growers.

Physico-chemical Properties and Composition of Cassia Oil

Crude cassia oil has the following physico-chemical properties (Guenther, 1950):

Sp. gravity (15 °C)	: 1.055–1.070
Optical rotation	: 1°0' to + 6°0'
Refractive index (20 °C)	: 1.600–1.606
Acid number	: 6–15, in exceptional cases up to 20
Cinnamic aldehyde content	: 75–90%
Solubility	: Readily soluble in 80% alcohol
Boiling Point	: 240–260 °C

Crude cassia oil is subjected to rectification by the importing countries (mainly the USA). The rectification is done to remove adulterants such as rosin, kerosene and lead. Only rectified cassia oil is permitted to go into the market. The rectified cassia oil has the following physicochemical properties:

Chinese Cassia 169

Table 6.4 Insect pests of cassia cinnamon

No.	Scientific name	Habitat	Affected part of trees
	I. Icoptera		
1.	*Macrotesmes* sp.	Cinnamon forest	Trunk, base
	II. Thysanoptera		
2.	*Liothrips* sp.	Nursery	
	III. Lepidoptera		
	Fam. Pyralidae		
3.	*Onaphanocrocis medinacis*	Nursery	Leaves
	Fam. Tortricidae		
4.	*Cophoprona* sp.	Cinnamon forest	
	IV. Orthoptera		
	Fam. Gryllidae		
5.	*Brachytrupes portentosus*	Nursery	Young tree
	Fam. Grillotapidea		
6.	*Grillotubpa africana*	Nursery	Young tree
	Fam. Acrididae		
7.	*Caryandra diminuta*	Nursery	Leaves
	V. Heteroptera		
	Fam. Pantatomicdae		
8.	*Nezera viridula*	Nursery	Leaves
9.	*Cletus* sp.	Cinnamon forest	Leaves
	VI. Plecoptera		
	Fam. Jassidae		
10.	*Empoasca flavescan*	Nursery	Leaves
11.	*Nephotettix* sp.	Nursery	Leaves
12.	*Anatkima insessa*	Nursery	Leaves
	Fam. Margarodidae		
13.	*Icerga* sp.	Cinnamon forest	Leaves
	Tetranychidae		
14.	*Schizotetranichus* sp.	Cinnamon forest	Leaves

Specific gravity (at 25 °C)	: 1.046–1.059
Optical rotation	: 0°40' to + 0'30'
Refractive index (20°)	: 1.6045–1.6135
Aldehyde content	: 88–99%
Solubility	: Soluble in 70% alcohol

Rectified cassia oil is a mobile, yellow to slightly brownish, highly refractive oil possessing a characteristic sweet and burning flavour that is typical of cassia bark.

Cassia oil chemistry has attracted the attention of many workers. Early information available on cassia oil chemistry was summarised by Guenther (1950). Adulteration of cassia oil is a common practice and hence efficient methods have also been evolved (Zhu *et al.*, 1996) by importing countries to detect such adulteration (mostly with the addition of synthetic cinnamaldehyde). Cassia oil resembles the Ceylon cinnamon bark oil in its composition, the chief constituent being cinnamaldehyde. Guenther (1950) identified the following substances in Chinese cassia oil:

Cinnamic aldehyde (cinnamaldehyde) – (80–95%), cinnamyl acetate, phenyl propyl acetate, salicylaldehyde, cinnamic acid, salicylic acid and benzoic acids, higher fatty

170　*Nguyen Kim Dao*

acids, coumarin, benzaldehyde, O-methoxy benzaldehyde and methyl-O-coumaraldehyde.

In the natural oil the auto-oxidation of cinnamic aldehyde to cinnamic acid is prevented by the presence of cinnamyl acetate, which acts as a very efficient preservative. In 1972 Ter Heide examined the chemical composition of a crude essential oil from China using GC, TLC, NMR, MS and IR, in addition to using authentic reference compounds and GC retention indices. He identified benzaldehyde, salicylaldehyde, 2-methoxy cinnamaldehyde, 2-hydroxy acetophenone, methyl benzoate, phenyl ethyl acetate, 3-phenyl propyl acetate, trans-cinnamyl acetate, trans-2-methoxy-cinnamyl acetate, phenyl ethyl alcohol, trans-cinnamyl alcohol, coumarin, phenol, O-cresol, guaicol, 2-vinyl phenol, chavicol, 4-ethyl guaneol and eugenol. The following acids were also identified: hexanoic, heptanoic, octanoic, benzoic, cinnamic, methoxy cinnamic, dihydro cinnamic and 2-methoxy dihydroxy cinnamic acids.

In 1977 Senanayake compared the chemical composition of lab-distilled leaf and bark oils with that of a commercial oil. The results of the study indicated a cinnamaldehyde content of 69.6%, 87.0% and 77.2% in leaf, bark and commercial oils, respectively. The coumarin content of the oil was found to be very high (8.06%, 0.28%, 15.3% respectively in leaf, bark and commercial oils (Tables 6.5 and 6.6).

Lawrence (2001) recently summarised a series of studies on Chinese cassia oil. Zhu *et al.* (1993) analysed the cassia leaf oil from China and recorded the compounds listed in Table 6.5.

Vernin *et al.* (1990) carried out a detailed compositional analysis of cassia bark oil (Table 6.6). In 1994 Vernin *et al.* analysed cassia bark powder using head space GC/MS and recorded the substances given in Tables 6.7 and 6.8.

Further reports on the chemical composition of cassia oil were published by Jayatilaka *et al.* (1995), Ehlers *et al.* (1995) Miller *et al.* (1995) and Kwon *et al.* (1996). Kwon *et al.* (1996) isolated salicylaldehyde from bark extract and reported that this compound inhibited farnesyl protein transferase activity. In 1998 Li *et al.* made a detailed comparison of leaf and bark oils from Chinese cassia grown in Yunnan province using GC/MS (Table 6.9). Li and Yuan (1999) reported 36 compounds in their analysis of a cassia oil sample from China by GC and GC/MS (Table 6.10). Gong *et al.* (2001) developed a complex interactive optimisation procedure to resolve co-eluting peaks in a GC/MS analysis of cassia bark grown in four different geographical regions of China.

Kondou *et al.* (1999) analysed the commercial cassia barks (different grades such as Xijang and Donxing bark from China and Yen Bai and Mien Nam barks from Vietnam) in order to bring out the salient differences among them based on the contents of coumarin, cinnamaldehyde, cinnamic acid, cinnamyl alcohol, cinnamic acetate and eugenol (Table 6.11).

In addition to the components present in the essential oil, a number of diterpenes named cinncassioles have been isolated from cassia bark. They include cinncassiols A, B and C_1, and their glucosides, cinncassiols C_2 and C_3, cinncassiols D_1, D_2 and D_3, and their glucosides, cinncassiol E, cinnzeylanol, cinnzeylanin, anhydrocinnzeylanol and anhydrocinnzeylanin (Yagi *et al.*, 1980; Nohara *et al.*, 1980a,b,c, 1981, 1982, 1985; Kashiwada *et al.*, 1981). The following compounds were also isolated from the stem bark of cassia:

Chinese Cassia 171

Table 6.5 Composition of leaf essential oils from *C. cassia* from two sources, China and Australia

1.1 Leaf oil:

From China		From Australia	
%	Compound	%	Compound
74.1	Cinnamic aldehyde	77.2	Cinnamic aldehyde
10.5	2-Methoxycinnamaldehyde	15.3	Coumarin
6.6	Cinnamyl acetate	3.6	Cinnamyl acetate
1.2	Coumarin	1.2	Benzaldehyde
1.1	Benzaldehyde	0.8	4-Ethylguaiacol
0.7	2-Phenylethyl acetate	0.4	Ethyl cinnamate
0.6	2-Phenylethanol	0.2	2-Phenylethyl acetate
0.6	2-Methoxybenzaldehyde	0.2	2-Phenylethanol
0.2	2-Methylbenzofuran	0.1	α-Terpineol
0.2	3-Phenylpropanal	0.1	Eugenyl acetate
0.2	Salicylaldehyde	0.1	Benzyl benzoate
0.2	Cinnamyl alcohol	0.1	Terpinen-4-ol
0.2	Nerolidol (unknown isomer)	0.1	Nonanal
0.1	Acetophenone	Trace	1-Hexanol
0.1	α-Pinene	Trace	Chavicol
		Trace	Camphene
		Trace	1,8-Cineole
		Trace	Cinnamyl alcohol
		Trace	Cuminaldehyde
		Trace	Eugenol
		Trace	Farnesol (unknown isomer)
		Trace	Geranial
		Trace	Guaiacol
		Trace	α-Humulene
		Trace	Isoeugenol
		Trace	Linalool
		Trace	Methyl benzoate
		Trace	3-Phenylpropanal
		Trace	α-Pinene
		Trace	β-Pinene
		Trace	Terpinolene
		Trace	2-Vinylphenol
96.5	Total	99.4	Total

Source: Zhu *et al.*, 1993.   Source: Senanayake, 1977.

Lyoniresinol 2α-O-β-D-glucopyranoside, 3,4,5-trimethoxy phenol-O-β-D-apiofura-nosyl-(1 → 6)-β-D-glucopyranoside, syringaresinol, two epicatechins-5,7,3′,4′-trimethyl-epicatechin and 5,7,0-dimethyl-3′,4′-di-0-methylene epicatechin, and two cinnamic aldehyde cyclic glycerol 1,3-acetals; 3-(2-hydroxyphenyl) propanoic acid and its O-β-D-glucopyranoside, cassioside, cinnamoside and 3,4,5-trimethoxyphenol-β-D-apiofuranyl-(1 → 6)-β-D-glucopyranoside (Tanaka *et al.*, 1989; Shiraga *et al.*, 1988).

In addition to the above, coumarin, cinnamic acid, β-sitosterol, choline, protocate-chuic acid, vanillic acid and syringic acid were isolated from the aqueous extract of cassia bark (Yuan and Jiang, 1982; Akira *et al.*, 1986).

Table 6.6 Composition of oils from cassia bark

From China		From Australia	
%	Compound	%	Compound
65.5	(E)-cinnamic aldehyde	87.0	Cinnamic aldehyde
8.7	Coumarin	4.7	Benzaldehyde
3.6	Cinnamyl acetate	2.5	2-Phenylethanol
2.7	2-Methoxycinnamaldehyde	2.0	3-Phenylpropanal
0.9	Benzaldehyde	0.7	1,8-Cineole
0.7	2-Methoxybenzaldehyde	0.5	4-Ethylguaiacol
0.6	Benzyl benzoate	0.5	Ethyl cinnamate
0.2	Cinnamyl alcohol	0.4	Cuminaldehyde
0.2	2-Phenylethyl acetate	0.3	Chavicol
0.2	Eugenyl acetate	0.3	Coumarin
0.1	(Z)-cinnamic aldehyde	0.1	Benzyl benzoate
0.1	2-Phenylethyl benzoate	0.1	Linalool
0.1	3-Phenylpropanal	0.1	Cinnamyl acetate
Trace	Caryophyllene oxide	0.1	Nonanal
Trace	α-Pinene	0.1	Eugenol
Trace	Camphene	0.1	α-Pinene
Trace	β-Pinene	Trace	Eugenyl acetate
Trace	Mycrene	Trace	Terpinen-4-ol
Trace	α-Phellandrene	Trace	Camphene
Trace	δ-3-Carene	Trace	Cinnamyl alcohol
Trace	α-Terpinene	Trace	Farnesol (unknown isomer)
Trace	Para-cymene	Trace	Fenchone
Trace	Limonene	Trace	Geranial
Trace	β-Phellandrene	Trace	Geraniol
Trace	(Z)-β-ocimene	Trace	1-Hexanol
Trace	(E)-β-ocimene	Trace	3-Hexenol-1
Trace	Terpinolene	Trace	α-Humulene
Trace	Linalool	Trace	Isoeugenol
Trace	Borneol	Trace	Methyl cinnamate
Trace	Terpinen-4-ol	Trace	2-Phenylethyl acetate
Trace	α-Terpinene	Trace	β-Pinene
Trace	Terpineolene	Trace	Sabinene
Trace	Linalool	Trace	Safrole
Trace	Borneol	Trace	β-Selinene
Trace	Terpinen-4-ol	Trace	α-Terpinene
Trace	α-Terpineol	Trace	α-Terpineol
Trace	Nerol	Trace	Terpinolene
Trace	Geraniol		
Trace	Linalyl acetate		
Trace	Bornyl acetate		
Trace	β-Caryophyllene		
Trace	α-Humulene		
Trace	α-Muurolene		
Trace	δ-Cadinene		
Trace	α-Copaene		
Trace	p-Cymene-8-ol		
Trace	Isoamyl isovalerate		
Trace	Spathulenol		
Trace	Eugenol		
Trace	Carvacrol		
Trace	β-Elemene		

Trace	Methyl eugenol		
Trace	Benzoic acid		
Trace	Isocaryophyllene		
Trace	γ-Muurolene		
Trace	Aromadendrene		
Trace	Anisaldehyde		
Trace	α-Cadinol		
Trace	Guaiacol		
Trace	ar-Curumene		
Trace	(E)-β-farnesene		
Trace	Safrole		
Trace	Acetic acid		
Trace	Vanillin		
Trace	Chavicol		
Trace	10-epi-α-Cadinol		
Trace	Methyl cinnamate		
Trace	Calamenene		
Trace	Benzyl alcohol		
Trace	trans-Linalool oxide (5) (furanoid)		
Trace	cis-Linalool oxide (5) (furanoid)		
Trace	α,p-Dimethylstyrene		
Trace	Ethyl cinnamate		
Trace	2-Phenylethanol		
Trace	Salicylaldehyde		
Trace	Acetophenone		
Trace	Carvotanacetone		
Trace	Palustrol		
Trace	α-Elemene		
Trace	Styrene		
Trace	Menthene		
Trace	Butyl 2-methylbutyrate		
Trace	(Z)-isoeugeonl		
Trace	α-Himachalene		
Trace	3-Phenylpropyl acetate		
Trace	Isoamyl benzoate		
Trace	2-Phenylethyl formate		
Trace	2-Vinylbenzaldehyde		
Trace	4-Vinylbenzaldehyde		
Trace	p-Tolylacetaldehyde		
Trace	(E)-2-methoxycinnamaldehyde		
Trace	Dimethoxycinnamaldehyde		
Trace	Dimethoxy allyl phenol		
Trace	3-Phenylpropanoic acid		
Trace	o-Methoxycinnamic alcohol		
Trace	(E)-cinnamic acid		
84.4	Total	100.0	Total

Source: Vernin *et al.*, 1990.

Some information is available on the leaf oil composition of *C. loureirii*, mostly from trees grown in Japan. Steam distilled leaves and twigs yielded 0.2% light brown oil, containing 40% linalool, 27% aldehyde (mainly citral) and cineole (Nitta, 1984). Steam distilled bark yielded 1.2% light brown oil. This oil contained mainly cinnamladehyde (80%), camphene, linalool and cineole as minor components (Asakawa, 1971; Weiss, 1997).

174   *Nguyen Kim Dao*

Table 6.7 Composition of cassia bark oil

%	Compound
92.5	(E)-cinnamic aldehyde
0.8	3-Phenylpropanal
0.6	(Z)-cinnamic aldehyde
0.6	Coumarin
0.3	Benzaldehyde
0.1	Eugenol
0.1	β-Caryophyllene
0.1	Benzyl benzoate
0.1	Camphor
0.1	1,8-Cineole
0.1	Linalool
0.1	β-Phellandrene
0.1	Salicylaldehyde
0.1	α-Terpineole
96.6	Total

Source: Vernin *et al.*, 1994.

Table 6.8 Chemical constituents of cassia bark recorded in the head space analysis using GC/MS

α-Pinene (0.10%)	Borneol + α-terpineol (3.80%)
Limonene (0.28%)	Hydrocinnamaldehyde (0.69%)
p-Cymene (0.82%)	2-Phenethyl acetate (0.96%)
α-Copaene (0.67%)	(Z)-cinnamaldehyde (0.21%)
Benzaldehyde (23.80%)	2-Phenethyl alcohol (0.24%)
Linalool (0.05%)	2-Methoxybenzaldehyde (0.82%)
Bornyl acetate (0.62%)	(E)-cinnamaldehyde (39.50%)
Linalyl acetate (0.44%)	(E)-cinnamyl acetate (0.24%)
Salicylaldehyde (1.27%)	

Source: Vernin *et al.*, 1994.

Properties and Biological Effects of Cassia Bark (*Cortex cinnamomi – Rou gui*)

C. cassia bark, is often referred to as *Cortex cinnamomi* and known as *Rou gui* in Chinese traditional medicine. The dried twig of *C. cassia*, known as *Ramulus cinnamomi* (*Guo zhi* in Chinese), is another drug used in Chinese traditional medicine. These two drugs are used differently and for different purposes.

Rou gui mixed with *Fuzi* (*Radix aconiti lateralis preparata*) produced a significant hypotensive effect in rats with adrenocortical hypertension due to the burning of one of the adrenal glands. However, the mixture had no effect on renal hypertensive rats. In rats with adrenocortical hypertension, the mixture affected the adrenal cholesterol metabolism, thereby intensifying the activity of the cauterised adrenal gland but not the intact one. It appears that the hypotensive action is due to the promotion and normalisation of the activity of the depressed adrenal gland (Wang, 1983). The decoction of the stem bark significantly increased coronary flow in the isolated heart of guinea pigs. A daily oral dose of 1.2 g/kg of decoction for six days improved the acute myocardial ischemia in rabbits due to putuitrin. Intravenous administration of 2 mg/kg of the

Table 6.9 Comparative percentages, composition of the leaf and bark oils of *Cinnamomum cassia*

Compound	Leaf oil	Bark oil
α-Pinene	0.05–0.36	0.10–0.25
Camphene	0.04–0.05	0.05–0.10
β-Pinene	0.04–0.15	0.14–0.22
Myrcene	0.02–0.03	t–0.10
α-Phellandrene	0.01–0.06	t–0.13
Limonene	0.13–0.24	0.14–0.29
1,8-Cineole	0.05–0.08	0.06–1.07
δ-3-Carene	0.03–0.05	t–0.07
p-Cymene	0.11–0.19	0.04–0.18
Camphor	0.07–0.15	0–0.08
Benzaldehyde	1.42–1.48	0.50–1.10
Linalool	0.11–0.23	0.08–0.16
Terpinolene	t	0–0.04
β-Caryophyllene	0.16–0.20	t–0.27
α-Humulene	t–0.03	0–0.15
β-Elemene	–	t–0.06
Isoborneol	0–0.20	0–0.27
Borneol	0.15–0.41	0.06–1.27
α-Terpineol	t–0.10	0.07–2.05
Geraniol	t	0.08–0.31
Carvone	0.57–0.64	0–0.34
2-Methoxybenzaldehyde	0.08	0–0.12
Safrole	–	t–0.20
γ-Elemene	0–t	0–0.41
δ-Cadinene	t	t–0.13
β-Cadinene	–	t–0.10
Hydrocinnamaldehyde	0.88–0.89	0–0.24
Phenylacetaldehyde	0.07–0.16	t–0.27
Methyl eugenol	0.14–0.15	t–0.05
(E)-cinnamaldehyde	64.10–68.30	80.40–88.50
α-Copaene	0.41–0.49	0.23–0.68
Vanillin	t	t–0.10
Salicylaldehyde	0.05–0.42	0.04–0.85
2-Phenethyl alcohol	0.11–0.27	t–0.16
Benzyl alcohol	t–0.05	–
Acetophenone	t–0.1	0–0.6
Eugenol	0.04–0.06	0.03–1.08
(Z)-isoeugenol	0.14–0.28	0.12–0.66
(E)-cinnamyl acetate	4.50–12.50	0.60–5.10
γ-Muurolene	t	t–0.50
Anisaldehyde	0.58–1.02	t
2-Phenethyl acetate	t–1.55	–
β-Bisabolene	t–0.06	t–0.18
β-Bisabolol	t	t–0.35
α-Muurolol	0–0.08	0–0.24
Coumarin	0.03–0.08	t–0.45
(E)-cinnamic acid	0.80–2.48	0.12–3.10
(E)-2-methoxycinnamaldehyde	8.40–10.50	t–2.50
Hydrocinnamic acid	0.18–0.51	0–0.24
4-Hydroxy-2-phenethyl alcohol	0–0.12	0–0.10
Caryophyllene oxide	0.15–0.17	0–0.10
Patchoulene	0.06–0.07	0–0.04

Table 6.9 (Continued)

Compound	Leaf oil	Bark oil
Octanoic acid	t	0–t
3-Phenylpropyl acetate	0.21–0.43	0.05–0.22
Nonanoic acid	t–0.10	0–t
Guaicol	t	0–0.08
(E)-cinnamyl alcohol	0.15	0.05–0.13
(E)-ethyl cinnamate	0.11–0.27	t–0.14
Benzyl benzoate	0.07–0.15	t–0.38
Methyl alaninate	t–0.05	–
Guaicyl cinnamate	t	t
Decanoic acid	t	0–t
Undecanoic acid	0–0.05	0–0.11
Dodecanoic acid	t–0.04	0–t
Benzoic acid	0.07–0.11	0.07–0.10
Salicylic acid	t–0.10	0.10–0.20

Source: Li *et al.*, 1998.

Note

t: trace.

decoction to anesthetised dogs did not affect blood pressure in the first 1–2 minutes, but markedly increased the blood flow in the coronory sinus and brain. Three to five minutes after administration the blood pressure dropped and the cerebral blood flow was slightly reduced. After five minutes the blood flow increased with gradual elevation of blood pressure and the heart rate was slightly reduced (Wang, 1983).

Intraperitonial administration of 100 mg/kg of aqueous extract of cassia bark prevented the occurrence of stress ulcers in rats due to exposure to a cold atmosphere

Table 6.10 Composition of a cassia oil sample produced in China

Styrene (0.07%)	(E)-cinnamaldehyde (67.12%)
Benzaldehyde (0.03%)	(E)-cinnamyl alcohol (1.02%)
p-Cymene (0.98%)	Eugenol (0.99%)
Benzyl alcohol (0.02%)	α-Copaene (0.23%)
Salicylaldehyde (1.01%)	α-Cubebene (0.11%)
Phenylacetaldehyde (0.34%)	Vanillin (0.30%)
Benzyl formate (0.08%)	β-Caryophyllene (5.58%)
Acetophenone (0.06%)	Benzofuran (0.34%)
2-Phenethyl alcohol (0.14%)	(E)-cinnamyl acetate (3.47%)
Hydrocinnamyl alcohol (0.09%)	Ethyl cinnamate (0.27%)
Menthol (0.10%)	β-Cubebene (0.07%)
2-Phenethyl acetate (0.15%)	(E)-cinnamic acid (0.23%)
Terpinen-4-ol (0.09%)	γ-Cadinene (0.07%)
Methyl salicylate (6.17%)	δ-Cadinene (0.05%)
2-Methylbenzofuran (0.13%)	(E)-2-methoxycinnamaldehyde (7.40%)
2-Phenylpropanal (0.05%)	(Z)-nerolidol (0.25%)
Benzylidenemalonaldehyde (0.12%)	(9H)-fluren-9-ol (0.06%)
2-Methoxybenzaldehyde (0.93%)	Adamantane (0.22%)

Source: Li and Yuan, 1999.

(3–5 °C) or due to water immersion. It also inhibited serotonin induced gastric ulcers in rats. Antiulcerative properties were noted for the following compounds from cassia bark: 3-(2-hydroxyphenyl)-propanoic acid and its O-β-D-glucopyranoside, cassioside, cinnamoside, and 3,4,5-trimethoxyphenol-β-D-apiofuranosyl-(1 → 6)-β-D-glucopyranoside (Shiraga et al., 1988).

Kubo et al. (1996) have shown that 70% methanolic extract of C. cassia bark inhibited the rise in vascular permeability induced by acetic acid and the increase of paw oedema induced by carrageenan in mice. It was ineffective against oedema induced by histamine and bradykinin, and exhibited a weak inhibitory effect against oedema induced by serotonin. The extract also showed an inhibitory effect against prekallikrein enzyme activity and ear oedema induced by arachidonic acid. It also had an inhibitory effect on cotton pellett-induced granuloma but showed no atrophying action against the adrenal or thymus glands. Little effect was shown on secondary lesions in the development of adjuvant-induced arthritis.

The daily administration of 0.35 mg of sodium cinnamate to dogs for five days increased peripheral leukocytes by 150–200%. Intraperitoneal administration of 0.35 mg of sodium cinnamate to mice and sc administration of 5.44 mg/kg of sodium cinnamate to dogs six hours, one, two, five and six days after irradiation with a lethal dose of ^{60}Co-rays increased the survival rate of the animals.

In anesthetised dogs and guinea pigs cinnamaldehyde produced a hypotensive effect mainly due to peripheral vasodilation. In isolated ileum of guinea pigs and mice cinnamaldehyde exhibited a papaverine-like activity. An increase in cardiac contractile force and beating rate was exerted by cinnamaldehyde in isolated guinea pig heart preparations. A repeated administration of cinnamaldehyde, however, led to a progressive decrease in such effects and to cardiac inhibition (Harada and Yano, 1975). Cinnamaldehyde induced the release of catecholamines from the adrenal glands of dogs. The positive ionotropic and chronotropic effects of cinnamaldehyde on perfused isolated guinea pig heart were presumably due to the release of endogenous catecholamines (Harada and Saito, 1978).

Oral administration of 250–500 mg/kg of cinnamaldehyde reduced the spontaneous motor activity in mice and antagonised the locomotor activity induced by methamphetamine. Prolongation of hexobarbital-induced hypnosis was also observed. Intraperitonial doses of 125 and 250 mg/kg produced similar effects. At an intraperitonial dose of 500 mg/kg, cinnamaldehyde delayed tetanic convulsion and death induced by strychnine (Wang, 1983).

Shimada et al. (2000) studied the protective effect of aqueous extract of cassia bark on glutamate-induced neuronal death and its action on Ca^{++} influx using cultured rat cerebellar granule cells. In a dose-dependent manner this extract significantly protected cells from glutamate-induced death and also inhibited glutamate-induced calcium ion influx. These results suggest that the cassia bark has a protective effect on glutamate induced neuronal death through the inhibition of calcium ion influx.

2′-hydroxycinnamaldehyde present in cassia bark inhibited farnesyl-protein transferase (FPTase). FPTase is an enzyme that catalyses the transfer of the farnesyl group from farnesyl pyrophosphate on to cysteine 186 at the c-terminal of the Ras-Protein, a mandatory process before anchoring to plasma membrane which is critical for triggering the 'ras' ocogene toward tumour formation (Kwon et al., 1996). Kim et al. (2000a,b) have shown that cinnamomi cortex extract exhibited strong antiallergic activity in the mouse system. The extract inhibited anaphylactic shock and inhibited the Arthus reaction. Among the

178 *Nguyen Kim Dao*

Table 6.11 Composition of the commercial Chinese and Vietnamese cinnamon barks (% dry weight)

Bark grade		Coumarin	Cinnam-aldehyde	Cinnamic acid	Cinnamyl alcohol	Cinnamyl acetate	Eugenol
Chinese cinnamon							
Xijiang cassia barks							
Qibian cassia bark		0.15	2.03	0.09	–	–	–
Peeled cassia bark		0.26	2.54	0.06	–	–	–
Flat cassia bark		0.01	2.03	0.07	–	–	–
Oily tube cassia bark							
	(1)	tr	0.92	0.04	tr	–	–
	(2)	0.03	1.68	0.07	tr	–	–
Non-pressed cassia							
Bark-whole	(1)	0.05	1.50	0.03	tr	–	–
	(2)	tr	2.65	0.06	tr	–	–
Pressed cassia							
Bark-whole	(1)	0.04	1.72	0.06	0.001	0.01	–
	(2)	0.07	1.17	0.08	tr	–	–
	(3)	0.08	1.66	0.11	0.02	0.05	–
	(4)	0.09	2.82	0.15	tr	tr	–
Cassia branch	(1)	0.03	0.69	0.05	–	–	–
	(2)	0.03	0.30	0.07	–	–	–
Donxing cassia barks							
Oily tube cassia bark		0.06	2.19	0.06	0.01	tr	–
Non-pressed cassia							
Bark-whole		0.02	3.23	0.13	–	–	–
Pressed cassia							
Bark-whole	(1)	0.03	1.72	0.16	–	–	–
	(2)	0.49	2.55	0.10	0.01	tr	–
Cassia bark-broken		0.04	1.61	0.30	–	–	–
Vietnamese cassia barks							
Yen Bai (YB) cinnamon							
YB I		0.58	3.70	0.13	0.03	0.01	tr
YB II		0.47	3.89	0.12	0.16	0.06	–
YB III		0.73	2.73	0.07	0.04	0.09	–
YB IV		0.40	2.70	0.06	0.02	0.01	–
YB V		0.32	2.70	0.08	0.05	0.06	–
YB Guiyisui		0.33	1.93	0.06	0.02	–	–
Mien Nam (MN) cinnamon							
MN I		0.57	3.57	0.05	0.23	0.34	0.01
MN II		0.53	4.08	0.07	0.08	0.15	0.03
MNK III		0.32	3.02	0.05	0.08	0.15	0.01
MN V		0.51	3.61	0.09	0.12	0.18	0.01
MN Guiyisui		0.51	3.52	0.10	0.09	0.28	0.02

Source: Kondou *et al.*, 1999.

Note
tr: trace.

fractions obtained in the successive fractionation with n-hexane, butanol and acetone, the butanol insoluble portion was shown to have the activities. In a detailed study on the effect of Chinese prescriptions and crude drugs on 1,1-diphenyl-2-picrylhyrazyl radical, cinnamomi cortex ranked third (after *Rhus* and *Rheum*) in its inhibitory effect. Cinnamomi cortex exhibited strong free radical scavenging activity, indicating that it was effective in curing diseases related to free radical reactions (Yokozawa *et al.*, 1998). The decoction of cassia bark is inhibitory to fungi *in vitro*. The essential oil is bactericidal, which is more potent against Gram positive than against Gram negative bacteria.

Side effects, toxicity

Dizziness, blurred vision, increase of intraocular pressure, cough, obliguina, thirst and rapid pulse occurred in patients taking 36 g of powdered bark once. The iv LD_{50} of the decoction was found to be 18.48 ± 1.80 g/kg. The iv LD_{50} values of cinnamic aldehyde were 132, 610 and 2225 mg/kg in mice by iv, ip and oral administrations, respectively. Cinnamic aldehyde at low dosage caused inhibition of the activity of animals and at high dosage caused strong spasms, molar disturbances, shortness of breath and finally death from respiratory paralysis (Wang, 1983).

Ramulus Cinnamomi – Gui zhi

In Chinese Materia Medica the crude drug *Gui zhi* is comprised of the dried twig of *C. cassia*. The drug is collected in spring-summer and dried for use. The dried twig contains 1–2% essential oil, the main ingredients of which are cinnamaldehyde and cinnamic acid.

Gui zhi in itself is a weak diaphoretic agent, but with its peripheral vasodilation effect, it can potentiate the diaphoretic effect of *Ma Huang (Herba Ephedra)*. The decoction of the herb can reduce body temperature in mice and artifical fever in rabbits. Cinnamaldehyde also exhibits a mild analgesic activity against acetic acid-induced writhing reaction in mice.

In nephritis, the aqueous extract of *Gui zhi* showed an inhibitory effect on the complement activity in the heterophile antibody reaction (Wang, 1983). This extract also prevented the increase of protein levels in the urine of rats with nephritis when orally administered. But it had no effect on the recovery of anemia induced by an acute blood loss and on the total leucocyte and lymphocyte counts. Immunosuppressive properties have been ascribed to this extract (Wang, 1983).

Recently Kim *et al.* (2000a,b) have shown that a 80% methanolic extract of *Ramulus cinnamomi* exhibited inhibitory effects on HMG-Co A reductase and DDPH-free radical scavenging effect *in vitro*. The extract was also shown to have hypolipaemic effects in rats with hyperlipidaemia induced by Triton WR 1339. The extract has significantly suppressed the elevated serum total cholesterol and triglyceride levels in corn-oil induced hyperlipidaemic rats. The chloroform fraction also exhibited remarkable inhibitory effects on HMG-Co A reductase activity.

Uses

Dried cassia bark is the source of an important spice. Immature fruits are the source of Cassia buds. The essential oil called cassia cinnamon oil (oleum cinnamon) obtained by steam distillation from leaves and twigs is used as a flavouring agent. The bark collected from old trees is a well-known medicine for diarrhoea. The drug is pungent and is a special tonic for the lumbar region. It is prescribed for cold hands and feet, for belly-ache and dysuria but it is not a medicine for pregnant woman (Chi, 1961). The bark collected from the fresh branchlets after drying in the sun is pungent, warming, with a sweet aftertaste, and is good for inducing perspiration and promoting circulation. It is prescribed as a febrifuge for colds, and to relieve pain in the joints and abdomen (Chi, 1961). In older works, cassia cinnamon is indicated as a stomachic astringent, a tonic stimulant (Rei, 1946), and a salivative and remedial for post-partum disorders (Stuart, 1911). Rei (1946) says that it is a general stimulant, which improves the circulation and respiration and most of the secretions and peristaltic movements of the digestive tract. The twigs are used as an emmenagogue and a diaphoretic. In Indo-China, cassia is one of many ingredients in remedies with antidiarrhoeic, antibilious and antifebrile properties and it is used to treat diarrhoea, jaundice and chronic malaria with splenomegaly. Cassia bark oils are also used in soaps and perfumes. Cassia leaf oil (usually referred to as cassia oil) is obtained by distilling twigs and leaves and is used for similar purpose as cassia bark oils in perfumery and flavouring, but it is of special importance in cola-type drinks (Dao *et al.*, 1999). Its timber is also useful.

In Chinese Materia Medica (Chi, 1961) dried cassia bark, (*Rou-gui*), has "pungent and sweet tastes and a hot property acting on the kidney, spleen, heart and liver channels". It has the functions of:

1. supplementing the body fire, reinforcing *Yang* and leading the fire back to the kidney; it is used in impotence, frigidity, feeling of coldness and pain in the loin and knees, dyspnea in kidney, deficiency syndrome, dizziness, inflammation of the eyes and sore throat due to *Yang* deficiency;
2. dispelling cold, relieving pain, it is used in precordial and abdominal pain with cold sensation; and,
3. warming and cleaning the channels, it is used in amenorrhea and dysmenorrhea.

Gui zhi, (dried twig of cassia cinnamon) is collected in spring and summer, freed from leaves and then dried in the sun or in the shade. This crude drug is mainly produced in the provinces of Guangdong and Guangxi in China. The decoction of the drug (containing cinnamaldehyde or sodium cinnamate) can reduce the normal body temperature of mice and artificial fever in rabbits. *Gui zhi* has pungent and sweet tastes and a warm property, which acts on the lung, heart and urinary bladder channels. In Chinese medicine *Gui zhi* is indicated as having the following uses and functions (Loi, 1996):

1. inducing perspiration and dispelling pathogenic factors from the exterior of the body, it is used in the common cold due to wind, cold infection manifested by headache, fever and aversion to cold;
2. warming the channels to relieve pain, used in arthralgia due to wind-cold dampness manifested by aching joints of the shoulders and limbs; and,
3. reinforcing *Yang* to promote the flow of *Qi*, it is used in irregular menstruation, amenorrheal abdominal pain and mass in the abdomen due to blood cold in women, and vague pain in the epigastric region which can be ameliorated by warming and pressing.

Gui zhi is used as an analgesic and antipyretic against the common cold, myalgia orthralgia and amenorrhea with abdominal cramp. Together with *Shoayao* (*Radix paeoniae*) and *Chai Hu* (*Radix bupleuri*) it has been used in the treatment of epilepsy.

In Vietnam, cassia bark is used widely. Bark is used as a spice and as a medicine and it is among the 114 recipes "*Thuong han luan*" of Truong Trong Canh. In fact barks from different parts of the cassia tree have different uses. Some of the uses of cassia cinnamon preparations prevalent in Vietnam, China and the adjoining areas are given below (Duc, 1997).

* Bark of cassia twig (bark taken from branch): Peppery-sweet, warm. Used to treat colds and to relieve, perspiration, rheumatism and hand and foot aches Use 5–10 g with other medicines.
* Quetam: to treat heart disease with 4–8 g combined with other medicines.
* Quethong: to treat cold internal organs, and to increase blood circulation.
* Quenhuc: sweet-peppery, burning. Improves circulation of the blood and improves respiration. It also increases secretion, uterus contraction and increases bowels peristaltism. Used to treat colds, apoplexy, swelling, bradycardia, cardiovascullar collapsus, severe epidemics. Ground with water, 2–3 g each time, or keep 12 g in simmered water for a drink, or combine with other medicines (Duc, 1997). In the ancient medical literature, cinnamon and cassia are described as sweet, rather hot and toxic to the liver and kidney. They have tonic effects on vital points and increase heat. They are used for the treatment of cramped hand and foot, tired back and knee, spermatorrhoea, sexual impotence, dysmenorrhoea, dysuria. They induce a hot sensation in the upper part of the body and a cold sensation in the lower part of the body. They induce the waning of the *Yin* and the thriving of the *Yang*. In oriental medicine, cinnamon and cassia are used for the treatment of conjunctivitis, coughs, as a tonic for postpartum women, and as an emollient after severe cholera. However, attention should be paid to the use of the bark of cinnamon and cassia according to physicians instructions. Overdoses can be dangerous for eyes and may cause epistaxis. It is contraindicated for pregnant women (Loi, 1996).
* For the treatment of running nose: branchlets of cinnamon – 8 g; *Abrus precatorius* L. (Fabaceae) – 6 g; *Paeonia lactiflora* Pall (Ranunculaceae) – 6 g; *Zingiber officinale* (Wild.) Roscoe (Zingiberaceae) – 6 g; *Zyzyphus jujuba* Mill. (Rhamnaceae) – four fruits, cooked down with 600 ml of water. The decoction is taken three times a day (Loi, 1996).
* For the treatment of cold, syncope: use cinnamon up to 12 g, keep it in simmering water. The decoction is taken gradually afterwards.
* Cinnamon is used in 'eight ingredients pills' as a heat tonic in the following conditions:

  • for the treatment of people with bad *Yin* and thriving *Yang*;
  • for the treatment of people suffering from thrilling, neurasthenia, tired back and knees, male spermatorrhoea and male sexual impotence;
  • for the treatment of people with damaged *Yin*, who are usually suffering from coldness, neurasthenia, tired knees, male spermatorrhea, impotence and sterility.

The ingredients for remedies are: Cassia cinnamon – 12 g; *Aconitum sinense* Paxt – 8 g; *Rehmannia glutinosa* (Gaertn.) Libosch (Scrophulariaceae) – 24–32 g; *Aconitum sinense* Paxt. (Ranunculaceae) – 8 g; *Cornus officinalis* Sieb et Zucc. (Cornaceae), *Dioscorea persimilis* Pr. & Bur. (Dioscoreaceae), *Poria cocos* Wolf. (Polyporaceae) – 16 g each; *Paeonia suffriticosa* Andr. (Ranunculaceae) – 8 g; *Alisma plantago-aquatica* L. (Alismataceae) – 8 g; when ground down and made up into pills, the daily dose is from 30 to 40 g.

If there is a kidney failure with chronic edema *Plantago asiatica* L. (Plantaginaceae) – 16 g and *Achiranthes bidentata* Blume (Amaranthaceae) – 16 g are added to the above ingredients. It is called "eight ingredients pill for the treatment of kidney".

182 *Nguyen Kim Dao*

For Cough or asthma: *Ophiopogon japonicus* (Liliaceae) and *Schizandra chinensis* Baill (Schizandraceae) are added to the above ingredients and prepared through decoction (Duc, 1997).

Prospects

The demand for Chinese cassia has always been satisfactory and the prospects are promising. Consumption increases with population growth, and the prospects for cassia leaf oil is expanding due to the growth in the soft drinks market. The use of cassia bark is steadily increasing mainly due to the US market. However, little attention is given to the use of cinnamon wood. This wood is useful for special decorative purposes and there may be scope for multipurpose plantations including timber production. This is especially important in the Vietnam region as the whole tree is harvested off the bark and the timber is a by-product which is wasted.

The exploitation of related species in the region has not been seriously attempted, and this is an area that needs to be looked into.

References

Akira, T., Tanaka, S., Tabata, M. (1986) Pharmacological studies on the antiulcerogenic activity of Chinese cinnamon. *Planta Medica*, **52**, 440–443.

Allen, C.K. (1939) Cinnamomum In. Cinnamomum and Neocinnamomum, *J. Arnold Arboretum*, **20**, 52–53.

Anon (2002) *Pharmacopoea Vietnamica* (3rd ed.). Health Ministry, Hanoi, p. 490.

Asakawa, Y. (1971) Chemical components of the benzene extract of *C. loureirii. Flav. Ind.*, **2**, 114–119 (cited from Weiss 1997).

Chi, C.Y. (1961) [*New Chinese Materia Medica*], Vol. 3.

Coppen, J.J.W. (1995) *Flavour and Fragrances of Plant Origin*. Non-wood Forest Products 1. FAO, Rome, Italy.

Dao, N.K., Hop, T. and Siemonsma, J.S. (1999) *Plant Resources of South East Asia. 13. Spices.* Backhuy's Publishers. Leiden.

Duc, L.T. (1997) *Medicine Plants of Vietnam*. Agricultural Publishing, Hanoi.

Dung, V.V. (1996) *Vietnam Forest Trees*. Agricultural Publishing House, Hanoi.

Ehlers, D., Hilmer, S. and Bartholomae, S. (1995) Quoted from Lawrence (2001).

Gong, F., Liang, Y.L., Xu, Q.S. and Chau, F.T. (2001) Gas chromatography-mass spectrometry and chromometric resolution applied to the determination of essential oils in cortex cinnamomi. *J. Chromatogr.*, **905**, 193–205.

Guenther, E. (1950) *The Essential Oils*, Vol. 4. Robert E. Krieger Pub. Co., Florida, USA, pp. 241–256.

Harada, M. and Yano, S. (1975) Pharmacological studies on Chinese cinnamon. II. Effects of cinnamaldehyde on the cardiovascular and digestive systems. *Chem. and Pharmaceu. Bull.*, **23**, 941–947.

Harada, M. and Saito, A. (1978) Pharmacological Studies on Chinese Cinnamon. IV. Effects of cinnamaldehyde on the isolated heart of guinea pigs and its catecholamine releasing effect from the adrenal gland of dogs. *J. Pharmaco-biodynamics*, **1**, 89–97.

IEBR (1996–1999) *Study Results of Insects of Kwangnan Province* (The document of Institute of Ecology and Biological Resources), Hanoi.

Jayatilake, A., Poole, S.K., Poole, C.F. and Chichila, T.M.P. (1995) Simultaneous microsteam distillation – solvent-extraction for the isolation of semivolatile flavour compounds from Cinnamomum and their separation by series coupled – column gas chromatography. *Ann. Chem. Acta.*, **30**, 147–162.

Kashiwada, Y., Nohara, T., Tomimatsu, T. and Nishioka, I. (1981) Constituents of Cinnamomi cortex. IV. Structure of cinncassiols C_1 glucoside, C_2 and C_3. *Chem. Pharmaceu. Bull.*, **29**, 2686–2688.

Khoi, H. (1991) *The Cinnamon and Cassia of Tra Mi District, Quangnam Province.* (The provincial document).

Kim-N.J., Jung-E.A., Kim-D.H. and Lee-S.I. (2000a) Studies on the development of anti-hyperlipidemic drugs from oriental herbal medicines (11) – antihyperlipidemic effects of Oriental herbal medicines. *Korean J. Pharmacognosy*, **31**, 190–195.

Kim-R.M., Kim-Y.H. and Lee-E.B. (2000b) Anti-allergy activity of cinnamomi cortex. *Natural Product Sci.*, **6**, 49–51.

Kondou, T., Kawamura, T., Noro, Y., Tanaka, T. and Inoue, K. (1999) Physical and chemical features of Vietnamese and Chinese cinnamon barks on the market. *Natural Medicines*, **53**, 178–182.

Kubo, M., ShiPing, M., JianXian, W., Matsuda, M., Ha, S.P. and Wu, J.X. (1996) Anti-inflammatory activities of 70% methanolic extract from cinnamomi cortex. *Biol. Pharmaceu. Bull.*, **19**, 1041–1045.

Kwon, B.M., Cho, Y.K., Lee, S.H., Nam, J.Y., Bok, S.H., Chun, S.K., Kim, J.A. and Lee, I.R. (1996) 2-hydroxy-cinnamaldehyde from stem bark of *Cinnamomum cassia. Planta Med.*, **62**, 183–184.

Lawrence, B.M. (2001) Progress in Essential oils. *Perfumer & Flavorist*, **26**, 44–57.

Lecomte, H. (1913) Lauraceae de Chine et Indochine Nouv. *Arch. Mus. Hist. Nat. (Paris)*, 5(5) pp (cited from Li and Li, 1998).

Li, H.W. and Li, J. (1998) *The Lauraceae Flora*, Kunming, Yunnan.

Li, H.W. (1978) Material ad florum *Lauraceae sinecorum. Acta Phytotax. Sinica.* 16(4) (cited from Li and Li 1998).

Li, H.W. (1984) *Flora Republicae Popularis Sinicae*, **31**, 214.

Li, Z.-Q., Luo, L., Huang, R. and Xia, Y.-Q. (1998) Chemical studies of cinnamon, true plants from Yunnan province. *Yunnan Daxue, Xuebao ziram kexueban*, **20**(suppl.), 337–379.

Li, L.-L. and Yuan,W.J. (1999) Cinnamon oil analysis by GC and GC/MS. *Fujian Feuxi Ceshi*, **8**, 1121–1125.

Loi, D.T. (1996) *Medicinal Plants and Medicinal Taste of Vietnam.* Science and Technological Publishing House, Hanoi.

Loureiro, J. (1793) *Flora Cochinchinensis*, Wildenow C.L. (ed.). *Trans. Amer. Phil. Soc.* (cited from Merrill, 1935).

Luc, P.V. (1999) *Study results on witches broom disease in Cinnamon and Cassia of South Vietnam, and prevention measures.* Vietnam National Centre for Natural Science & Technology, Hanoi.

Merrill (1935) *Trans. American Phil. Soc.*, **24**(2), 281–282 (quoted from Allen, 1939).

Miller, K.G., Poole, C.F. and Chicila, T.M.P. (1995) Solvent assisted supercritical fluid extraction for the isolation of semi-volatile flavor compounds from the cinnamons of commerce and their separation by series coupled column gas chromatography. *J. High Resol. Chromatogr.*, **18**, 461–471.

Nitta, A. (1984) Cinnamon bark from Vietnam. *Yakugaku Zasshi*, **104**(3), 261–274 (cited from Weiss, 1997).

Nohara, T., Kashiwada, Y., Yomimatsu, T., Kido, C., Tokobuchi, M. and Nishioka, N.I. (1980a) Cinncassiol D_1 and its glucoside, novel pentacyclic diterpenes from cinnamomi cortex. *Tetrahedron Lett.*, **21**, 2647–2648.

Nohara, T., Nishioka, I., Tokubuchi, N., Miyahara, K. and Kawasaki, T. (1980b) The Constituents of Cinnamomi Cortex. II. Cinncassiol C, a novel type of diterpene from cinnamomi cortex. *Chem. Pharmaceu. Bull.*, **28**, 1969–1970.

Nohara, T., Tokubuchi, N., Kuroiwa, M. and Nishioka, I. (1980c) The constituents of cinnamomi cortex. III. Structures of cinncassiol B and its glucoside. *Chem. Pharmaceu. Bull.*, **28**, 2682–2686.

Nohara, T., Kashiwada, Y., Murakami, T., Tomimatsu, T., Kido, M. and Yagi, A. (1981) Constituents of cinnamomi cortex. V. Structures of five novel diterpenes, cinncassiols D_1, D_1 glucoside, D_2, D_2 glucoside, and D_3. *Chemi. Pharmaceu. Bull.*, **29**, 2451–2459.

Nohara, T., Kashiwada, Y., Tomimatsu, T. and Nishioka, I. (1982) Studies on the constituents of *Cinnamomum* cortex. VII. Two novel diterpenes from bark of *Cinnamomum cassia*. *Phytochemistry*, **21**, 2130–2132.

Nohara, T., Kashiwada, Y. and Nishioka, I. (1985) Cinncassiol E, a diterpene from the bark of *Cinnamomum cassia*. *Phytochemistry*, **24**, 1849–1850.

Purseglove, J.W., Brown, I.G., Green, C.L. and Robbins, S.R.J. (1981) *Spices Vol. 1*, Longman, U.K, pp. 100–173.

Rei, J. (1946) *Atlas de Plantes Medicinales Chinenses*. Paris, p. 125.

Senanayake, U.M. (1977) *The nature, description and biosynthesis of volatiles in* Cinnamomum *spp*. Ph.D. thesis, Univ. New South Wales, Australia.

Shimada, Y., Goto, H., Kogure, T., Kohta, K., Shintani T., Itoh, T. and Terasawa, K. (2000) Extract prepared from the bark by *Cinnamomum cassia* Blume prevents gluamate-induced neuronal degeneration. *Phytotherapy Res.*, **114**, 466–468.

Shiraga, Y., Okano, K., Akira, T., Fukaya, C., Yokoyama, K. and Taneka, S. (1988). Structures of potent antiulcerogenic compounds from *Cinnamomum cassia*. *Tetrahedron*, **44**, 4703–4711.

Stuart, G.A. (1911) *Chinese Materea Medica. Vegetable Kingdom*. Shanghai, p. 558.

Tanaka, S., Yoon, Y.H., Fukui, H., Tabata, M., Akira, T. and Okan, K. (1989) Antiulcerogenic compounds isolated from Chinese cinnamon. *Planta Medica*, **55**, 245–248.

Ter Heide, R. (1972) Qualitative analysis of the essential oil of cassia (*Cinnamomum cassia* Blume). *J. Agric. Food Chem.*, **20**, 747–751.

Vernin, C., Vernin, G., Metzger, J. and Puigol, I. (1990) La canelle, Premire partie. Analyse CPG/Sm Banuqe SPECMA d'huile essentielle de canelle de ceylan etde chine *Parfumes, cosmetiques, Aromes*, **93**, 85–90.

Vernin, G., Vernin, C., Metzger, J., Pujol, L. and Porkanyi, C. (1994). GC/MS analysis of cinnamon and cassia essential oils : a comparative study. In G. Charalambous (ed.) *Spices, Herbs and Edible Fungi*, Elsever. pp. 411–425.

Wang, Y.S. (1983) *Pharmacology and Applications of Chinese Materia Medica*. People's Health Pub., Beijing, pp. 442–446, 862–866.

Weiss, E.A. (1997) *Essential Oil Crops*. CAB International, U.K.

Yagi, A., Tokubuchi, N., Nohara, T., Nonaka, G., Nishioka, I. and Koda, A. (1980) The constiutuents of cinnamomi cortex. I. Structure of cinncassiol A and its glucoside. *Chem. Pharmaceu. Bull.*, **28**, 1432–1436.

Yuan, A.X., Tan, L. and Jiang, D.G. (1982) Studies on chemical constituents of Rou Gui (*Cinnamomum cassia* Presl.). *Bull. Chinese Materia Medica*, **7**, 26–28.

Yokozawa, T., CuiPing, C. and ZhongWu, L. (1998) Effect of traditional Chinese prescriptions and their main crude drugs on 1,1-diphenyl-2-piperylhydryl radical. *Phytotherapy Res.*, **12**, 94–97.

Zhu, Z. and Zhang, M. (1993) Pharamacological study on spleen-stomach warming and analgesic action of *Cinnamomum cassia* Presl. *China J. Chinese Mat. Medica*, **18**, 553–557, 574–575.

Zhu, L., Ding, D. and Lawrence, B.M. (1994) The *Cinnamomum* species in China. *Perfum. Flavour.* **19**, 17–22.

Zhu, M.S., Liu, S., Las, R.J. and Bu, Y.-L. (1996) GC, MS detection of synthetic cinnamic aldehyde added to cassia oil. *Acta pharm. Sinica*, **31**, 461–465.

Zhu, I.F., Li, Y.H., Li, B.L., Lu, B.Y. and Xia, N.H. (1993) *Aromatic Plants and their Essential Constituents*. South China Inst.Bot., Chinese Academy of Sciences, Peace Book Co., Hong Kong.

Zhu, L.F., Li, Y.H., Li, B.L., Lu, B.Y. and Zhang, W.L. (1995) *Aromatic Plants and their Essential Constituents*. Supplement I. South China Inst. Bot., Chinese Academy of Sciences, Peace Book Company, Hong Kong.

7 Indonesian Cassia (Indonesian Cinnamon)

M. Hasanah, Y. Nuryani, A. Djisbar, E. Mulyono, E. Wikardi and A. Asman

Introduction

C. burmannii Nees – Indonesian cinnamon, Indonesian cassia, Java cassia, Fagot cassia, Padang cinnamon, Batavia cassia, Korintji cassia, cassia vera.

Indonesian cassia or Indonesian cinnamon is the dried bark of *C. burmannii* which is grown in the Malaysia-Indonesia regions and commercially cultivated in the Indonesian islands. It is grown most extensively in the Sumatera, Java and Jambi Islands and extends up to Timor, growing from sea level to about 2000 m. The main centre of cultivation is the Padang area of Sumatera, at altitudes of 500–1300 m. A variant of *C. burmannii*, which has red young leaves, is grown at a higher elevation in the region of Mount Korintji (Kerinci). This cassia is of better quality and is traded in the international market as Korintji (or Kerinci) cassia. The form having green young leaves is grown at lower elevations, and is referred to in the international market as Padang cassia, Batavia cassia or cassia vera. In a small scale it is also cultivated in Phillippines.

The main centres of cultivation are Jambi and west Sumatera, which have around 59,490 ha and 28,893 ha areas respectively, producing around 20,185 t and 18,525 t of cassia bark, respectively. In 1999 there was 123,979 ha of cassia cinnamon that produced 42,590 t of bark. Most of the cassia bark produced is exported and domestic consumption is very little. In 1998, Indonesia exported 36,202 t of cassia bark valued at US$31.7 million. The main importing countries are the USA, Germany and the Netherlands. Almost 85–90% of the product exported from Indonesia comes from west Sumatera.

Habit

C. burmannii is a small evergreen tree, up to 15 m tall, having subopposite leaves. The petiole is 0.5–1 cm long, with a blade that is oblong–elliptical to lanceolate, 4–14 cm × 1.5–6 cm; pale red and finely hairy when young. Older leaves are glabrous, glossy green above and glaucous pruinose below. Inflorescence is a short axillary panicle (Fig. 7.1). Flowers are borne on 4–12 mm long pedicel, perianth 4–5 mm long and after anthesis the lobes tear off transversely about half way. Stamens about 4 mm long, staminodes 2 mm, fruit (berry) is ovoid, about 1 cm long. (Dao *et al.*, 1999) (see Chapter 2 for details on nomenclature and botanical aspects).

0-415-31755-X/04/$0.00 + $1.50
© 2004 by CRC Press LLC

Figure 7.1 A twig of *C. burmannii* with panicle.

Kostermans (1964) lists the following varieties of *C. burmannii*:
var. *angustifolium* Meissner
var. *chinense* (Bl.) Meissner
var. *kiamis* (Nees) Meissner
var. *lanceolatum* Miquel
var. *microphyllum* Miquel
var. *sumatrense* Teijsm & Binnenad
var. *suleavene* Miquel

Production

Light, rich sandy loam soil is best suited for the cultivation of Indonesian cinnamon for the production of the high quality bark. Growth is reported to be good in andosol, latosol and organosol soil types (Siswoputranto, 1976). But many small holders' plantations are on steep hillsides where the soil is stony, lateritie and less suitable for the production of high quality bark. The annual rainfall of the hilly regions of Padang is about 2000–2500 mm with short dry periods in May and September. The growth is slower at higher altitudes, while at low elevations the trees grow faster, but the bark is thinner and of lower quality and value.

Indonesian cassia is cultivated very similarly to that of Sri Lankan cinnamon except in the matter of harvesting (see Chapter 4). Cassia plants are raised from seeds. Vegetative propagation is possible through cutting and layering but it is not practiced as such plants produce thinner bark of lesser quality. Ripe fruits are collected from selected mother trees having thick bark and good aroma. Such trees are covered with nets in order to protect the fruits from birds, which are attracted to cinnamon fruits. Fruits are harvested at full ripening (when they become bluish black in colour), heaped for two or three days to allow the pericarp to rot, and are then washed in water to remove the fruit wall. Seeds, freed from the pericarp, are dried in the shade and sown immediately in seedbeds. The viability of seeds is lost rapidly, and storing even for a few days may result in drastic reduction in germination.

However, seeds stored at 15–20 °C maintain viability a little longer. By this treatment the seed moisture content decreases slowly, reducing pathogen activity and slowing down the metabolic processes. Osmoconditioning of seeds is one way to increase seed viability by soaking seeds in an osmoticum solution (KNO_3, H_2O, PEG 6000) before planting (Nursandi *et al.*, 1990). Darwati and Hasanah (1987) carried out seed germination studies and reported that fruits harvested when their pericarps turned black produced seeds with higher dry weight and gave higher germination percentage, greater seedling vigour and more uniformity in growth.

Nursery and planting

Raised nursery beds, 1 m wide are prepared in fertile, shaded areas. Artificial shade is provided if necessary, for which plaited palm leaves can be used. Seeds are sown at a spacing of 5 × 5 cm and at a depth of 1 cm. Seeds germinate in 5–15 days and when seedlings are about two-months old they are transplanted to polybags in a soil and dung mixture. Alternatively seeds can be sown at a wider spacing of 20–25 cm and plants can be left in the bed until they are ready for field planting after 10–12 months. Seedlings are hardened by the gradual removal of shade. The polybag seedlings have a definite advantage and give much higher field establishment.

Spacing adopted in field planting varies, but the common practice is to leave 1 m spacing between plants. Spacing is a compromise between two factors. Plants should be close enough to encourage formation of tall, straight trunks with as few branches as possible. At the same time, planting too close may lead to reduction in bark thickness, and hence spacing should be wide enough to encourage thicker bark formation (Dao *et al.*, 1999). Pits of 30 × 30 × 30 cm are usually used, in which the polybag seedlings are planted. Seedlings growing on beds should be lifted with a ball of earth around the roots and planted in pits. Damage to the root system predisposes plants to infection by *Phytophthora cinnamomi*. In the Sumatera region *Tephrosia candida* is recommended as a nurse crop. This may be sown six months ahead of planting cassia in rows about 1 m apart. Cassia seedlings are planted in *Tephrosia* rows after digging pits. Tephrosia is cut later and spread in the inter spaces. Intercropping with groundnut, ginger etc. is practiced during the first one or two years.

Studies carried out by Usman (1999) indicated that spacing treatments had little effect on the yield of bark except in one particular year when wider spacing produced a higher bark yield (Table 7.1). This was consistent with girth size that was increased by wider spacing. The highest total yield per plant was 1.21 and 3.40 kg per tree, respectively, by employing the cut and peel methods of bark extraction. This yield was equivalent to 1.45 and 4.14 t dry bark/ha (population base of 1200 trees/ha) obtained from the cut and peel methods, respectively. There is no significant effect due to fertilizer

188   M. *Hasanah* et al.

Table 7.1 Mean height and girth sizes as affected by spacing and fertilizers with the peel method

Treatments	1997		1981		1985
	Height (m)	Girth (cm)	Height (m)	Girth (cm)	Girth (cm)
Spacing (m²) 2.5 × 3.3	4.23 a	24.0 a	8.86 a	52.3 a	76.3 a
	4.11 a	23.9 a	8.29 a	48.7 a	70.6 ab
	4.21 a	23.8 a	8.39 a	49.2 a	68.5 b
Fertilizer control	4.19 a	23.6 a	8.19 a	46.4 a	68.3 a
	4.30 a	24.5 a	8.53 a	50.9 b	72.5 a
NPK (15 × 15 × 15) + 12.5 g urea	4.04 a	23.7 a	8.41 a	52.3 b	74.5 a

Source: Usman, 1999.

Note
Values followed by the same letters are not statistically significant.

treatments on yield. The addition of higher rates of fertilizer tended to reduce bark yields (peel method) (Table 7.2).

Plantations of Indonesian cassia are maintained like forest plantations, without any cultural operations or fertilizer applications. The only operation is the removal of the lower branches from the main shoots, which are removed by cutting close to the trunk. The recommended fertilizer ratio is a 15:15:15 mixture of urea, rock phosphate and potash (40–100 kg/ha depending on the age), but, as mentioned above, growers seldom apply fertilizer.

Bark yields obtained from the peel method are about three-fold higher than that of the cut method. The total yield of bark obtained from the cut and peel methods varies from 1.04 to 1.25 and 3.02 to 3.65 kg/tree, respectively. Rusli and Hamid (1990) reported a yield of 2.9 kg and 1.45 kg dry bark, respectively, from the main trunk and branches per tree.

Ridwansyah (1998) carried out an economic analysis of cinnamon cultivation in the Kerinci National Park region of Jambi Province, Indonesia. Here, cinnamon farming is the only source of income for about half the farming community, and more than 40% have less than 1 ha of land. The mean cultivated area is around 1.8 ha per household.

Table 7.2 Mean yields as affected by spacing, fertilizer applications and method of harvesting

Treatment	Yield (kg/tree)							
	Cut				Peeled			
	1997	1981	1985	Total	1977	1981	1985	Total
Spacing (m²)								
2.5 × 3.3	0.35 a	0.53 a	0.33 a	1.21	0.13 a	0.49 a	2.84 a	3.46
2.5 × 2.5	0.36 a	0.49 ab	0.31 a	1.16	0.13 a	0.45 a	2.69 a	3.27
2.5 × 2.0	0.34 a	0.37 b	0.29 a	1.00	0.12 a	0.45 a	2.57 a	3.12
Fertilizer control	0.31 a	0.43 a	0.30 a	1.04	0.13 a	0.44 a	2.65 a	3.22
3.75RY	0.38 a	0.44 a	0.36 a	1.18	0.14 a	0.49 a	3.02 a	3.65
NPK (15 × 15 × 15) + 12.5 g urea	0.35 a	0.53 a	0.37 a	1.25	0.12 a	0.44 a	2.46 a	3.02

Source: Usman, 1999.

Notes
RY = Rustica Yellow; values followed by the same letters are not statistically significant.

Table 7.3 Production of stem bark (SB), branch bark (BB) and twig bark (TB) in eight selected groups of Indonesian cassia

No.	Group	Production of quill kg/plant		
		SB	BB	TB
1	BG-1	1.67	0.64	0.39
2	BG-2	1.78	1.05	0.57
3	BG-3	2.92	1.04	0.98
4	BG-4	2.21	0.73	0.75
5	BG-5	4.57	1.12	1.53
6	BG-6	5.75	1.16	1.65
7	BG-7	6.97	1.73	2.41
8	BG-8	10.83	3.27	3.63
HSD 5%		5.10	1.10	1.33
		6.93	1.49	1.80

Source: Djisbar, 1994.

The mean number of cinnamon trees per farm unit is 2443 with a mean stand age of eight years. The government has introduced a cinnamon protection programme for protecting the crop and its sustainable exploitation.

Little crop improvement work has been developed for this important crop. Preliminary studies on collection and evaluation were carried out at the Bukit Gompong Research Station, west Sumatera (Djisbar, 1994). High variability was observed for stem diameter and eight groups of trees have since been identified based on their stem diameter (Table 7.3) but their oil content remained almost the same.

Oil cells in the leaves are always present in both the palisade and the spongy parenchyma. They are oblong-ovoid in the palisade parenchyma and more or less globular in spongy parenchyma (Bakker *et al.*, 1992). The oil content from the leaves is 1.12% (Rusli and Abdullah, 1988), less than that from the stem's bark (3.45%) and than from the branch bark (2.38%) (Rusli *et al.*, 1990).

Harvesting and Post-Harvest Handling

The periodical thinning of trees is often undertaken. The first thinning is usually in the third year and it produces some inferior bark. The first harvesting by thinning (or selective felling) is done in the fifth year when bark of fair to average quality is obtained. Thereafter some selective harvesting is done annually and continues for about 15 years. Trees are harvested at the beginning of the rainy season when the bark can be peeled easily. The trunk is first scraped with a blunt knife to remove moss, lichens and outer cork tissue. The bark is harvested from the lower part of the trunk in strips about 1 m in length and 7.5–10 cm wide (Fig. 7.2). The tree is then felled leaving a stump of 20–30 cm, and the bark is stripped from the upper part of the trunk and larger branches. The stump then regenerates and one or two strong shoots are allowed to develop into new stems (Purseglove *et al.*, 1981). Thus a field will be continuously harvested for a considerable period of time. Both peel and cut methods are used for the extraction of bark, the former gives higher yield and better quality bark. Bark is also separated by beating the harvested stem with a mallet (Fig. 7.3).

Figure 7.2 Harvesting of bark by peeling.

The extracted bark is then dried by spreading it on mats or wire netting either in the sun or in partial shade. In some areas the harvested bark is heaped for a few days and allowed to undergo fermentation. Then the bark is washed and dried. For preparing higher quality quills the bark is tied around bamboo poles, and on drying such bark assumes an attractive curved appearance (Dao *et al.*, 1999). Rusli and Hamid (1990) reported that from eight-year old trees about 14 t dry bark per hectare can be achieved.

On drying, the bark curls into quills, and is then ready for marketing. The dried quill is reddish brown in colour. The yield is highly variable. On an average a tree gives about 3 kg of stem bark and 1.5 kg of branch bark. In a crop cycle of ten years the total yield from 1 ha is about 2 t/ha of bark (Dao *et al.*, 1999).

The dried bark is graded based on type (scraped, unscraped, quills, quillings, featherings, chips), appearance (length, colour) and volatile oil. The Korintji and Padang forms are graded by appearance into A, B, C and D types according to length, colour and quality, and are sold on their content of essential oil. The following grades have been recognised in the international trade circles for Korintji cassia:

Figure 7.3 Separation of bark by beating.

Vera C/W	AA 2–3/4
Korintji A	3.00 ML/SVO
Korintji A	2.75 ML/SVO
Korintji B	2.50 ML/SVO
Korintji B	2.25 ML/SVO
Korintji C	1.50 ML/SVO
Korintji C	1.00 ML/SVO

The USA is the main importer of Indonesian cassia, and experience there shows that volatile oil content usually varies between 1.3% for Korintji C and 4% for Korintji A and between 1% for Batavia C and 2.7% for Batavia A. The grade AA is used primarily for packing in glass bottles where appearance is more important than oil content. The type cassia vera, produced from Java, Celebes, Phillippines, etc. are exported mainly to Germany. Rusli and Hamid (1990) reported that KA (Korintji A) contains 0.86% oil, KB 0.47% oil and KC 0.35% oil.

192 M. Hasanah et al.

Pests, Diseases

Many diseases and pests of *C. verum* also attack *C. burmannii* (Deinum, 1949; Mardinus *et al.*, 1974). The most serious disease is the stripe canker (cancer) caused by *Phytophthora cinnamomi* that often leads to widespread damage to plants (Djafaruddin and Hanafiah, 1975). Other diseases include pink disease (*Corticium salmonicolor* syn. *C. javanicum*), white rot (*Fomes lignosus*), rust (*Accidium cinnamomi*) and anthracnose (*Glomerella cingulata*). A list of diseases occurring in Indonesia is given below (see also Chapter 10).

As a tropical country Indonesia's climate is conducive to the development of many pathogens. Usually, cinnamon grows in the wet climate with a relative humidity of 80–95%, which predisposes the plant to many pathogens. However, data on yield loss caused by pathogens in Indonesia is still very limited.

Diseases commonly found on cinnamon plants in Indonesia are summarised below.

Red rust (Cephaleuros sp.)

Symptoms: Greyish orange or brownish orange spots on the upper leaf surface. On top of the spots can be seen the spores or conidia. On the lower leaf surface the spots appear brown.

Black rust (Puccinia cinnamomi)

Symptoms: Wet greyish brown irregular spots on the lower leaf surface. The spores of the fungi are clearly seen.

Grey leaf spots (Pestalotia cinnamomi)

Symptoms: Grey necrotic spots on the leaves. The spots enlarge between the leaf ribs. Sometimes the fungus also infects the twigs near the petiole.

Anthracnose (Colletotrichum sp.)

Symptoms: Blackish brown leaf spots, especially on the young leaves. In the more severe infection the old leaves and twigs are also attacked. Young leaves role up and become black. It can cause dieback and plant death.

Swollen and rotten leaf disease (Aecidium cinnamomi)

Symptoms: Lower leaf surface is swollen like blister blight, and becomes greyish to brownish grey. Tissues rot and show white to black spots. The leaves become concave and convex. The blister blight form is irregular; sometimes in one leaf there are lot of blisters and sometimes only one or two.

Pink disease (Cortitium salmonicolor)

Symptoms: The parts of the plant attacked are twigs, branches and stem. The affected twigs and branches appear pink due to mycelial growth on the upper and lower part of the branch. The mycelium layer penetrates the bark and wood. Small branches die off due to the fungus infection. On the big branches that are infected the leaves get yellow and gradually die when the upper branches are infected, and further infection quickly spreads to the lower branches.

Split canker disease (Endothia sp.)

Symptoms: The disease especially attacks young plants. Split and striped cankers are seen. The swollen part looks like grafting. The grafting like part is the dead bark.

For striped canker disease, part of the infected stem is removed with a knife and then painted with fungicides. The heavily infected plants are cut down and destroyed.

Crown-gall disease

Symptoms: Infects plant stems especially the young ones (about five-years old). The leaves become yellow, and sometimes the plant dies. The bark becomes dry, thick, and forms a hollow space, rough and brittle. The size of the canker is approximately 5–8 cm. On older plants (five- to ten-years old), the symptom is intermittent stripes. Brown ooze flows out from the stripes. The length of the stripes is more or less 2 m, and depends on the age of the plants.

Gall caused by the bite of Eriophyes doctersi

Symptoms: On the upper surface of young leaves; gall look convex and black. The leaves are still smooth. The lower surface is more concave with more black spots and sometimes becomes hairy.

Control measures

Generally, farmers do not apply any fungicides to control the disease. Some sanitation and eradication of the infected plants is recommended to contain the disease.

Insect pests

Insect pests affecting the Indonesian cassia include caterpillars of cinnamon butterfly (*Chilasa clytia*), leaf miners (*Aerocercops* spp.), caterpillars of leaf webber (*Sorolopha archimedias*) and mole crickets (*Gryllotalpa* spp.), all of which are damaging to young seedlings (Dao *et al.*, 1999) (Fig. 7.4). Insecticide application is not practiced.

Chemistry

Chemically, Indonesian cassia is similar to that of Chinese cassia. The bark oil has a composition similar to Ceylon cinnamon and Chinese cassia oils. Cinnamaldehyde is the major component in bark and leaf oils. Bark oil is obtained by steam or hydrodistillation and the yields range from 0.5 to over 2.0%. It is a colourless to brownish-yellow liquid having an odour similar, but less delicate, to that of the Ceylon cinnamon bark oil (Purseglove *et al.*, 1981). Reports indicated a cinnamaldehyde content of about 80–95% (Guenther, 1950) and later studies revealed the presence of alpha-terpineol, coumarin and benzaldehyde in

Figure 7.4 Insect pest damage in *C. burmannii*, (a) *Pompelon marginata* (Inset: Close up of *P. marginata*), (b) *Eriophyes boisi*, (c) seedling damage by *Scoliptidae* (beetle).

batavia and Korintji cassia (Lawrence, 1967). Eugenol is absent in the oil. The leaf on distillation gives about 0.4–0.5% oil, having about 45–62% cinnamaldehyde and about 10% phenols. The properties of Indonesian cassia bark and leaf oils are regarded similarly to those of Chinese cassia bark and leaf oils (Table 7.4). However, later reports indicate a chemical composition that is distinctly different. Xiao-duo *et al.* (1991) reported 1,8-cineol as the major component in bark and leaf oils (Table 7.5).

Yu-Jing *et al.* (1987) investigated the chemical composition of a physiological type rich in borneol (Mei Pan tree) and detected 34 constituents in leaf oil (Table 7.6).

Table 7.4 The yield and characteristic of Indonesian cassia oil

Specification		
Yield of oil (%)	–	1.32
Specific gravity (25 °C)	–	0.9593
Refractive index (25 °C)	–	1.5215
Optical rotation	–	+19°, 5′
Solubility in ethanol	–	Solution in 1:1 alcohol
Aldehyde (total %)	–	27.15
Cinnamaldehyde	–	–

Table 7.5 Composition of bark and leaf oils of *C. burmannii*

Bark component	%	Leaf component	%
1,8-Cineole	51.4	1,8-Cineole	28.5
α-Terpineol	12.5	Borneol	16.5
Camphor	9.0	α-Terpineol	6.4
Terpinen-4-ol	8.5	para-Cymene	6.1
Borneol	1.8	Spathulenol	5.8
α-Pinene	1.6	Terpinen-4-ol	4.1
β-Caryophyllene	1.6	Bornyl acetate	3.1
para-Cymene	1.0	β-Caryophyllene	2.9
β-Eudesmol	0.5	α-Pinene	1.9
Camphene	0.5	β-Pinene	1.7
Elemol	0.4	Cinnamyl acetate	1.5
Myristicin	0.4	Myristicin	1.2
β-Pinene	0.4	Elemol	0.6
α-Humulene	0.3	α-Humulene	0.4
Bornylacetate	0.1	Linalool	0.4
		Camphene	0.2
		β-Eudesmol	0.1

Source: Xiao-duo *et al.*, 1991.

Moestafa and Badeges (1986) studied the distillation of cassia cinnamon using the cohobation method and tried to identify the components. They reported that 70 °C is the optimum temperature for distillation and the yields of oil when trapped at 25 °C, 70 °C and 85 °C, were 0.57%, 1.62% and 0.31% respectively. They also found that the major components of bark oil are cinnamic aldehyde and eugenol. Chen *et al.* (1997) analysed *C. burmannii f. heyneanum* leaves, which yielded 0.54–0.85% oil on steam distillation. The oil contained 96.28–99.7% safrole, and is thus an excellent source of safrole.

Bark and leaf oils of Indonesian cassia have little commercial significance, as the entire cassia oil production for commerce is derived almost entirely from Chinese cassia and cinnamon oil from Ceylon cinnamon.

Other Aspects: Cassia Vera for Soil Conservation

Research during the past five years on cassia vera at locations 240 m above sea level which receive an average rainfall of 370 mm/month have indicated a reduction in soil

196   *M. Hasanah* et al.

Table 7.6 Chemical composition of Mei Pan tree (*C. burmannii*) physiological type

Constituent	%
d-Borneol	70.81
1,8-Cineole	10.73
Bornyl acetate	5.02
Carene-4	2.09
Myrcene	1.59
α-Phellandrene	1.47
α-Pinene	1.24
β-Pinene	1.11
α-Terpineol	1.22
γ-Elemene	0.14
Nerolidol	0.11
β-Phellandrene	0.92
Linalool	0.96
α-Terpinyl acetate	0.10
α-Thujene	0.03
Limonene	0.05
P-Cymene	tr
2-Bornene	tr
3,7-Dimethyl 1,3,b-octariene	0.03
Terpinen-4-ol	0.02
Perillaldehyde	0.01
Fenchylalcohol	0.01
Camphor	0.07
Citronellol	0.03
Geraniol	0.07
Eugenol	0.38
Geranyl acetate	0.01
Neryl acetate	0.03
4,11,11-Trimethyl-8-methylene bicyclo-(7,2,0) undec-4-ene	0.03
Isoeugenol	tr
4-Ethyl cyclohexane methanol	0.08

Source: Yu-jing *et al.*, 1987; De Guzman and Siemonsma, 1999.

Note
tr: trace.

erosion compared with grass cover. The prediction of erosian under the cassia vera were 14.25 t/ha/year, while on grass vegetation it was 24.46 t/ha/year. So cassia vera can be used as a conservation plant. This effect will be prolonged if yields are harvested in the form of peeled bark rather than cutting trees (Syakir and Hermanto, 1995). Dijisbar and Dhalimi (1996) indicated that in a sloping area, such as in Singkarak lake, west Sumatera, alley cropping was recommended using Sloping Agriculture Land Technology. Contour lines and seed beds $\pm$ 20 cm wide and 10–15 cm high are prepared and cassia vera plants are planted between two contour lines as a main crop along with annual crops such as corn, dry land rice and legumes as intercrops. If sloping increases up to 15–30%, then each contour line consists of one swallow ditch, one line of king grass and one line of vetiver. With 30% slope, each contour line consists of one swallow ditch, one line of vetier and two lines of king grass. Cassia plants are planted between two contours. This has been proven to be an efficient method for cultivating sloppy lands.

Conclusion

In spite of the fact that *C. burmannii* is an important export crop, little research and development has gone into its production and improvement. Even efforts on collection-conservation and evaluation of germplasm have been initiated only very recently. In Sumatera these activities were started during 1991–1992. The Research Institute for Spices and Medicinal Crops has initiated this work at its research centre in Sumatera where a germplasm collection has been assembled. This collection consists of progenies of 71 trees selected from the field, 232 random collections, and 35 elite collections. There are also four blocks of half a ha each of Bukit Gompong cinnamon collection conserved there. No superior line has so far been released for commercial growing, through a couple of superior lines have been identified.

References

Bakker, M.E., Gerritsen, A.F. and Van Der Schaaf, P.J. (1992). Leaf anatomy of *Cinnamomum* Schaeffer (Lauraceae) with special reference to oil and mucilage cells. *Blumea*, **37**, 1–30.

Dao, N.K., Hop, T. and Siemonsma, J.S. (1999) *Cinnammomum* Schaeffer. In C.C. De Guzman and J.S. Siemonsma (eds), *Plant Resources of South East Asia*, Vol. 13, *Spices*. Backheys Pub., Laden, pp. 94–99.

Darwati, I. and Hasanah, M. (1987) Perkecambahan benih kayumanis dari bergagai tingkat kemasakan dan periode penyimpanan. *Pebrt. LITTRI*, XIII (1–2), 5–11.

Denium, H.K. (1949), Kaneel: In de Lanbouw in die Indische Archipel, deel IIB, pp. 746–762.

De Guzman, C.C. and Siemonsma, J.S. (eds) (1999) *Plant Resources of South East Asia*, Vol. 13, *Spices*. Backheys Pub. London.

Djafaruddin, M. and Hanafiah, A. (1975) *Research on cinnamon diseases especially canker* (Phytophthora cinnamomi) *in West Sumatera*. Ris Mc, Bogor, Indonesia, pp. 76.

Djisbar, A. (1994) Usulan pelepasan varietas kayumanis yang cocok untuk daerah-daerah Sumatera Barat, *Jambi dan Sumatera*. p. 24.

Djisbar, A. and Dhalimi, A. (1996) Pengembangan kayu manis di sekitar danau Singkarak dalam rangka konservasi dan peningkatan pendapatan petani. Prosiding seminar dan temu lapang. Teknologi konservasi air berwawasan agribisnis pada ekosistem wilayah Sumatera Barat. *Singkarak* 21–22 December 1995, *Bilittro. hal*, 18–28.

Guenther, E. (1950) *The Essential Oils*. Vol. 2. Van Nostrand Co., New York, p. 516.

Kostermans, A.J.G.H. (1964) *Bibliographia Lauracearum*, National Herbareum, Bogor.

Lawrence, B.M. (1967) A review of some commercial aspects of cinnamon, *Perf. Ess. Oil Rec.* April 236–241.

Mardinus, D., Rifai, F. and Kiman, Z. (1974) Pest and diseases and their effects on the development of technique and marketing of Cassia vera. Presented in the Workshop of Development of Culture Technique and Marketing of Cassia vera. Faculty of Agriculture. Andalas University, Padang. p. 15.

Moestafa, A. and Badeges, F. (1986). Penyulingan minyak daun kayu manis *Cinnamomum burmannii, C. zeylanicum* dan, *C. cassia* secara kohobasi dan identifikasi komponen minyak yang dinasilkan. *Warta IHP*, **3**(1), 22–25.

Nursandi, F., Murniati, E. and Suwarto (1990) Pengaruh priming pada benih kedelai terhadap nilai vigor kecambah dan vigor tanaman. *Proceeding seminar sehari kebijaksanaan perbenihan nasional*. Forum komunikasi antar peminat dan ahli benih. Keluarga Benih. I (2: 11–21).

Purseglove, J.W., Brown, E.G., Green, C.I. and Robbins, S.R.J. (1981) *Spices* Vol. 1, longman, London, pp. 100–173.

198   *M. Hasanah* et al.

Ridwansyah, M. (1998) *Economic analysis of protecting cinnamon resources in Kerinci-Seblat National Park, Kabupaten Kerinci, Jambi province, Indonesia.* M.S. thesis, College of Agriculture, Laguna, Philippines, p. 164.

Rusli, S. and Abdullah, A. (1988) Prospek penfembangan kayu manis di Indonesia. *Journal Penelitian dan Pengembangan Pertanian*, VII (3), 75–79, Badan Litbag Pertanian. Department Pertanian.

Rusli, S. and Ma'mun (1990) Karakteristik Tiga Jenis Minyak Kulit Kayu Manis Komersial. *Media Komunikasi Penelitian dan Pengembangan Tanaman Industri*, No. 6, 62–64.

Rusli, S. and Hamid, A. (1990) Cinnamon spp. *Balai Penelitian Tenaman Remph-dan-obat (Indonesia). Edisikhususy*, 6(1), 45–53.

Siswoputranto, P.S. (1976) *Komoditi ekspor Indonesia.* Penerbit P.T. Gramedia, Jakarta, p. 310.

Syakir, M. and Hermanto (1995). Tingkat erodibilitas tanah dan bahaya erosi pada kayu manis (*Cinnamomum burmanii*) sebagai tanaman konservasi. Seminar Nasional dan Temu Lapang Teknologi Konservasi Lahan Kritis Berwawasan Agribisnis Pada Ekosistem Wilayah Sumatera Barat Padang. 21–22 December 1995. Tidak diterbitkan.

Usman (1999). Effect of spacing fertilizers and methods of harvesting on growth and yield of cassia vera (*Cinnamomum burmannii* BL). *Indonesian Journal of Crop Science*, 14(2), 1–6.

Yu-Jing, L., Liang-feng, Z., Bi-yao, L., Langtian, M., Zhao-lun, L. and Liang-zhi, J. (1987) Studies on mei pian tree (*Cinnamomum burmannii* physiological type) as a new source of natural D-borneol. *Acta Brt. Sinica*, 29, 373–375.

Xiao-duo, J., Quan-Long, P., Garraffo, H.M. and Pannell, L.K. (1991) Essential oil of the leaf barkl and branch of *Cinnamomum burmannii* Blume. *J. Essent. Oil Res.*, 3, 373–375.

8 Indian Cassia

Akhil Baruah and Subhan C. Nath

Introduction

Indian cassia (known as *tejpat* in Hindi), is a small to moderately sized evergreen tree, known botanically as *Cinnamomum tamala* Nees (Fig. 8.1). The leaf of this tree is a spice having a warm, clove-like taste and a faintly pepper-like odour. It is very popular among the people of northern India and since antiquity has been used as an essential flavouring agent in the preparation of many vegetarian and non-vegetarian dishes. It holds in Indian cookery the same status as that of 'bay leaves' (*Laurus nobilis*) in Europe.

C. tamala, occurring mostly in the tropical and sub-tropical Himalayas and extending to north-east India up to an altitude of 2000 m, is the main source of the spice *tejpat* which is commercialised. It also grows in Nepal, Bangladesh and Myanmar. However, whilst conducting an ethnofloristic census of *Cinnamomum* species in north-east India during 1994–1997, Baruah and Nath (1998) came across the leaves of a total of five taxa of *Cinnamomum*, (*C. tamala, C. bejolghota, C. impressinervium, C. sulphuratum*, and one unknown *Cinnamomum* sp.) all used as the spice *tejpat* by the people in the region (Baruah and Nath, 1998; Baruah *et al.*, 2000). The essential oils of *C. tamala* collected from different geographical locations have been investigated. Apart from the common occurrence of eugenol type essential oil (Gulati, 1982), linalool, cinnamaldehyde and linalool-cinnamaldehyde, predominating type oils from the leaves of the species have also been reported (Gulati *et al.*, 1977; Sood *et al.*, 1979; Bradu and Sobti, 1988; Nath *et al.*, 1994b). Likewise, cinnamaldehyde and eugenol type essential oils have been reported from *C. impressinervium* from different geographical origins (Kya and Min, 1970; Nath *et al.*, 1994c). In contrast, linalool, citral-geraniol and methyl cinnamate type oils have been reported from *C. sulphuratum*, another species native to north-east India that is also used as *tejpat* (Nath *et al.*, 1994a; Baruah *et al.*, 1999, 2001). Only linalool type essential oil has been reported so far from the leaf of *C. bejolghota*, another species traded as *tejpat* (Baruah *et al.*, 1997; Choudhury *et al.*, 1998). The present chapter will give brief accounts on *C. tamala* and other species used as Indian cassia or *tejpat* in north and north-east India (for botanical aspects refer to Chapter 2).

C. tamala Nees and Eberm

See Chapter 2 for nomenclatural citation and botanical studies.

Based on the morphology of leaves, the population of *tejpat* growing in north-east India has been classified into four types (Baruah *et al.*, 2000), the details of which are presented below. In addition to *C. tamala*, a few other species are also traded and consumed as *tejpat*. Brief descriptions of these species are also provided.

0-415-31755-X/04/$0.00 + $1.50
© 2004 by CRC Press LLC

Figure 8.1 A *Cinnamomum tamala* tree.

C. tamala *type 1*

This is a medium-sized evergreen tree, up to 7.5 m high with zigzag branching; trunk up to 95 cm girth; bark rather rough, dark grey to reddish-brown, darkening on exposure; leaves alternate, sub-opposite or opposite on the same twig, coriaceous, glabrous, shining above, dark green, pale below, pink when young, strictly ovate to ovate-lanceolate, apex acute to sharply acuminate, base acute to obtusely acute, variable in size, 2–4.5 × 6.5–11.5 cm, triplinerved, lateral nerves not reaching the tip, insertion basal to suprabasal, midrib moderate to stout, 2° veins distinct, sub-parallel. Petiole up to 1.3 cm long, concave above; panicle sub-terminal to axillary, equal to the leaves or slightly exceeding them, stout, pale yellow, silky pubescent, glabrate with age; flowers 5–7 mm long, pedicel 3 mm long, perianth 3 + 3, elliptic-ovate-lanceolate, sub-equal, outer 2.5 mm long, inner 2–2.5 mm long, silky pubescent on both surfaces, longitudinally brown-ribbed. Stamens 3 + 3 + 3, 1.5–2 m long, anther four-locular, pollen dehiscence occurs through the opening of valve, introrse, extrorse in whorl III, glands of whorl III attached 1/3 of the base of the filament, glands and anther pale yellow, filaments silky tomentose, pale yellow, head sagittate; pistil 1.5–2 mm long, pale yellowish-green, silky minutely puberulous, style filiform, ovary elliptic-oblong; fruit (drupe) black when ripe, 10–14 mm long, 5–6 mm in across, ovoid to oblong-elliptic, supported by a thickened peduncle and an enlarged truncate toothed base of the perianth, peduncle 3–4 mm long.

Significant foliar epidermal and venation characteristics: Epidermal cells pentagonal to polygonal and sinuous, hypostomatic, stomata sunken, stomata/mm^2 589, stomatal index 19.34, areoles tetragonal to polygonal, veinlet entering present, average frequency of areole/mm^2 14.93 (Baruah, 2000; Baruah and Nath, 1998).

Significant essential oil characteristics: Leaves are aromatic, spicy and yield volatile oil. Oil yield 1.07% (FWB), golden-yellow in colour, refractive Index (25 °C) 1.5262. On GC analysis, 23 components representing 99.01% of the total oil of the leaves are identified (Nath *et al.*, 1999) where eugenol alone constitutes 77.50%. The other components above 1% concentration in the oil are α-pinene (1.65%), α-phellandrene (10.47%), p-cymene (2.23%) and caryophyllene oxide (1.20%).

Phenology: Flowers from February–May; Fruits from June–September.

Occurrence and distribution: Found only in cultivated stands mostly in Arunachal Pradesh, Assam, Manipur, Nagaland and Tripura.

C. tamala *type 2*

This type of *C. tamala* can be easily distinguished from the others by its leaf morphology, having as elliptic-lanceolate to ovate-lanceolate shape, apex acuminate, base acute, being comparatively larger in size, 2.3–4.5 × 8–18 cm; the floral and other characteristics including phenology, distribution and ecology are found to be similar with that of type 1.

Significant foliar epidermal and venation characteristics: Epidermal cells pentagonal to polygonal and sinuous, hypostomatic, stomata sunken, stomata/mm^2 546, stomatal index 18.45, areoles tetragonal to polygonal, veinlet entering present, average frequency of areole/mm^2 8.47 (Baruah and Nath, 1997, 1998).

Significant essential oil characteristics: Leaves are aromatic, spicy and yield volatile oil. Oil yield 1.35% (FWB), pale yellowish-brown in colour, refractive index (25 °C) 1.5248. On GC analysis, 20 components representing 99.91% of the total oil of the leaves can be identified (Nath *et al.*, 1999), where eugenol alone constitutes 68.10% of the oil. The other components of above 1% concentration in the oil are α-pinene (2.25%), α-phellandrene (14.50%), p-cymene (4.00%), 1,8-cineole (2.35%), linalool (1.20%), α-terpineol (1.30%) and eugenyl acetatete (1.60%).

C. tamala *type 3*

A moderately sized evergreen tree, attaining a height of about 6 m trunk up to 70 cm girth, branches slender; bark rather rough, dark grey, aromatic, reddish-brown, darkening on exposure; leaves opposite, sub-opposite to alternate on the same twig, thinly coriaceous, aromatic, green and shining above, glabrous, pale below with sparsely distributed microscopic unicellular hairs, margins entire but undulate, broadly elliptic-lanceolate, apex acute, base decurrently acute to acute, variable in size, 2.1–7.8 × 5.5–20 cm, triplinerved, lateral nerves not continued to the tip, suprabasal perfect to imperfect, midrib moderate, nervules prominent on both surfaces: petiole concave above, up to 2 cm long. Inflorescence paniculate cyme, sub-terminal to axillary, not stout, pale yellowish-green, sub-quadrangular, silky pubescent, glabrate with age, shorter or equal to the leaves, up to 11 cm long (2 ° peduncle up to 4 cm long), flowers 4–5 cm long, perianth 3 + 3, outer broadly elliptic-lanceolate, 2.5–3 mm long, inner elliptic-lanceolate, 2 mm long, silky puberulous on both surfaces; stamens 3 + 3 + 3, 1.5 mm long, minutely puberulous to tomentose at base, pale yellowish-green, anther four-locular, introrse, whorl III extrorse, glands white, attached at 1/3 of the base of the filament; staminode 3, 1 mm long, white, minutely puberulous, head sagittate; pistil 1.5 mm long, minutely puberulous, stigma capitate, ovary elliptic to globose.

Significant foliar epidermal and venation characteristics: Epidermal cells pentagonal to polygonal and sinuous, hypostomatic, stomata sunken, stomata/mm^2 472, stomatal index 16.23, areoles tetragonal to polygonal, veinlet entering present, average frequency of areole/mm^2 7.27 (Baruah, 2000; Baruah and Nath, 1998).

Significant essential oil characteristics: Leaves are aromatic, spicy and yield about 0.7% volatile oil (FWB), golden-yellow in colour, refractive index (25 °C) 1.4980. On GC analysis, 17 components representing 94.04% of the total oil of the leaves can be identified (Nath *et al.*, 1999), where eugenol alone constitutes 57.90% of the oil. The other components of above 1% concentration in the oil are α-phellandrene (5.45%), p-cymene (2.68%), 1,8-cineole (4.35%), terpinen-4-ol (2.25%), eugenyl acetate (8.73%), α-farnesene (7.90%) and caryophyllene oxide (2.90%).

Phenology: Flowers from March to May; Fruits from June to August.

Occurrence and distribution: Found in the lower Assam zone, but rare in occurrence.

Use: Leaves are known as *tejpat* in the zone. Rarely sold commercially.

C. tamala *type 4*

Excepting the leaf morphology, this type is similar to type 1. In this type, the leaves are comparatively smaller in size (1.5–3 × 6–15 cm) and its shape varies from elliptic to oblong-lanceolate, apex acuminate, base cuneately acute.

Significant foliar epidermal and venation characteristics: Epidermal cells pentagonal to polygonal and sinuous, hypostomatic, stomata sunken, stomata/mm^2 611, stomatal index 19.88, areoles tetragonal to polygonal, veinlet entering present, average frequency of areole/mm^2 8.90 (Baruah, 2000; Baruah and Nath, 1998).

Significant essential oil characteristics: Leaves are strongly aromatic, spicy and yield about 1.5% volatile oil (FWB), pale yellow in colour, refractive index (25 °C) 1.5248. On GC analysis, 12 components representing 97.7% of the total oil of the leaves can be identified (Nath *et al.*, 1999), where eugenol alone constitutes 82.50% of the oil. The other components of above 1% concentration in the oil are α-phellandrene (6.38%), p-cymene (1.09%), caryophyllene (1.47%), and eugenyl acetate (4.36%).

Ecology and distribution: Found abundantly in both wild and cultivated conditions in Meghalaya up to an altitude of 1250 m, and in the North Cachar Hills of Assam up to an altitude of 1050 m.

Use: The leaves are commonly known as *Teji-bol* or *Dieng latyrpet* in Meghalaya. It is the main component of the *tejpat* market of Shillong from where it is transported to other places.

Other Species Traded as *Tejpat*

C. impressinervium *Meissn*

This is a mid-sized evergreen tree; 6–7.5 m *tall*; branchlets terete and slender; bark rough, aromatic, brown, inside creamish-brown, on exposure turning brown, 6–10 mm thick; leaf buds silky; leaves alternate, sub-opposite or opposite on the same twig, coriaceous, smell like *tejpat* leaves, glabrous, shining above, dark green, pale below, elliptic-oblong to elliptic-lanceolate, apex acute to acuminate, base decurrently acute, variable in size, 2.5–3.8 × 7–14 cm, triplinerved, lateral nerves reaching near the base of the acumen, suprabasal, midrib stout, 2° nerves sub-horizontal, nervules not so distinct; petiole stout, slightly concave above, 0.8–1.1 cm long; panicle sub-terminal to axillary, shorter than leaves, up to 6.5 cm long, glabrate, perianth 3 + 3, sub-equal, minutely puberulous on truncate cup-shaped fruiting tepals, pedicel obconic, pedicel with fruiting tepal up to 8 mm long.

Significant foliar epidermal and venation characteristics: Epidermal cells pentagonal to polygonal and highly sinuous, hypostomatic, stomata sunken, stomata/mm^2 550, stomatal index 19.52, areoles tetragonal to polygonal, vein endings simple, average frequency of areole/mm^2 7.88 (Baruah and Nath, 1997, 1998).

Significant essential oil characteristics: Leaves are aromatic, similar to that of *C. tamala* leaves and yield 2.00% (FWB) volatile oil, oil yellowish-brown in colour. Refractive Index (25 °C) 1.5320. On GC analysis, ten components representing 96.80% of the total oil of the leaves can be identified (Nath *et al.*, 1999) where eugenol alone constitutes 88.3% of the oil. The other components of above 1% concentration in the oil are δ-3-carene (1.6%), limonene (4.1%) and eugenyl acetate (1.1%).

Phenology: Flowers from February–April; Fruits from May–August.

Occurrence and distribution: Found in the North Cachar Hills and Cachar districts of Assam at an altitude between 800–1050 m, but is rare in other areas of north-east India. Also seen under cultivated stands in those areas.

Use: Leaves are used as 'best quality *tejpat*' among the people of the North Cachar Hills and Cachar districts of Assam and is even sold in the local markets under the same name.

C. sulphuratum *Nees*

A moderately sized evergreen tree; bark dark brown; leaves alternate, sub-opposite or opposite on the same twig, coriaceous, aromatic, glabrous, shining above, dark green, pale beneath, elliptic-lanceolate to narrowly ovate-lanceolate, apex sharply acute to shortly acuminate, base acute to cuneately acute, variable in size, 2.6–4.8 × 9.5–18 cm, triplinerved, lateral nerves not reaching the tip of apex, basal to suprabasal perfect or imperfect, midrib moderate, 2° veins distinct, sub-parallel, bent at middle; petiole up to 1.1 cm long, concave above; panicle terminal to sub-terminal or solitary axillary, shorter than leaves, up to 8.5 cm long, lax and rather few flowered, minutely tomentose, branchlets few, slender, flowers densely minutely pilose, pale yellow, pedicel up to 5 mm long, perianth 3 + 3, thickish, ovate, acutish, 2–4 mm long, anther oblong, four-locular, pollen dehiscence valvular, introrse, whorl III extrorse, filament pilose, glands adnate to the basal part of the filament of slightly higher up; staminode 3, 1 mm long, pilose, head narrowly sagittate; ovary ellipsoid, as long as the style, stigma minute, peltate; fruiting peduncle stout, 4–5 mm long at the time of fruit initiation.

Significant foliar epidermal and venation characteristics: Epidermal cells pentagonal to polygonal and moderately sinuous, hypostomatic, stomata sunken, stomata/mm^2 511, stomatal index 16.46, areoles trigonal to polygonal, veinlet entering present, average frequency of areole/mm^2 23.13 (Baruah, 2000; Baruah and Nath, 1998).

Significant essential oil characteristics: Leaves are aromatic and yield volatile oil. Oil yield 0.70% (FWB). Colourless. Refractive Index (28 °C) 1.4721. On GC analysis 13 components representing 93.44% of the total oil of the leaves can be identified (Nath *et al.*, 1999) where linalool alone constitutes 60.73% of the oil. Other components of above 1% concentration in the oil are benzaldehyde (1.40%), α-pinene (10.54%), camphene (3.06%), β-pinene (10.42%), limonene (3.21%) and geraniol (2.24%).

Phenology: Flowers from September–November; Fruits from December–February.

Occurrence and distribution: Found in the North Cachar Hills district of Assam, sporadic in occurrence between altitudes of 800–1230 m.

Use: Leaves are commonly used locally as *tejpat* and are also sold in the local markets.

Cultivation and Processing

C. tamala is the main source of Indian cassia or *tejpat* and is cultivated in certain parts of Khasi, Jaintia, the Garo Hills of Meghalaya and the North Cachar Hills of Assam for commercial purposes. Shillong is the main market for *tejpat*, from where it is transported to other places. Additionally, throughout the north-east region of India, local people cultivate this tree in their own gardens to meet their domestic needs.

Tejpat trees are planted at a spacing of 3 m × 2 m apart in regular plantations. Seedlings are raised in beds and planted out permanently when the plants are four to five years old. The trees take six to nine years to grow and harvesting of the leaves is done when trees are eight to ten years old, continuing for a century. No special care is required for cultivation. Mature leaves are collected during October to December until March, i.e. between the monsoons, as rains effect the aroma and quality of leaves. Leaf harvesting is carried out every year for young, vigorous plants and in alternate years for old and weak trees. Leaves are collected (small branches with leaves are also tied into bundles), dried in the sun and marketed. A single tree yields about 9–19 kg of leaves every year. The cultivation of *tejpat* forms part of an agro-forestry system in north-east India.

Chemical Composition

Leaves and bark are mildly aromatic, and yield on distillation an essential oil of about 0.13–2% concentration. The oil resembles cinnamon leaf oil and contains

Table 8.1 Physico-chemical characteristics of leaf oil

Characteristics	Mandi District	Palampur	Local
Colour	Light yellow	Light yellow	Light yellow
Specific gravity	0.9349 (30 °C)	0.9252 (20 °C)	0.9481 (20 °C)
Refractive index	1.4090 (30 °C)	1.4902 (20 °C)	1.5120 (20 °C)
Optical rotation	+1.30'	+2.26'	+2.12'
Acid value	4.3	3.9	4.9
Ester value	45–49	36–38	–
Ester value after acetylation	152.7	160.0	143.2
Cinnamic aldehyde (%)	38.43	29.23	60.35
Phenol content	4.5–5.2	3–4.5	1–1.5

cinnamaldehyde as its major component. The oil has roughly the following character-istics: Sp. gravity 1.025; $[\alpha]_D + 16°37'$; $nd^{20°}$ 1.526; soluble in 1.2 vol. of 70% alcohol. Bradu and Sobti (1988) analysed the leaf oils from three distinct sources. The properties of the oil samples are given in Table 8.1.

The chemical composition also exhibits variations. Bradu and Sobti (1988) analysed the chemical composition of oil from different sources (Table 8.2). In these

Table 8.2 Chemical composition % of *C. tamala* leaf oil samples

Sl. no.	Compound	I (Mandi District)	II (Kangra District)	III (Telikot Kumaon)	IV (Market sample)
1.	α-Pinene	0.87	0.63	0.30	0.26
2.	β-Pinene	0.41	0.46	0.22	0.05
3.	Benzaldehyde	1.071	3.88	–	–
4.	Unidentified	5.24	5.04	0.52	0.09
5.	Unidentified	0.42	2.25	0.46	–
6.	Linalool	28.26	22.75	15.67	15.28
7.	Benzylacetate	0.65	1.01	0.68	0.77
8.	Linalool acetate	3.40	5.15	–	–
9.	α-Terpineol	1.87	1.28	1.77	1.54
10.	Geraniol	1.68	1.783	0.67	1.20
11.	Limonene	1.14	2.68	0.50	0.10
12.	Cinnamic aldehyde	35.21	37.72	41.20	55.19
13.	Eugenol	9.98	11.57	–	–
14.	Benzyl cinnamate	0.95	0.19	–	–
15.	α-Phellandrene	–	–	0.10	3.95
16.	p-Cymene	–	–	0.82	0.64
17.	Ocimene	–	–	0.14	traces
18.	α-Terpinene	–	–	0.22	traces
19.	Camphor	–	–	3.19	0.90
20.	Borneol	–	–	1.18	1.07
21.	β-Caryophyllene	–	–	4.00	7.26
22.	Cadinene	–	–	3.06	–
23.	Eugenolacetate	–	–	12.45	2.06
24.	Unidentified compound	7.0	5.0	8.0	5.0

Source: Bradu and Sobti, 1988.

206 Akhil Baruah and Subhan C. Nath

Table 8.3 Composition of essential oil from a sample of *C. tamala*

Component	%
α-Pinene	10.54
Camphene	3.00
β-Pinene	10.42
Benzaldehyde	1.40
Myrcene	0.08
Limonene	3.21
p-Cymene	0.02
α-Linalool	60.73
Benzylacetate	0.11
α-Terpineol	0.24
Cinnamic aldehyde	0.24
Geraniol	2.24
Lindylacetate	0.30
Eugenol	0.85
Total	93.44

Source: Nath *et al.*, 1994b.

samples cinnamaldehyde and linalool were found to be the chief constituents. Nath *et al.* (1994b) identified 14 compounds that constituted 93.44% of the oil (Table 8.3). Linalool was the major component (60.73%), and eugenol and cinnamic aldehyde were less than 1%. Physico-chemical constants of this oil were: $n_D^{28} = 1.4791$; d^{28} 0.09034; $[\alpha]$ D28 $= +6$. In composition this particular sample is a quite distinct chemotype of *C. tamala*.

Upadhyaya *et al.* (1994) analysed the leaf oil *C. tamala* from Nepal using GC.MS and reported the following constituents: linalool (54.66%), alpha-pinene (9.67%), p-cymene (6.43%) beta pinene (4.45%), limonene (2.64%) and 16 minor components (less than 2%). Cinnamaldehyde concentration was only 1.16%, while engenol was absent.

Nath *et al.* (1999) carried out a comparative chemical study of four tamala types (morpho types) from the north-east regions of India. The oil content of the four types were 0.7% (type 3), 1.0% (type 1), 1.3% (type 2) and 1.5% (type 4). The chemical composition of the leaf oil of the four morphotypes is given in Table 8.4. The major constituent in all the four was eugenol accounting for 77.5%, 68.1%, 51.9% and 82.5%, respectively, in types 1–4. In type 1, the other important components were α-phellandrene, p-cymene, α-pinene and caryophyllene oxide. α-phellandrene, α-pinene, p-cymene, 1,8-cineole, eugenyl acetate, linalool and α-terpineol were the other important compounds in the oil of type 2. In type 3, α-phellandrene, 1,8-cineole, p-cymene, α-pinene, linalool, α-terpineol and eugenyl acetate occurred in above 1% concentration. α-phellandrene, eugenyl acetate, β-caryophyllene and p-cymene were the other important compounds in type 4 (Table 8.4).

Related Aspects and Properties

Joshi and Tandon (1989, 1990, 1991) reported the isolation and growth factor requirements of leaf gall induced by a mite on *C. tamala*. The growth regulator

Table 8.4 Composition (%) of essential oils of the leaves of four morphotypes of *Cinnamomum tamala*

Components	C. tamala			
	Variant I	Variant II	Variant III	Variant IV
Benzaldehyde	0.60	0.80		0.15
α-Pinene	1.65	2.25	t	0.25
Camphene	t	0.34	t	t
β-Pinene	0.20	0.50	t	0.06
Myrcene	0.42	0.60	t	0.14
α-Phellandrene	10.47	14.50	5.45	6.36
p-Cymene	2.23	4.00	2.68	1.09
1,8-Cineole	0.90	2.35	4.35	0.70
β-Phellandrene	0.20	0.30	0.68	0.08
γ-Terpinene	0.14	0.30	t	0.06
δ-3-Carene	–	t	–	–
Limonene	–	t	t	–
Terpinolene	t	t	t	t
Linalool	0.72	1.20	1.68	0.36
Camphor	0.10	t	t	t
Borneol	0.10	0.10	0.80	t
Terpinen-4-ol	–	0.14	2.25	0.20
Benzyl acetate	–	t	t	–
Guaiacol	–	t	t	–
α-Terpineol	0.60	1.30	1.12	0.24
Geraniol		t	t	t
Linalyl acetate	–	–	t	–
3-Phynyl propanal	0.20	t		
(E)-cinnamaldehyde	0.20	t	t	
Eugenol	77.50	68.10	51.90	82.50
(E)-methyl-cinnamate	0.24	t	1.08	0.07
Methyl eugenol	–	t	0.36	0.25
β-Caryophyllene	0.50	0.50	1.08	1.47
Iso-eugenol	0.14	t	–	0.17
Eugenyl acetate	0.75	1.60	8.73	4.36
(E)-ethyl cinnamate	–	–	0.05	0.86
(Z)-methyl isoeugenol	–	–	1.05	
α-Humulene	–	–	1.12	0.20
α-Farnesene	–	–	7.90	t
Caryophyllene oxide	1.20	0.65	2.90	0.16
Total	99.01	94.99	95.14	99.71

Source: Nath *et al.*, 1999.

Note
t = <0.05%.

required by gall tissue did not differ fundamentally from those required by healthy tissue. Optimum callusing of explants occurred on a medium containing 2,4-D (10 mg/l), kinetin (0.1 mg/l) and beta-mercaptoethanol (1–2 mM). Callus grew better in a medium fortified with 2, 4-d (4 mg/l) and kinetin (0.4 mg/l). Gall tissue grew faster than healthy tissue. Healthy tissue failed to grow in an auxin-free medium. One-year old cultures of both healthy and gall tissues grew on a medium devoid of cytokinin for a long time. The addition of caffeic acid and catechol (pyrocatechol) enhanced the growth of both normal and gall tissues, but these phenolies had no effect in the absence of auxin. Joshi and Tandon (1991) also reported that both normal and gall tissues showed indolepyruvic acid pathway of auxin biosynthesis. A direct correlation between tryptophan and auxin contents was recorded suggesting a substrate dependent regulation of IAA. Joshi and Tandon (1989) also observed a gradient of auxin protection activity in leaf galls (from young to the brown stage). Three auxin protectors with molecular weights of about 200, 8 and 2 Kda respectively, were isolated from gall tissue using Sephadex gel filtration. These protectors appeared to be oligomers or polymers of lower molecular weight phenolic substances.

The essential oil of the leaves of *C. tamala* exhibited fungal toxicity against *A. flavus* and *A. parasiticus* at 3000 ppm and 1000 ppm, respectively. The fungi-toxic property was not affected by temperature, autoclaving or storage. The active constituent in the oil was identified as eugenol. Alcohol extract of *C. tamala* was also shown to have above 50% schizomatocidal activity against Malarial parasites (*Praomys natalensis* and *Plasmodium berghei*) both *in vivo* and *in vitro* when tested at a dose of 1 g/kg × four days and 100mg/ml respectively (Misra *et al.*, 1991).

The oil of *C. tamala* exhibited absolute antidermatophytic activity against two ringworm fungi, *Microsporum audouini* and *Trichophyton metagrophytes* at 500 ppm. The ointment containing essential oil, prepared in polyethylene glycol (0.5 ml of oil in 50 ml base), showed promising efficacy as a herbal antifungal agent in treating dermatomycosis of guinea pigs, with zero positive culture recovery after 21 days following twice daily application of 2 ml of the ointment. *C. tamala* oil was assessed for its oral toxicity in mice and its LD 50 was recorded as 5.36 ml/kg (Dubey *et al.*, 1998; Yadav *et al.*, 1999).

Chughtai *et al.* (1998) reported that the aqueous extract of *C. tamala* significantly increased the rate of gene conversion and reverse mutation in diploid yeast (*Saccharomyces cerevisiae*, strain D_7) and also caused cell death and the inhibition of cell division.

C. tamala has also been found to have hypoglycaemic and hypolipidemic effects. Oral administration of a 50% ethanolic extract of leaves (single dose of 250 mg/kg) significantly lowered the plasma glucose levels in normoglycemic and streptozotocin-induced hyperglycemic rats. The extract also exhibited antihypercholesterolaemic and antihypertriglyceridaemic effects in streptozotocin-induced hyperglycemic rats (Sharma *et al.*, 1996).

End Uses

Besides flavouring, *tejpat* is used as a clarifier along with the products of *Emblica officinalis* fruits, and for tanning and dyeing leather (Anon, 1950). It is reported to have hypoglycemic, stimulant and carminative effects, and is used in Indian systems of traditional medicines to treat colic, coughs, diarrhoea, gonorrhoea, rheumatism, irritation, boils, conjunctivitis and itching (Chopra *et al.*, 1956; Chatterjee and Prakashi, 1991; Hussain *et al.*, 1992). In Kashmir, the leaves are used as a substitute for pan or betel leaves. The strong flavour of *tejpat* is due to an alcohol – soluble essential oil – Indian Cassia Lignea oil – (Anon, 1950), which is rich in eugenol, widely used in pharmaceutical preparations, perfumes for soap, cosmetics and as a flavouring agent in many kinds of foods, meats and sauces (Zutshi, 1982).

References

Anonymous (1950) *The Wealth of India, Raw Materials.* Vol. II. CSIR Publication, New Delhi.

Baruah, A. (2000) *Cinnamomum species associated with the livelihood of people in North-East India: A systematic census with emphasis to ethnobotany.* Ph.D. Thesis, Gauhati University, Assam.

Baruah, A. and Nath, S.C. (1997) Foliar epidermal characters in twelve species of *Cinnamomum* Schaeffer (Lauraceae) from Northeastern India. *Phytomorphology*, 47, 127–134.

Baruah, A. and Nath, S.C. (1998) Diversity of *Cinnamomum* species in North-East India: A micromorphological study with emphasis to venation patterns. In: A.K. Goel, V.K. Jain and A.K. Nayak (ed.) *Modern Trends in Biodiversity*, Jaishree Prakashan, Muzaffarnagar, U.P., pp. 147–167.

Baruah, A., Nath, S.C. and Boissya, C.L. (2000) Systematics and diversities of *Cinnamomum* species used as "tejpat" in Northeast India. *J. Econ. Tax. Bot.*, 41, 361–374.

Baruah, A., Nath, S.C., Hazarika, A.K. and Sarma, T.C. (1997). Essential oils of leaf, stem bark and panicle of *Cinnamomum bejolghota* (Buch-Ham) Sweet. *J.Essent. Oil Res.*, 9, 293–295.

Baruah, A., Nath, S.C. and Leclark, P.A. (1999). Leaf and stem bark oils of *Cinnamomum sulphuratum* Nees. From Northeast India. *J. Essent. Oil Res.*, 11, 194–196.

Baruah A., Nath, S.C. and Hazarika, A.K. (2001) Methyl cinnamate, the major component of the leaf and stem bark oils of *Cinnamomum sulphuratum* Nees. *Indian Perfumer*, 45, 39–41.

Bradu, B.L. and Sobti, S.N. (1988) *Cinnamomum tamala* in North West Himalayas; evaluation of various chemical types for perfumery value. *Indian Perfumer*, 32, 334–340.

Chatterjee, A. and Prakashi, S.C. (1991) *The Treatise on Indian Medicinal Plants* Vol. 1. Publication & Information Directorate, New Delhi.

Chopra, R.N., Nayar, S.L. and Chopra, I.C. (1956) *Glossary of Indian Medicinal Plants.* Council of Scientific & Industrial Research, New Delhi.

Choudhury, S.N., Ahmed, R.Z., Borthel, A. and Leclark, P.A. (1998) Essential oil composition of *Cinnamomum bejolghota* (Buch-Ham) Sweet – A secondary muga (*Antherara assama* W/W) food plant from Assam, India. *Sericologia*, 38, 473–478.

Chughtai, S.R., Dhmad, M.A., Khalid, N. and Mohamed, A.S. (1998) Genotoxicity testing of some spices in diploid yeast. *Pakisthan J. Bot.*, 30, 33–38.

Dubey, N.K., Yadav, P., Joshy, V.K. and Yadav, P. (1998) Screening of some essential oils against dermatophytes. *Philippine J. Sci.*, 127, 137–147.

Gulati, B.C. (1982) Essential oils of *Cinnamomum* species In: C.K. Atal and B.M. Kapur (ed.) *Cultivation and Utilization of Aromatic Plants*, Regional Research Laboratory (CSIR), Jammu-Tawi. pp. 607–619.

Gulati, B.C., Agarwal, S.G., Thappa, R.K. and Dhar, K.L. (1977) Essential oil of Tejpat (Kumaon) from *Cinnamomum tamala. Indian Perfumer*, 21, 15–20.

Hussain, A., Virmani, O.P., Popli, S.P., Mishra, L.N., Gupta, M.M., Srivastava, G.N., Abraham, Z. and Singh, A.K. (1992) *Dictionary of the Indian Medicinal plants.* Central Institute of Medicinal & Aromatic Plants, Lucknow, India.

Joshi, S.C. and Tandon, P. (1989) Association of auxin protectors in *Cinnamomum tamala* Fr. Nees leaf gall formation. *Indian J. Exp. Biol.*, 12, 1020–1023.

Joshi, S.C. and Tandon, P. (1990) Isolation and maintenance of normal and mite-incited leaf gall tissue of *Cinnamomum tamala* in culture. *Indian J. Exp. Biol.*, 28, 838–841.

Joshi, S.C. and Tandon, P. (1991) Possible cause of hyperauxinity in *Cinnamomum tamala* leaf gall. *Indian J. Exp. Bio.*, 29, 192–194.

Kya, P. and Min, N.C. (1970) Studies on some local *Cinnamomum* species. *Union Burma J. Life Sciences*, 3, 197–204.

Misra, P., Pal, N.L., Guru, P.Y., Katiyar, J.C. and Tandon, J.S. (1991). Antimalerial activity of traditional plants against erythrocytic stages of *Plasmodium berghei. Indian J. pharmacog.*, 29, 19–23.

Nath, S.C., Hazarika, A.K., Baruah, R.N., Singh, R.S. and Ghosh, A.C. (1994a) Major components of the leaf oil of *Cinnamomum sulphuratum* Nees. *J. Essent. Oil Res.*, 6, 77–78.

Nath, S.C., Hazarika, A.K. and Singh, R.S. (1994b) Essential oil of leaves of *Cinnamomum tamala* Nees & Eberm. From North East India. *J. Spices & Aromatic Crops*, 3, 33–35.

Nath, S.C. and Baruah, A.K. (1994c) Eugenol as the major component of the leaf oil of *Cinnamomum impressinervium* Meissn. *J. Essent. Oil Res.*, 6, 211–212.

Nath, S.C., Baruah, A. and Hazarika, A.K. (1999) Essential oils of the leaves of *Cinnamomum* Schaefer members. *Indian Perfumer*, 43, 473–478.

Sharma, S.R., Dwivedi, S.K. and Swarup, D. (1996) Hypoglycemic and hypolipidemic effects of *Cinnamomum tamala* Nees leaves. *Indian J. Exp. Biol.*, 34, 372–374.

Sood, S.P., Padha, C.D., Talwar, Y.P., Jamwal, R.K., Chopra, M.M. and Rao, P.R. (1979) Essential oils from the leaves of *Cinnamomum tamala* Nees & Eberm. growing in Himachal Pradesh. *Indian Perfumer*, 23, 75–78.

Upadyaya, S.P., Kirihata, M. and Ichimoto, I. (1994) Cinnamon leaf oil from *Cinnamomum tamala* grown in *Nepal. J. Jap. Soc. Food Sci. Technol.*, 41, 512–514.

Yadav, P., Dubey, N.K., Joshi, V.K., Chansouria, J.P.N. and Yadav, P. (1999) Antidermatophytic activity of essential oil of *Cinnamomum. J. Med. Aromatic plant. Sci.*, 21, 347–351.

Zutshi, N.L. (1982) Essential Oils – Isolates and Semisynthesis. In: Atal, C.K. and Kapur, B.M. (eds) *Cultivation and Utilization of Aromatic Plants*, Regional Research Laboratory (CSIR), Jammu-Tawi, pp. 38–89.

9 Camphor Tree

K. Nirmal Babu, P.N. Ravindran and M. Shylaja

Introduction

The camphor tree (*Cinnamomum camphora*) grows naturally in China, Japan, Taiwan and in the adjoining regions of South-East Asia, its natural habitat extending up to the sub-Himalayan regions of India. It is also cultivated in many tropical and sub-tropical countries. The camphor tree has become naturalised in Australia, where it is regarded as a tree weed. It was introduced into Europe possibly in the seventeenth century. In 1676, the tree was planted in Holland, and subsequently in other West European countries (Morton, 1977). In the decades that followed the camphor tree found its way to the Philippines, Eastern and Southern Africa, Queensland, central America, Cuba, Trinidad and the USA (Florida, Texas, California). It is common in the gardens in Ceylon, India and Malaysia. It grows best between 1000–1800 m elevation, under a rainfall of 120–400 cm per year. The camphor tree is commercially cultivated in Taiwan, Japan and eastern China, and to some extent in India.

Before the First World War, Taiwan (formerly Formosa) was the major producer and exporter of camphor. China was another producer, but the produce was consumed internally. Due to unrestricted exploitation, natural stands of camphor trees have dwindled, and subsequently plantations were established, initially in Taiwan and later in Japan. Since 1990 large plantations have also been established in China.

Little information is available on the production figures of camphor and camphor oil. Taiwan, Japan and China are the major producers. Japan used to produce several thousand tonnes of camphor annually, but the whole Japanese camphor industry declined as a result of the entry of synthetic camphor into the world market. Eventually camphor oil (the residual liquid left after crystallisation of camphor) became the most valuable product from the camphor tree, as it contains most of the volatile aromatic compounds.

Taxonomical Features

Windadri and Rahayu (1999) provide the following description of the camphor tree. Camphor trees are evergreen, aromatic, medium to large trees reaching 15 to 30 m height. (Fig. 9.1). The tree has an extensive shallow root system; short and stout trunk, bark deeply furrowed, crown spreading, twigs brown, yellowish or pinkish when young; twigs are glabrous, buds stout, ovoid, pubescent, with many imbricate scales. Leaves alternate, aromatic, petiole slender, 1.5–3 cm long. Leaf blade broadly ovate-elliptic to oblong-lanceolate, base obtuse, margin slightly undulate, apex acute or acuminate, chartaceous, deep green, shiny, glabrous above, glabrous or sparsely hairy

0-415-31755-X/04/$0.00 + $1.50
© 2004 by CRC Press LLC

Figure 9.1 Camphor tree.

beneath, with three main veins, impressed glands in vein axils, major veins prominent on both sides. Inflorescence an axillary, many flowered panicle, up to 7 cm long, pedicel 1–1.5 mm long, glabrous, flowers bisexual, small; perianth tubular, six-lobbed, membraneous, partly persistant in fruit; lobes ovate, 2.5−3 mm × 1 mm, obtuse, yellowish green, glabrous outside, pubescent inside, transversely tearing off near the base; with nine fertile stamens, in three whorls, pubescent; first and second whorls glandular, anthers oblong, 0.5 mm long, introrse, third whorl with subsessile, ovate glands at the base and extrorse anthers; fourth whorl consists of three glandular staminodes, ovoid with short filaments; anthers open upwards by flaps, ovary superior, ovoid, subsessile, glabrous, style up to 2 mm long (Fig. 9.2). Fruit is a compressed − globose berry, 7–10 mm in diameter, violet-black when ripe, one seeded. Seed 6–7 mm in diameter (Windadri and Rahayu, 1999). For the detailed nomenclature of *C. camphora* see Kostermans (1964).

Figure 9.2 1. Flowering twig of camphor; 2. Open flower; 3. L.S. of flower; 4. Stamen of third whorl showing valvular opening and basal glands; 5. Fruit.

Sub-specific division of C. camphora

C. camphora is a complex species in which a tremendous amount of intraspecific variability has been recorded. Such variability exists both in morphology and in chemical constitution. Kostermans (1964) listed the following varieties that were identified based mostly on morphological characteristics:

var. *cuneata* Blume ex Meissner
var. *glaucescens* A Braun
var. *linaloolifera* Fujita (here the major component of the oil is linalool)
var. *nominalis* Hayata
var. *normale* Hayata ex Ito
var. *parviflora* Miquel

var. *procera* Blume ex meissner
var. *quintuplinervia* Miquel
var. *rotundata* Meissner
var. *rotundifolia* Makino ex Honda
var. *typica* Petyaev

Hirota (1951) suggested the following subdivision based on chemical characteristics

1. Subspecies *Eucamphor* Hirota
2. Subspecies *Formosana* Hirota

Var. *occidentalis* Hirota
  Sub var. *eucamphor*
  Sub var. *cineola*
  Sub var. *safrola*
  Sub var. *sesquiterpenia*
  Sub var. *linaloola*
  Sub var. *linaloid*

Var. *Orientalis* Hirota
  Sub var. *eucamphor*
  Sub var. *cineola*
  Sub var. *safrola*
  Sub var. *sesquiterpenia*
  Sub var. *linaloola*
  Sub var. *linaloid*
  Sub var. *borneola*

3. Subspecies *Newzealandia* Hirota
    var. *eucamphor*
    var. *cineola*

A great number of variations have been noticed even in the segregating progenies of camphor trees. Processes such as natural crossing and segregation might have led to the great diversity in morphological and chemical characteristics present in this species.

Ecology, propagation and husbandry

The natural habitat of *C. camphora* is primary forest, but it also grows in open fields up to 3000 m altitude. It flourishes in sub-tropical climates and at higher elevations of the tropics. High rainfall is favoured by camphor trees; in its natural habitat the rainfall ranges from 1000–3500 mm but it does not tolerate water logging (Windadri and Rahayu, 1999).

The plant has been successfully cultivated in many countries. In India commercial cultivation has existed in Dehra Dun, Saharanpur, Calcutta, Nilgiris and Mysore. In Ceylon the camphor grows well up to an elevation of 1500 m; while in Nilgiris (south India) the tree thrives even up to 2000 m. Being a shade tolerant species it has colonised well under the pine forests of the sub-Himalayan zones as a secondary succession species (Singh and Negi, 1997). However, it has been shown that unshaded trees have higher essential oil content. Trees growing in the shade and even shaded leaves of a specific tree usually have lower essential oil and camphor contents. For cultivation, fertile and well drained sandy loam is most suitable. Soil type also effects essential oil content and its composition. Trees grown on lighter soils tend to have higher essential oil contents. Neutral or slightly alkaline soils are preferred for plantations (Windadri and Rahayu, 1999).

Propagation of camphor tree is through seed, though vegetative propagation by shoot or root cuttings and root suckers is also possible. Cuttings from trees rich in camphor, however, are difficult to root (Burkill, 1935). Seeds from later crops are used for sowing. The viability of seeds is rather short, and the germination rate is low. Fresh seeds freed of pulp should be air-dried for four to five days and sown in nursery beds of light, sandy loam soil enriched with manure. Germination takes three to four weeks. In parts of China where the winter is cold, seeds are kept until the following spring. Cleaning and soaking the seeds in water for 24 hours improves and hastens germination. Seedlings are ready for transplanting to field after 12–24 months. Before transplanting, seedling plants are pulled out with the onset of the rainy season, and the plants are cutback to 5–10 cm. In India, root pruning is also practiced and planting density is 2000–2500 trees per ha. When the trees are grown for leaf oil the chemotype of the planting material is checked (usually the farmers crush and smell the leaf). Micropropagation protocols have been developed for the multiplication of camphor trees (Nirmal Babu *et al.*, 1997, 2003; Huang *et al.*, 1998) (Fig. 9.3).

Figure 9.3 Micropropagation of camphor tree.

Camphor tree plantations are managed like forest tree plantations. Some initial weeding is undertaken, (up to five to six years), after which the tree canopy grows and prevents weed growth. When grown for leaves trees are topped at about 2 m height and coppiced to encourage a bushy growth.

Diseases and insect pest problems are minimal, though some of the insect pests and pathogens affecting cinnamon and cassia also affect the camphor tree. Plant protection measures are seldom employed.

Harvesting and Camphor Processing

In the major producing countries such as Japan and Taiwan, camphor is obtained by the distillation of wood from trees that are 50-years old or older, while in other camphor growing countries such as America and India the practice is to utilise twigs and leaves for camphor production. This latter practice has the advantage of an early yield, as plants of about five-years old can be harvested for leaves and used for distillation. In such cases where the leaves are harvested for distillation, camphor trees are maintained as bushes. Camphor tree coppices grow vigorously and shoots grow rapidly. New leaves take about two months to grow to full size and cropping at two-month intervals is possible. Usually leaves are harvested three to four times a year in India as well as in Sri Lanka.

The Japanese camphor industry

An account of the Japanese camphor industry in its heyday is given by Hiraizumi (1950), on which the following discussion is mainly based. Traditionally, Japan has been a major producer of camphor. Camphor trees existed in abundance in south Japan, the main areas being the islands of Kyu-Shu, Shikoku and the southern part of Hon-Shu. In earlier times the camphor industry was a government monopoly. Due to unrestricted exploitation the camphor tree population dwindled very much and the government of Japan embarked on a massive replanting programme in the years following the Second World War.

In Japan camphor and camphor oil are distilled from the wood of trees that are 25-years old or more. Before a camphor tree is felled, the soil is removed from its base to expose the root system, and all the roots, as much as possible, are removed. Then the tree is cut close to the ground. Roots, stump, trunk and branches are all cut into a convenient size and transported to the distillation factory. In the factory, the wood is reduced to small chips using mechanical chippers and fed into a distillation still. Essentially a distillation unit consists of a boiler, a retort (vat or still) and several condensers. The boiler is an iron pan in which water is brought to boiling, and its size varies from 1.2 m to 1.5 m depending on the size of the retort. The fire box beneath the boiler is constructed of stone or fire-brick. The retort (still), known as *koshiki*, consists of a wooden tank without a bottom and rests upon the boiler, but the two are separated by a pine wood grid that consists of many holes for the passing of steam. Camphor wood chips are charged into the retort and covered with a heavy lid. Once the distillation is completed, the exhausted waste is taken out through a man hole at the bottom of the retort, just above the grid. The capacity of stills vary from 600 kg to 800 kg. A pipe about 10 cm in diameter runs from the upper part of the still to the condenser. Usually a series of condensers are connected, and all connections are made air-tight.

The condenser is the most bulky and troublesome unit of the conventional camphor distillation unit. Since camphor oil partly crystallises at ordinary temperatures, the conventional spiral condensers cannot be used for the distillation of camphor. The low price of camphor prevents the growers from high investment in distillation units. Hence relatively crude condensers are still used for camphor distillation. The vapours coming out from the distillation chamber (*Koshiki*) goes through a pipe to three wooden barrels or boxes, connected in a series. The first condenser serves for preliminary cooling, where the heavier high boiling fractions are condensed. This condenser is smaller than the other two. The second condenser is the main part where most condensation takes place. The barrel type condensers are covered with a V-shaped copper basin, which is fed with running water throughout the operation. Prior to distillation the barrel is half filled with water. In the box type (called the *Tosa* type) the condenser consists of a big box constructed of heavy wooden planks (about 1.8 m long, 1.2 m wide and 0.85 m deep). The inside chamber is divided into interconnected compartments. Prior to distillation the box is half filled with water. The lid of the box consists of a shallow tray of the same length and width and about 10 cm deep, and the bottom is made of copper. Water flows through this shallow lid during the distillation period to keep the lower box cool. The third condenser is similar to the second one, but smaller.

Distillation

The flow chart of camphor extraction is given in Fig. 9.4. This chart indicates the various products (camphor and the different types of camphor oil) obtained in the industrial distillation of camphor wood and fractionation of the distillate.

Chipped wood material is charged tightly into the wooden retort (*koshiki*), through the opening at the top, and water in the kettle is brought to a boil. Steam passes through wood chips carrying vapours of camphor and camphor oil and passes to the condenser where the vapours condense. Camphor separates in the form of white crystals along the walls of the condensers. The distillation water is returned into the retort and redistilled (cohobated) during distillation. The duration of the distillation process depends upon the quantity of chips charged and the type of wood. Roots require more time than wood (15 hours compared to 8–10 hours). The optimum rate of distillation is attained when 700–800 ml of distillate per minute is produced from a 500 kg charge of wood material. The ratio of oil and water in the distillation is 1:25. On completion of the distillation, the exhausted chips are discharged through the man-hole in the side of the retort and are used as fuel.

The camphor and camphor oil accumulating in the condensers are usually removed once a month, after about 40 distillations. The semi-crystalline camphor is first separated from the camphor oil by crude filtering. The oil is poured into cans, the camphor packed into barrels and the two products are then carried to the collection centres of camphor manufacturing companies. Final purification of camphor oil by fractionation is done by such companies.

Advanced camphor distillation factories were established in Japan, Taiwan and in China. In such factories, an integrated production process for camphor and fractionated camphor oil is employed. Continuous vacuum fractionation is used for separating the various commercially useful fractions. A continuous vacuum fractionation unit consists of a series of distillation towers where distillation and fractionation take place as a continuous process. Such units separate camphor and linalool containing fractions and produce a rectified Ito oil that contains more than 85% linalool.

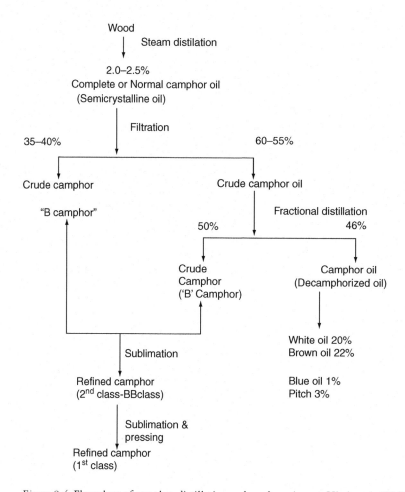

Figure 9.4 Flow chart of camphor distillation and products (source Hiraizumi, 1950).

Yield of oil

The yield of camphor and camphor oil from trees depends upon many factors:

1. Naturally growing trees contain more camphor and camphor oil than reforested ones.
2. Isolated trees contain more camphor and camphor oil than trees growing in dense stands.
3. Healthy and strong trees contain more camphor and camphor oils than weak ones.
4. Old trees contain more camphor than young trees.
5. Trees growing on sandy soil contain much more camphor than trees growing in other types of soil.

The relation between the yield of camphor, camphor oil and the age of trees is given in Table 9.1. As a camphor tree grows older it develops fewer leaves and the root system enlarges. The various parts of a camphor tree yield the average weight percentages of camphor and camphor oil as given in Table 9.2.

Table 9.1 Relation between age of trees and yield of camphor and camphor oil

Age of trees (yrs)	Growth of tree			Camphor and camphor oil		
	Height (cm)	Diameter at breast height	Volume (cu. m)	Camphor %	Camphor oil %	Total %
5	1.75	1.19	0.0006	–	1.00	1.00
10	4.68	5.92	0.0095	0.13	1.08	1.21
15	6.95	10.65	0.0503	0.27	1.15	1.42
20	8.78	15.37	0.1170	0.50	1.50	2.00
25	10.27	20.10	0.2160	0.63	1.50	2.13
30	11.52	24.82	0.3473	0.82	1.40	2.22
35	12.57	29.55	0.5164	0.88	1.40	2.28
40	13.48	34.22	0.7255	1.00	1.30	2.30
45	14.26	39.00	0.9770	1.05	1.30	2.35
50	14.94	43.72	1.2760	1.10	1.30	2.40
55	15.55	48.45	1.6210	1.14	1.30	2.44
60	–	50.90	–	1.28	1.20	2.48
65	–	56.20	–	1.35	1.20	2.55
70	–	60.00	–	1.35	1.20	2.55
75	–	62.80	–	1.42	1.20	2.62
80	–	65.70	–	1.52	1.10	2.62
100	–	78.10	–	1.60	1.10	2.70
120	–	85.70	–	1.70	1.10	2.80

Source: Hiraizumi, 1950.

Table 9.2 Yield of camphor and camphor oil from different plant parts

	Leaves	Branches	Trunk	Stump	Roots
Camphor (%)	1.0	0.3	0.8	1.3	0.8
Camphor oil (%)	0.3	0.6	1.4	1.8	2.5
Total yield (%)	1.3	0.9	2.2	3.1	3.3

The Taiwan (Formosan) camphor industry

Taiwan (previously Formosa) was earlier a major producer and exporter of camphor and camphor oil. According to tradition, the process of camphor recovery was introduced into Formosa from Japan between 1664 and 1683 by Chang Chang Kong (Hiraizumi, 1950), or from nearby mainland China. When Formosa was conquered by China in 1683, camphor trees came under the protection of the state. In the nineteenth century the camphor industry began to prosper and Formosan camphor became world famous. During the civil war there was such a demand for camphor in the United States that it bought Formosa's entire camphor yield. The US even made an effort to purchase Formosa to monopolise the camphor trade (Morton, 1977). In 1895 Japan annexed Formosa and established monopoly over the camphor industry. Following the Second World War Formosa became the Nationalist Republic of China under Chiang-Kaishak. The camphor industry became a state industry and it continues to be so even now.

220 K. Nirmal Babu, P.N. Ravindran and M. Shylaja

Taiwan (Formosan) camphor trees

There are four morphologically distinct varieties as well as species of camphor trees in Taiwan. One of these (*Kusunoki*) is subdivided into three physiological forms.

1. Kusunoki (*Cinnamomum camphora*)
    (i)   *Hon-Sho*
    (ii) *Ho-Sho*
    (iii) *Yu-Sho*
2. *Rau-Kusu* (*C. camphora* var. *nominale* Hayata)
3. *Sho-Gyu* (*C. kanahirai* Hayata)
4. *Obha-Kusu* or *Pha-Chium* (*C. micranthum* Hayata)

The *Hon-Sho* (hon = true, sho = camphor tree) is the true camphor tree and the most important, and is identical with the Japanese camphor tree. The *Ho-Sho* (ho = fragrant) oil contains linalool as an important ingredient, in addition to camphor. The rectified oil from *Ho-Sho* became world famous as a source of linalool, which is used widely in the perfume industry. This variety is later classified as *C. camphora* var. *linaloolifera* by Fujita (1967, Fujita *et al.*, 1973).

The *Yu-Sho* (Yu – oil, Sho – camphor tree) variety is found in China, but rarely in Taiwan. The oil of this variety contains little camphor, hence the name camphor oil tree. The oil contains a high percentage of cineole. This oil has no commercial prospects now, and hence is not exploited. *Rau-Kusu* (*C. camphora* var. *nominale*) is morphologically a distinct variety found in the drier parts of Taiwan. The oil resembles the *Hon-Sho* and *Ho-Sho* varieties and is classified similarly. Its stumps regenerate profusely. Five distinct types of oils have been reported containing 78% camphor, 59% cineole, 87% safrole, 85% linalool and sesquinterpenes only. *Sho-Gyu* (*C. kanahirai*) resembles the true camphor tree in appearance, except that the leaves are blackish and its bark lusterous. The essential oil derived from the *Sho-Gyu* tree contains no camphor, its chief constituent is terpinen-4-ol. *Obha-Kusu* (*C. micranthum*) has much larger leaves, the essential oil contains mainly safrole.

Hirota (1951) has designated the Formosan camphor trees as *C. camphora* spp. *formosana* and distinguished seven chemical varieties under this (chemovars). They are: chvar. borneol; chvar.cineole; chvar. camphor; chvar. linalool; chvar. safrole; chvar. sesqunterpene and chvar. sesquiterpene alcohol (Tetenyi, 1974). The chvar. linalool further consists of two chemical forms; ch forma 86% and ch forma 71% (of linalool) (terminology proposed by Tetenyi, 1974).

The Formosan camphor industry developed almost in line with the Japanese camphor industry and employs similar distillation procedures. The average yield of camphor and camphor oil from the Formosan trees is given in Table 9.3. The average yield of camphor and camphor oil from trees of different height is given in Table 9.4. The content of camphor and camphor oil in the trunk decreases from ground level to the top. It has also been shown that camphor and camphor oil accumulate more towards the periphery of the trunk than in the centre.

The oils derived from *Hon-Sho* and the *Ho-Sho* trees yield the following main fractions (Table 9.5): white oil (contains 25% cineole), the *Ito* oil (80–85% linalool), the brown oil (50–60% safrole) and the terpineole (contains 70% α-terpineol).

Prior to the Second World War Formosa had established large, more sophisticated factories for the distillation of camphor wood. Production of synthetic camphor from

Camphor Tree 221

Table 9.3 Yield of camphor and oil from different parts of camphor tree types

	Roots		Trunk		Bough, twigs		Whole tree	
	Camphor %	Camphor oil %	Cam. %	Cam. oil %	Cam. %	Cam. oil %	Cam. %	Cam. oil %
Hon-Sho	1.02	1.87	0.67	1.53	0.47	1.07	0.80	1.60
Ho-Sho	–	3.43	–	2.26	–	0.91	–	2.40
Leaves of Hon-Sho	–	–	–	–	–	–	0.7	0.4

Table 9.4 Average yield of camphor and oil from trees of different heights

Ht. above ground (m)	Hon-Sho tree 120-years old			Ho-Sho tree 120-years old	
	Dia. (cm)	Camphor %	Camphor oil %	Diameter	Ho-Sho oil
0.30	107.3	1.73	2.24	109.1	2.68
3.03	80.6	1.11	1.61	74.5	2.19
6.67	71.8	0.82	1.26	67.9	1.72
12.12	54.5	0.48	0.81	60.9	1.55
15.76	40.0	0.17	0.58	51.8	1.47
19.39	24.2	0.06	0.34	42.4	1.26
21.21	23.0	0.02	0.22	39.7	0.98
23.03	21.2	–	–	34.5	0.89
26.66	11.2	–	–	26.1	0.56
30.30	–	–	–	17.3	0.27

Source: Guenther, 1950.

Table 9.5 Main fractions of Hon-Sho and Ho-Sho oils

	Camphor %	White oil %	Ho oil %	Brown oil %	Terpineol %	Blue oil %	Pitch %
Hon-Sho oil	50	22	–	20	2	2	2
Ho-Sho oil	40	18	20	13	2	2	2

pinene was also started using the following synthetic route: pinene → bornyl chloride → camphene → camphor. However, in the post-war years the Formosan camphor industry suffered a set back due to the large scale production of synthetic camphor by the USA. The Formosan camphor oil (Taiwan camphor oil) is still an important product of international trade.

The Chinese camphor industry

The Chinese camphor industry was dependent on natural stands of the 'Hon-Sho' variety. In China camphor trees occur south of the Yangtse river, in the provinces of Kiangsi, Chekiang, Fukren, Hunan, Kwangtung and Kwangsi. All three varieties – Hon-Sho, Ho-Sho and Yu-Sho – occur under natural stands, and for centuries only Hon-Sho trees were used for camphor production resulting in a very drastic reduction of this variety. Until relatively recently the Chinese camphor industry was interested only in the production of camphor, the camphor oil was never used. The producers have developed the Chinese method of camphor extraction (bowl method) and the improved barrel method, which is a simpler version of the Japanese method. In the first quarter of the last century, China exported large quantities of camphor. The Chinese camphor industry also suffered a lot during the world war years and during the post-war years the Chinese camphor industry continued to decline along with the other camphor producing countries. However, the introduction of leaf distillation for the production of camphor oil has given a new impetus to the camphor industry of China. With government support camphor oil production has gone up substantially in the past two decades. The Sichuan province is now the major camphor leaf oil production region where a new distillary at Yibui produces more than 1000 t of camphor leaf oil annually, about 70% of China's total output (Weiss, 1997).

Physico-chemical Properties of Camphor Oil

Original oil (non-rectified)

Oil of Kusunoki

Hon-Sho oil:

Specific gravity (25 °C)	– 0.9543
Optical rotation	– +32°24′
Refractive index (25 °C)	– 1.4806
Acid number	– 1.1
Saponification number	– 4.1
Saponification number after acetylation	– 23.3
Alcohol content (as $C_{10}H_{18}$)	– 6.5%
Camphor content	– 50.5%

Hon-Sho oil (from Taiwan):

Specific gravity at 25 °C	– 0.936
Optical rotation	– 21°58′
Refractive index (at 25 °C)	– 1.4702
Alcohol (content calculated as $C_{10}H_{18}O$)	– 6.5%
Camphor content	– 50.5%

Ho-Sho soil (from Taiwan):

Specific gravity at 25 °C	– 0.9306
Optical rotation	– 21°58′
Refractive index (at 25 °C)	– 1.4702
Alcohol (content calculated as $C_{10}H_{18}O$)	– 30.0%
Camphor content	– 44.0%

Ho-Sho leaf oil:

Specific gravity (25 °C)	– 0.8925
Optical rotation	– 11°2′
Refractive index (25 °C)	– 1.4659
Alcohol content (calculated as $C_{10}H_{18}O$)	– 71.2%

Yu-Sho oil:

Specific gravity	– 0.942 (at 13 °C), 0.9712 (16 °C)
Optical rotation	– +18°53′ to +29°51′
Refractive index	– 1.4746 (at 19 °C) to 1.4789 (20 °C)

Sho-Gyu oil (C. kanahirai):

Specific gravity (at 9 °C)	– 0.910
Optical rotation at 9 °C	– +23°15′
Refractive index at 20 °C	– 1.4748
Acid number	– 0.63
Ester number	– 0.65
Saponification number after acetylation	– 100.7
Alcohol content (calculated as $C_{10}H_{18}O$) – 30.05%	

Obha-Kusu (or *Pha-Chium*) oil (*C. micranthum*):

Specific gravity (20/4)	– 1.089
Optical rotation (26°)	– 0°14′
Refractive index (20 °C)	– 1.5457
Acid number	– 0.5
Ester number	– 3.5
Saponification number after acetylation – 5.7	
Safrole content	– 90% or more

Physico-chemical properties of fractionated camphor oil

White (light) camphor oil:

Boiling range	– 160–185 °C
Specific gravity	– 0.870–0.880
Optical rotation	– +15°0′ – 20°0′
Refractive index (25 °C)	– 1.4663
Camphor content	– Below 2.5%
Cineole content	– 20–25%

Brown camphor oil (red camphor oil):
It is reddish brown in colour.

Boiling range	– 210–250 °C
Specific gravity	– 1.000–1.035
Optical rotation	– 0° to +12 °C
Refractive index at 25 °C	– 1.4663
Camphor content	– Below 2.5%
Cineole content	– 20–25%

Blue camphor oil (green camphor oil):
This is a higher boiling fraction and exhibits a blue colour.

Boiling range	– 220–300 °C
Specific gravity (15 °C)	– Below 1.000
Refractive index (25 ° C)	– About 1.5050
Camphor content	– Below 2.5 °C

Ho-oil 'A' (extra Ho oil):

This is the best quality Ho-oil.

Specific gravity at 25 °C	– 0.860–0.865
Optical rotation	– Above –14°0'
Refractive index (25 °C)	– About 1.4613
Camphor content	– Below 0.7%
Linalool content	– 94% or more

Ho oil 'B' (Ho oil):

Specific gravity (25°)	– 0.860–0.865
Optical rotation	– Above –12°0'
Refractive index (25°)	– About 1.4621
Camphor content	– Below 1.5%
Linalool content	– 85–90%

Yellow camphor oil:

This is a by-product of brown camphor oil after the separation of safrole.

Boiling range	– 215–225 °C
Specific gravity (15°)	– 0.97–0.99
Optical rotation	– +1°0' to +5°0'
Refractive index at 20°	– About 1.5010

Camphor oil from India:

Specific gravity (25 °C)	– 0.9428–0.9436
Refractive index (25 °C)	– 1.4713–1.4727
Optical rotation	– +26° to –29°
Acid number	– 2.48–3.125
Ester number	– 4.0–5.0
Acetyl value	– 21.6

Chemical Composition

Gildemeister and Hoffmann (1925) and subsequently Hiraizumi (1950) made compilations on the chemical composition of camphor oils. The following chemicals have been listed by the latter worker.

Acids:

Formic acid, acetic acid, propionic acid, isobutyric acid, isovaleric acid, n-caproic acid, n-caprylic acid, n-enanthic acid, dl-citronellic acid, lauric acid, myristic acid, piperonytic acid.

Aldehydes:

Acetaldehyde, propionaldehyde, isovaleraldehyde, n-hexaldehyde, furfural, α-methyl furfural, α,β-hexanal, pentenylaldehyde, stearic aldehyde, decyclaldehyde, pentadecylaldehyde, myristic aldehyde, piperonylacrolein.

Ketones and oxides:

Camphor, piperitone, cineole, 1,4-cineole, safrole, linalool monoxide.

Alcohols:

l-Linalool, geraniol, citronellol, d-citronellol, d-borneol, cuminyl alcohol, dihydro-cuminyl alcohol, α-terpineol, dl-α-terpineol, l-α-terpineol, terpinen-1-ol, terpinen-4-ol.

Phenols and phenol ethers:

Cresol, ethyl phenol, ethyl guaicol methylethyl phenol, carvacrol, eugenol, methyl-eugenol, dihydroeugenol, propylpyrocatechol, elemicin, laurol, microl.

Terpenes:

α-Pinene, β-pinene, camphene, l-camphene, fenchene, phellandrene, dipentene, l-limonene, p-cymene, d-sabinene, α-terpinene, γ-terpinene, l-α-thujene.

Sesquiterpenes:

Bisabolene, cadinene, caryophyllene, camphazulone.

Diterpenes:

α-Camphorene, β-camphorene, ho-sho-diterpene.

Hirota (1956) and Hirota and Horoi (1967) have given the comparative figures of the composition of unrectified camphor oils found in Table 9.6.

Wan-Yang *et al.* (1989) conducted a detailed study of the chemical composition of camphor trees growing in China (Tables 9.7, 9.8). They have analysed 363 essential oil samples using GC and GC-MS and identified 34 chemical compounds. Depending upon the predominance of particular chemical compounds they recognised five chemo-types: camphor-type, linalool-type, cineole type, isonerolidol type and borneol type (Table 9.9).

Van Khein *et al.* (1998) studied the chemical segregation of progenies of camphor trees with high camphor and linalool contents. The leaf oil contents were analysed by

Table 9.6 Composition of unrectified camphor oils (%)

Compounds	Japan	Taiwan	China
Monoterpenes	14	7.2	18.9–18.5
Cineole	4.6	4.1	19.0–21.6
Camphor	45.6	39.6	36.8–33.3
Terpenealcohols	9.9	36.8	16.3–9.3
Safrole	18.1	7.8	2.9–0.8
Sesquiterpenes (plus alcohol)	6.2	2.0	6.3–3.0
Others	1.2	2.0	1.7–1.8
Resinous matter	0.4	0.5	1.1–1.7

Source: Hirota and Hiroi, 1967.

Table 9.7 Oil content and major compounds from *Cinnamomum camphora* leaves of different chemotypes

Type of leaf oil	No. of trees analysed	Oil content from leaves		Major compound content in oil	
		Max.	Min.	Max.	Min.
Camphor type	47	3.29	1.16	97.49	54.84
Linalool type	65	2.91	0.96	92.09	58.88
Cineol type	25	3.00	1.51	52.21	32.62
Isonerolidol type	25	0.94	0.16	57.67	16.48
Borneol type	2	1.93	1.53	81.78	67.06

Source: Wan-Yang *et al.*, 1989.

Table 9.8 Analytical results of different organs from cultivated young trees

Types	Organs	Content (%)					
		Oil	Camphor	Linalool	Cineol	α-Terpineol	Safrole
Camphor oil	Leaf	1.70	72.0	1.3	3.3	3.5	0.7
	Root	1.86	33.6	1.2	10.1	4.6	38.3
Linalool type	Leaf	2.05	0.7	80.2	4.8	2.1	0.8
	Root	1.15	5.8	8.9	0.5	0.5	78.8
Cineol type	Leaf	2.22	1.4	0.5	48.1	15.9	0.3
	Root	1.52	12.2	2.1	10.6	5.5	60.7

Source: Wan-Yang *et al.*, 1989.

Table 9.9 The compounds identified in the essential oil from *Cinnamomum camphora* leaves

Peak no.	Compounds	Content (%)				
		1	2	3	4	5
1	α-Thujene	0.14	0.07	–	0.18	
2	α-Pinene	1.29	0.12	1.65	2.01	0.08
3	Camphene	1.55	–	–	1.51	0.25
4	β-Pinene	0.77	0.05	6.92	0.81	0.06
5	Sabinene	0.25	0.05	0.25	0.33	0.06
6	Myrcene	0.35	0.08	–	0.93	0.10
7	α-Phellandrene	0.81	0.05	–	0.05	0.47
8	Limonene	0.03	0.01	–	1.62	0.17
9	1,8-cineole	1.73	0.21	50.00	1.63	0.92
10	Ocimene	–	0.09	–	0.07	0.03
11	γ-Terpinene	0.19	0.30	–	0.16	0.04
12	p-Cymene	0.02	0.01	1.12	0.16	0.03
13	Terpinolene	0.03	0.02	0.03	0.08	0.05
14	Camphor	83.87	0.47	0.25	2.95	0.31
15	Linalool	0.53	90.57	2.02	0.92	2.31
16	α-Copaene	0.45	0.11	0.23	0.72	0.14
17	Borneyl acetate	0.60	0.18	3.14	0.51	0.03
18	Carryophyllene	0.75	1.99	0.20	0.93	1.42
19	Terpen-4-ol	0.02	0.04	0.11	0.03	1.27

20	Citronellyl acetate	0.50	–	0.15	0.03	0.09
21	β-Bisabolene	0.13	–	1.73	0.63	–
22	Borneol	1.10	0.16	0.31	81.78	1.14
23	α-Terpineol	1.29	0.53	14.35	0.27	3.58
24	Humulene	0.16	0.17	0.13	0.12	0.39
25	Terpinyl acetate	0.11	0.18	0.22	0.06	0.12
26	β-Citronellol	0.11	0.25	0.12	–	–
27	Geranyl acetate	0.04	0.03	0.08	–	0.47
28	Iso-geraniol	0.05	0.19	0.17	–	0.04
29	Safrole	0.16	0.88	0.20	0.20	0.48
30	Methyl eugenol	0.50	0.11	0.12	0.03	0.29
31	Iso-nerolidol	0.51	0.46	0.65	0.11	57.67
32	Methyl iso-eugenol	0.03	0.05	0.33	0.38	0.98
33	β-Eudesmol	0.14	0.11	0.31	0.14	0.23
34	Guaiol	0.03	0.02	1.46	0.17	0.15

Source: Wan-Yang *et al.*, 1989.

Notes
1 – Camphor type; 2 – Linalool type; 3 – Cineole type; 4 – Borneol type; 5 – Isonerolidol type.

capillary GC and GC-MS. Four chemotypes could be distinguished among the 115 progenies of a single camphor-rich mother tree (Table 9.10). The first group was rich in camphor (75.8–80.6%), the second group contained 1,8-cineole and α-terpineol as major compounds (45–54% and 15–20%, respectively); the third group was rich in E-nerolidol (43–22%) and 9-oxo-nerolidol (22–25%) (Table 9.10). The fourth group contained a high proportion of safrole (36–58%), nerolidol (15–19%) and oxonerolidol (6–12%). Progenies of a linalool rich mother tree was found to have six chemotypes: (1) linalool type (90–93%); (2) camphor type (75–84%); (3) nerolidol (36–59%) and oxonerolidol (19–29%) type; (4) camphor (36–66%) + linalool (14–50%) type; (5) 1,8-cineole type (53–54%); and (6) linalool (37.5%) + 1,8-cineole (36.4%) type (Table 9.11).

Table 9.10 Chemical composition (%) of a camphor-rich mother tree and its four chemotypes

Compound	Mother tree	1	2	3	4
α-Thujene	0.1	–	–	–	–
α-Pinene	0.4	–	–	–	–
β-Pinene	0.2	–	–	–	–
α-Phellandrene	0.9	0.38–0.71	6.44–10.1	0.18–0.39	0.10–0.42
1,8-Cineole	4.9	2.02–3.11	45.0–53.6	0.24–0.54	0.14–0.21
Camphor	81.9	75.8–80.6	0.30–2.53	0.30–1.34	0.09–4.63
Terpinen-4-ol	0.9	0.87–1.22	1.07–1.61	0.53–1.28	0.35–1.30
α-Terpineol	1.3	2.02–2.65	15.9–19.9	1.82–3.34	1.06–9.64
Safrole	0.2	1.09–3.68	1.15–9.71	3.90–11.0	36.7–57.7
E-nerolidol	–	–	–	33.2–41.6	15.2–18.9
9-Oxonerolidol	–	–	–	22.2–24.7	6.65–11.3

Source: Van Khein *et al.*, 1998.

Notes
1 – Camphor type; 2 – 1,8-Cineole-α-terpineol type; 3 – Nerolidol-oxonerolidol type; 4 – Safrole-nerolidol type
% ranges are given.

228 K. Nirmal Babu, P.N. Ravindran and M. Shylaja

Table 9.11 Chemical composition (% range) of various chemotypes segregated from a linalool-rich camphor tree

Compound	1	2	3	4	5	6
α-Thujene	t	0.03–0.17	t	0.00–0.26	0.5–0.7	0.2
α-Pinene	t	1.81–2.59	t	0.17–1.49	3.0–4.6	1.5
Camphene	t	1.54–1.98	t	0.07–1.10	0.00–0.2	t
β-Pinene	0.0–0.17	1.07–1.82	t	0.02–0.72	16.0–24.1	11.0
Myrcene	t	1.03–1.60	t	0.00–2.01	0.00–2.0	1.0
α-Phellandrene	t	0.00–0.25	t	0.00–0.17	0.00–1.6	t
1,8-Cineole	0.10–0.64	2.73–4.13	0.24–0.86	1.96–4.37	53.9–54.2	36.4
Limonene	–	0.00–2.91	–	–	–	–
Linalool	90.7–93.3	0.10–1.36	0.72–4.56	15.9–50.4	0.00–3.5	37.5
Camphor	0.78–1.75	75.6–83.5	1.50–3.31	36.5–66.1	t – 3.6	0.6
Terpinen-4-ol	0.17–0.33	0.34–0.72	0.04–0.24	0.09–0.51	0.00–0.9	–
α-Terpineol	0.29–0.57	0.35–1.36	0.18–0.94	0.07–0.70	9.00–10.6	6.6
Safrole	0.0–0.10	0.00–0.23	–	0.00–0.61	0.00 – t	t
α-Copane	0.05–0.21	0.01–0.53	10.00–1.36	0.00–1.12	0.00–0.4	–
β-Caryophyllene	0.48–0.68	0.59–1.21	1.84–5.19	0.33–1.04	0.3–0.4	0.2
α-Humulene	0.57–0.82	0.65–1.15	2.11–7.52	0.52–1.71	0.5	0.3
(E)-nerolidol	0.43–0.81	t	36.2–58.9	0.06–0.34	t – 1.0	0.5
9-Oxonerolidol	0.30–0.66	t	19.6–28.9	t	0.00–0.2	0.3
α-Terpinene	–	–	–	–	0.00–0.1	–
p-Cymene	–	–	–	–	0.00–0.3	t
z-(β)-Ocimene	–	–	–	–	0.00–0.2	–
(E)-β-ocimene	–	–	–	–	0.00–0.3	–
γ-Terpinene	–	–	–	–	0.00–1.8	1.3
Terpinolene	–	–	–	–	0.00–1.8	–

Source: Van Khien *et al.*, 1998.

Notes
1 – Linalool type; 2 – Camphor type; 3 – Nerolidol-oxonerolidol type; 4 – Camphor-linalool type; 5 – β-pinene-1,8-cineol type; 6 – 1,8-cineol-linalooltype (single tree); t – trace.

The chemical composition of the camphor tree varies depending upon the chemotype (or chemovar). Zhu *et al.* (1993) have listed the chemical composition for cineole type, borneol type and isonerolidol type of camphor tree from China found in Table 9.12.

Pelissier *et al.* (1995) analysed the leaf, stem and bark of camphor trees growing in the Ivory Coast using GC-MS and reported 52 compounds. The main constituents were camphor (37.8–84.1%) and 1, 8-cineole (1.0–12.0%).

Pino and Fuentes (1998) analysed the composition of leaf oil of camphor trees growing in Cuba by GC-MS. They identified 39 compounds, the major one being camphor (71.2%). Chalchat and Valade (2000) analysed the chemical composition of camphor trees growing in Madagascar. The average chemical composition of five sample trees anlysed by them is given in Table 9.13. The Madagascar camphor population seems to be predominantly 1,8-cincole type, but is also unique in having sabinene (11.4–14.0%).

Camphor Cultivation in Other Countries

Camphor trees are also grown in many other tropical sub-tropical regions of the world. In India camphor trees have been successfully cultivated in Dehra Dun, Saharanpur,

Camphor Tree 229

Table 9.12 Chemical composition (%) of camphor oil of three chemovars from China

C. camphora cineole type	*C. camphora* borneol type	*C. camphora* iso-nerolidol type
50.0% 1,8-Cineole	81.8% Borneol	57.7% Iso-nerolidol
14.4% α-Terpineol	3.0% Camphor	3.6% α-Terpineol
6.9% β-Pinene	2.0% α-Pinene	2.3% Linalool
3.1% Bornyl acetate	1.6% 1,8-Cineole	1.3% Terpinen-4-ol
2.0% Linalool	1.6% Limonene	1.1% Borneol
1.7% β-Bisabolene	1.5% Camphene	1.1% β-Caryophyllene
1.7% α-Pinene	0.9% Myrcene	1.0% Methyl isoeugenol
1.1% para-Cymene	0.9% β-Caryophyllene	0.9% 1,8-Cineole
0.7% Iso-nerolidol (unknown isomer)	0.9% Linalool	0.5% Safrole
0.3% Methyl isoeugenol	0.8% β-Pinene	0.5% α-Phellandrene
0.3% Borneol	0.7% α-Copaene	0.5% Geranyl acetate
0.3% β-Eudesmol	0.6% β-Bisabolene	0.4% α-Humulene
0.3% Sabinene	0.5% Bornyl acetate	0.3% Camphor
0.3% Camphor	0.4% Methyl isoeugenol	0.3% Methyl eugenol
0.2% α-Copaene	0.3% Sabinene	0.3% Camphene
0.2% Terpinyl acetate (unknown isomer)	0.3% α-Terpineol	0.2% β-Eudesmol
0.2% β-Caryophyllene	0.2% Safrole	0.2% Limonene
0.2% Safrole	0.2% α-Thujene	0.2% Guaiol
0.2% Isogeraniol	0.2% Guaiol	0.1% α-Copaene
0.2% Citronellyl acetate	0.2% para-Cymene	0.1% Terpinyl acetate (unknown isomer)
0.1% α-Humulene	0.2% γ-Terpinene	0.1% α-Pinene
0.1% Citronellol	0.1% β-Eudesmol	0.1% Sabinene
0.1% Methyl eugenol	0.1% α-Humulene	0.1% β-Pinene
0.1% Terpinen-4-ol	0.1% iso-Nerolidol (unknown isomer)	0.1% Myrcene
0.1% Geranyl acetate	0.1% α-Phellandrene	0.1% Terpinolene
84.4% Total	0.1% Terpinolene	0.1% Citronellyl acetate
	0.1% β-Ocimene	73.1% Total
	0.1% Terpinyl acetate (unknown isomer)	
	99.6% Total	

Source: Zhu *et al.*, 1993.

north Bengal, Western Ghats (Nilgiris) and Mysore. Camphor trees are grown in Sri Lanka up to an altitude of 1600 m. In India only leaf oil distillation was previously practiced, but in recent years the distillation has become uneconomical and discontinued. In the past large quantities of natural camphor were imported into India from Japan, but the cheap, synthetic camphor has almost totally replaced natural camphor. Now natural camphor is used only for medicinal purposes.

The camphor industry in India was based on leaf distillation. Simple distillaries existed in many growing areas in Dehra Dun, Sahranpur, Mysore, Wynad and Nilgiris. The distillation apparatus used in these areas was a simple one consisting of a boiler, a still, a condenser and a press for extracting the oil. Leaves were packed tightly in a copper still provided with an aluminium condenser. Steam was led in from a boiler and the distillation was continued throughout the day. The water in the condensing

Table 9.13 Average chemical composition of five samples of leaf oils of *C. camphora* from Madagascar

Compound	%	Compound	%
α-Pinene	3.7–4.6	p-Cymene	0.9–1.8
α-Thujene	0.6–0.9	Terpinolene	t–0.3
Camphene	0.1–0.3	trans-p-Menth-2-en-1-ol	0.3–0.6
β-Pinene	2.7–3.3	Linalool	t
Sabinene	11.4–14.0	cis-p-Menth-2-en-1-ol	0.1–0.3
α-Phellandrene	t	β-Caryophyllene	0.1–1.3
Myrcene	0.6–1.2	Terpinen-4-ol	1.6–2.4
α-Terpinene	t–0.5	α-Humulene	0.4–1.7
Limonene	0.6–0.9	δ-Terpineol	t–0.7
1,8-Cineole	56.7–63.7	α-Terpineol	6.9–8.3
γ-Terpinene	t–0.8	Bicyclogermacrene	t
(E)-β-ocimene	t–0.4		

Source: Chalchat and Valade, 2000.

Note

t = trace (≤0.05%).

vessel was decanted and the residual camphor pressed first in a screw press and then in a hydraulic press. The yield of camphor was about 1% of the weight of the leaves. The camphor oil was never used.

Camphor production was also practiced in Sri Lanka, Algeria and also in the USA; in all these regions the produce was based on leaf distillation. The productivity was about 200–250 kg per ha in USA, 400–450 kg in Algeria, 90–100 kg in India and 80–100 kg in Sri Lanka (Anon, 1950).

The oil distilled from leaves of plants grown in Dehra Dun is pale yellow in colour, the main constituents, apart from camphor (22.2%), are d-α-pinene, dipentene, cineol, terpineol and caryophyllene. The oil distilled at Dehra Dun from twigs is deep brown in colour and contains camphor (20%), d-α-pinene (26%), dipentene (11%), terpineol (6%) and caryophyllene and cadinene (10%) (Baruah, 1975).

Bhandari *et al.* (1992) and Bhandari (1995) have investigated the yield and composition of oil from leaves harvested at different times of the year at Dehra Dun conditions. Interesting variations were noted in the chemical composition of the oil extracted at different times of the year (Table 9.14).

In 20-year old trees the oil content remained relatively constant while in two- and four-year old bushes there was fluctuation in oil content. The highest oil content was in August–September. Compositional variations indicated that camphor oil of different composition could be obtained by timing the harvesting of the leaves.

Singh and Negi (1997) investigated the biomass accumulation and distribution under the Dehra Dun conditions. In the Dehra Dun region camphor trees have been successfully established as a secondary succession species in the pine forests. The type of density is typical of that of a climax community formation as defined by Kimmins (1987). However, due to the homogeneous conditions created by this tree species, succession has been temporarily arrested in the Dehra Dun valley and the community has been designated as ser-climax (Singh and Negi, 1997). They have calculated regression equations for biomass estimation based on various relationships, the most reliable variables are diameter at breast height (DBH) and D^2H (D^2 = square of the diameter

Table 9.14 Yield and composition of oil in relation to harvesting time from 2-, 4- and 20-year old plants

Date of collection	Percentage of yield			Composition of essential oil		
	(2)	(4)	(20)	(2)	(4)	(20)
15th Dec.	2.77	3.25	2.94	β-pinene and cineole maximum, low camphor, good borneol	Low percentage of α-pinene, β-pinene, cineole, limonene, good % of camphor, borneol-acetate and borneol	Maximum cineole, low % of α-pinine, β-pinene camphor and borneol
1st Feb.	1.49	2.94	1.85	Good % of camphor other components in low quantity	Maximum % of isoborneol acetate good % of camphor	Maximum borneol, small % of other components
15th March	2.72	2.87	2.79	Limonene, iso-borneol acetate and camphor	β-Pinene cineole iso-borneol acetate and less camphor	Iso-Borneol maximum, good% β-pinene
1st May	2.98	3.57	2.63	Maximum % of camphor	Very good yield of camphor	Limonene maximum
15th June	2.67	3.57	2.53	Iso-borneol maximum	Camphor yield good	β-pinene, iso-borneol acetate
1st Aug.	4.34	4.94	2.71	Pinene maximum good amount of camphor	Maximum % of camphor	Good % of safrole
15th Sep.	3.16	4.97	2.71	Pinene, bornyl acetate good % of camphor	Limonene, bornyl-acetate maximum, good % of – pinene, β-pinene and camphor	Safrole – pinene maximum
1st Nov.	3.18	4.28	2.54	Good yield of – pinine, limonene, camphor, bornyl-acetate and borneol	Good yield of pinine, β-pinene camphor, borneol-acetate, iso-borneol acetate and borneol	Limonene – pinine, β-pinene, borneol, and safrole good %

Source: Bhandari, 1995.

at breast height in cm, and H = height of the tree). Singh and Negi (1997) thus calculated the standing biomass based on stem DBH.

The bole contributes considerably (62.9%) to the biomass, followed by twig + branch (27.5%), bark (6.1%), leaf (3.4%) and fruit (0.06%), respectively. Satoo (1986) estimated that in a 46-year old camphor tree stand the total biomass was around 196.1 t/ha from a plantation that contained 1250 trees/ha. Here bole + bark contributed 80%, branch + twig 17.8% and leaves 2% (Table 9.15).

Chopra *et al.* (1958) reported camphor and camphor oil content from trees growing in three agroecologically different locations in India (Table 9.16, 9.17). They also reported 55.66% camphor content in leaf oil of trees growing in Calcutta and 80.13% of camphor content in oil from trees growing in Dehra Dun. The physico-chemical properties of oil from different plant parts are given in Table 9.18 (Gulati, 1982).

232   K. Nirmal Babu, P.N. Ravindran and M. Shylaja

Table 9.15 Dry matter distribution in camphor trees (kg/ha)

Diameter class (cm)	Leaf	Actual above ground biomass			Fruit	Total biomass
		Twig + branch	Bark	Bole		
07–12	81.4	425.7	327.8	2302.8	–	3137.7
12–17	333.7	3700.1	1190.2	9584.8	–	14808.8
17–22	897.1	6766.2	1785.4	18699.7	14.9	28163.3
22–27	906.3	7824.9	1450.5	15913.9	18.0	26114.0
27–32	904.3	6552.3	1212.1	13805.8	16.4	22490.9
32–37	442.1	3422.9	440.8	5312.5	9.1	9627.4
Total	3565.2 (3.4)	28692.1 (27.5)	6406.9 (6.1)	65619.5 (62.9)	58.3 (0.06)	104342.0

Source: Singh and Negi, 1997.

Table 9.16 Camphor and camphor oil content of trees growing in different areas in India

Oil from	Dehra Dun (22-year old tree)		Calcutta		Bangalore (40-years old trees)	
	Camphor	Camphor Oil	Camphor	Camphor oil	Camphor	Camphor oil
Leaves	1.32	0.69	0.54	1.50	0.90	1.2
Twigs	Nil	0.12	–	–	0.33	0.69
Small branches	Nil	0.13	Nil	0.07	0.06	1.03
Large branches	0.50	0.25	Nil	0.17	0.13–0.30	0.93–2.0
Stumps	0.69	0.29	Nil	0.95	1.06–1.44	4.36–8.04
Roots	–	–	–	2.03	1.91	6.04

Source: Chopra *et al.*, 1958.

Table 9.17 Camphor and camphor oil contents in leaves from trees growing in different locations

	Total oil	Camphor	Camphor oil
Nilgiris	1.0	0.1–0.7	0.9–0.3
Madras	2.62	1.99	0.63
Burma	1.51	1.03	0.48
Cochin	2.33	2.01	0.32
Dehra Dun	4.04	0.38	3.66

Source: Chopra *et al.*, 1958.

Camphor leaf oil is no longer produced in India and in the other countries mentioned above as its production became uneconomical, leaving China, Taiwan and Japan as the only commercial producers of camphor oil.

Camphor seeds contain about 42% of a yellowish white crystalline aromatic fat (mp.21–23°) with a high laurin content, which has the following physical constants: sp-gravity (25°) 0.925; optical rotation (25°) 1.4442; acid value 0.6; saponification value 272.3; iodine value 4.0; unsaponifiable matter 0.7%. The oil is similar to coconut

Camphor Tree 233

Table 9.18 Physico-chemical characteristics of oil samples from various parts of the camphor tree

Characteristics	Oil samples from Calcutta			Chunabhati leaves
	Leaves	Branch wood	Trunk wood	
Colour	(Colourless)	(Light brown)	(Light brown)	(Pale yellow green)
Specific gravity (29.2 °C)	0.9280	0.9428	0.9436	0.9300 (30.1 °C)
Optical rotation	+30°60'	+26°21'	+28°48'	+32°19'
Refractive index (29.5 °C)	1.4786	1.4713 (29.2 °C)	1.4727	1.3807 (30 °C)
Acid number	2.238	2.98	3.125	3.217
Ester number	12.148	3.965	4.396	3.476
Acetyl value	29.8	–	216.1	–
Solubility	1:1 (90% alcohol)	1:2 (80% alcohol)	1:2 (80% alcohol)	1:1 (90% alcohol)

Source: Gulati, 1982.

oil in properties and is suitable for soap making. Zhang-Bin *et al.* (2001) reported that camphor fruit contains oil similar to the camphor wood oil, and the extraction of green seeds yielded 2.7–3.1% 'wood oil' on solvent extraction and about 2.8% oil by steam distillation.

Synthetic Camphor

Camphor is a bicyclic ketone. Due to its varied industrial uses a synthetic process for manufacture of camphor was developed. Camphor is now synthesised by the process involving conversion: Pinene $\rightarrow$ bornyl/isobornyl chloride $\rightarrow$ borneol/isoborneol $\rightarrow$ camphor. This is a very cheap process as the starting material pinene (turpentine oil) is abundantly available from conifer wood. The USA and UK synthesise large quantities of camphor for various industrial uses. With the introduction of synthetic camphor the importance of natural camphor declined and the camphor oil gained more importance as a source material for many aromatic chemicals. Now the camphor tree is valued not for its camphor but for its oil.

Properties and Uses

Camphor has been indicated for the treatment of a variety of illnesses and afflictions in Chinese, Japanese and Indian systems of medicines. The varied medicinal uses to which oil is put to use are summarised below.

Camphor possesses stimulant, carminative, and aphrodisiac properties, and is widely used in traditional medicine, both externally and internally. Its primary action is that of a diffusible stimulant and diaphoretic; its secondary action is that of a sedative, anodyne, and antispasmodic. In large doses it is an acro-narcotic poison. Camphor has been extensively used in the advanced stages of fevers and inflammation, insanity, asthma, angina pectoris, whooping-cough, palpitations connected with hypertrophy of the heart; affections of the genito-urinary system, comprising dysmenorrhoea, nymphomania, spermatorrhoea, cancer, and irritable states of the uterus; chordee, incontinence of urine, hysteria, rheumatism, gangrene and gout. It has also been employed as an antidote to strychnia, but with doubtful results. It is regarded as a medicine in impotence (Dymock *et al.*, 1890). The Hindus consider camphor to be aphrodisiacal, but the Muhammadans hold a contrary opinion; both regard it as a valuable application to the eyelids in inflammatory conditions of the eye (Watt, 1885; Morton, 1977).

The following extract on the use of camphor is from Waring's most useful little book, *Bazar Medicines*:

In asthma, camphor in 4-grain doses, with an equal quantity of asafoetida, in the form of a pill, repeated every second or third hour during a paroxysm, affords in some instances great relief. Turpentine stupes to the chest should be used at the same time. Many cases of difficulty of breathing are relieved by the same means. These pills also sometimes relieve violent palpitation of the heart. In the coughs of childhood, camphor liniment, previously warmed, well rubbed in over the chest at nights, often exercises a beneficial effect. For young children, the strength of the liniment should be reduced one-half or more by the addition of some bland oil.

In rheumatic and nervous headaches, a very useful application is one ounce of camphor dissolved in a pint of vinegar, and then diluted with one or two parts of water. Cloths saturated with it should be kept constantly to the part.

In spermatorrhoea, and in all involuntary seminal discharges, no medicine is more generally useful than camphor in doses of 4 grains with half a grain of opium taken each night at bed time. In gonorrhoea, to relieve that painful symptom, chordee, the same prescription is generally very effectual; but it may be necessary to increase the quantity of opium to one grain, and it is advisable to apply the camphor liniment along the under-surface of the penis as far as the anus. To relieve that distressing irritation of the generative organs which some women suffer from so severely, it will be found that 5 or 6 grains of camphor, taken in the form of a pill twice or three times a day, according to the severity of the symptoms, will sometimes afford great relief. In each of these cases it is important to keep the bowels freely open.

In painful affections of the uterus, camphor in 6- or 8-grain doses often affords much relief. The liniment should at the same time be well rubbed into the loins. In the convulsions attendant on child-birth, the following pills may be tried: Camphor and calomel, of each 5 grains. Beat into a mass with a little honey, and divide into two pills; to be followed an hour subsequently by a full dose of castor oil or other purgative.

In the advanced stages of fever, small-pox and measles, when the patient is low, weak, and exhausted, and when there are at the same time delirium, muttering, and sleeplessness, 3 grains of camphor, with an equal quantity of asafoetida, may be given every third hour; turpentine stupes or mustard poultices being applied at the same time to the feet or over the region of the heart. It should be discontinued if it causes headache or increased heat of the scalp. Its use requires much discrimination and caution.

To prevent bed-sores, it is advisable to make a strong solution of camphor in arrack or brandy, and with this night and morning to bathe, for a few minutes, the parts which, from continued pressure, are likely to become affected

(Watt, 1885).

The Lancet (31 May 1884) gives an account of a simple process of curing coryza by the inhalation of camphor vapour through a paper tube, the whole face and head being covered so as to secure the full action (Watt, 1885).

Camphor is esteemed as an analeptic in various cardiac depressions and has been used in the past in the treatment of myocarditis. It has a calming influence in hysteria and nervousness and is used in the treatment of diarrhoea (dosage 2–5 grains) (Anon, 1950).

Camphor is extensively employed in external applications as a counter-irritant in the treatment of muscular strains, rheumatic conditions and inflammations. In conjunction with menthol or phenol it relieves itching of the skin, however, its application with phenol may cause ulceration in the case of sensitive people.

Toxic doses taken internally result in convulsion usually accompanied with vertigo and mental confusion, which may lead to delerium and sometimes to coma and death due to respiratory failure (Anon, 1950). It is reported that camphor fumes during the distilling process cause a continuous flow of tears from the eyes of workers. Camphor poisoning has occurred in children and, with large doses, in adults. The effects include a feeling of warmth, nausea and vomiting, headache, confusion, excitement or occasionally depression, restlessness, delirium, hallucinations, unconsciousness, convulsions and death by respiratory failure (Watt and Breyer Brandwijk, 1962).

The Pharmacographia of India (Anon, 1966) has listed the following camphor preparations:

Ammoniated camphor liniment:
Camphor   – 125 g
Eucalyptus oil – 5 ml
Ammonia solution (strong) – 250 ml
Alcohol (90%) to make up – 1000 ml

Dissolve the camphor and Eucalyptus oil in 600 ml of alcohol (90%) and to this add the ammonia solution and shake. Finally make up with alcohol to 1 l. Store in a glass stoppered bottle.

Camphor liniment:
Camphor liniment contains 20.0% w/w of camphor.
Camphor 20 g. Arachis oil 800 g

Dissolve the camphor in oil in a closed vessel.

Camphor water
Camphor   – 1 g
Alcohol (90%) – 2 ml
Pure water   – 1000 ml

Dissolve the camphor in alcohol (90%) and then add the solution to water and shake well until camphor is dissolved.

In Cuba, camphor is used as an antiseptic, antispasmodic, anaphrodisiac, antiasthmatic and anthelmintic. It is given in the case of nervous and eruptive fevers. The leaves are crushed and steeped in alcohol, which is applied as a rub to relieve rheumatic pains (Roig and Mesa, 1945). In the south western United States, the American Indians melted camphor and pork fat together or mixed camphor with whisky for the same purpose (Curtin, 1947). Camphor in olive oil is popular in Mexico and South America as an application on bruises, concusions and neuralgia and camphor is much employed in treating rhinitis, asthma, pulmonary congestion, chronic bronchitis and ephysema (Manfred, 1947). In the past it was a common and nearly universal practice to wear around the neck a little bag containing a lump of camphor in the belief that the emanation would cure colds and related ills (Wren, 1970).

The industrial use of camphor is mainly for the manufacturing of explosives, disinfectants and insecticides. Camphor oil is used in perfumes, soap, detergents, deodorants, as a solvent in paints, varnishes and inks. White camphor oil, free of safrole, contributes to the formulation of vanilla and peppermint flavours, and enters into the flavouring of certain soft drinks, baked products and condiments (Furia and Bellanca, 1971).

Camphor is a basic ingredient in the manufacture of celluloid and films and of medicinal and pharmaceutical preparations such as 'vita camphor' (trans-π-oxo-camphor), camphenal (oxo-camphor) and bromo-camphor. Camphor is employed in mineral flotation, as a deodoriser and antiseptic and in agrochemicals. In India large quantities of camphor are used for burning during religious ceremonies and in Hindu worship.

All over the world the camphor trees were originally introduced as an ornamental shade tree and also for wind breaks and hedges. A variant known as the Majestic Beauty, which has an erect growing habit, large leaves and a rich green colour, is a valuable avenue tree. In China camphor wood is greatly prized as it does not warp or split, it is immune to insect attacks and repels moths. It is valued as a cabinet wood especially for the production of storage chests for clothes. This wood was once commonly used for panelling, but is today very expensive.

Indeed the versatile camphor tree is a rare gift of nature. Its varied uses, from ornamental and horticultural to medicinal and industrial, continue to be, exploited by mankind.

References

Anonymous (1950) *The Wealth of India, Raw materials* Vol. III. CSIR, New Delhi, pp. 173–183.

Anonymous (1966) *Pharmacopoeia of India*. Pub. Dept. Govt. of India (Ministry of Health).

Baruah, A.K.S. (1975) Volatile oil of *Cinnamomum camphora* grown in Jorhat, Assam. *Indian J. Pharm.*, **37**, 39–41.

Bhandari, J., Shiva, M.P., Mehra, S.N., Paliwal, G.S. and Jain, P.P. (1992). Prospects of camphor and camphor oil production from the leaves of pollarded bushes of *Cinnamomum camphora*: In S.P. Ray Chandhuri (ed.) *Recent Advances in Medicinal, Aromatic and Spice crops*, Vol. 2. Today and Tomorrow Printers & Pub., New Delhi, pp. 441–447.

Bhandari, J. (1995) Comparative chromatographic studies of *Cinnamomum camphora* (Linn.) Nees and Eberm. young pollarded bushes and trees of essential oil. *J. Eco. Tax. Bot.*, **19**, 287–292.

Burkill, I.H. (1935) *Dictionary of the Economic Products of the Malay Peninsula* Vol. 2. Crown agents for the colonies, London.

Chalchat, J.-C. and Valade, I. (2000) Chemical composition of leaf oils of *Cinnamomum* from Madagascar : *C. zeylanicum* Blume, *C. camphora* L., *C. fragrans* Baillon and *C. anguistifolium*. *J. Essential Oil Res.*, **12**, 537–540.

Chopra, R.N., Chopra, I.C., Handa, K.L. and Kapur, L.D. (1958). *Indigenous Drugs of India*. U. N. Dhur & Sons, Calcutta, pp.120–129.

Curtin, L.S.M. (1947) *Healing Herbs of the Upper Rio Grande*. Laboratory of Anthropology, Santa Fe, New Mexico, 28 pp.

Dung, N.X., Khien, P.V., Chien, H.T. and Leclercq, P.A. (1993) The essential oil of *Cinnamomum camphora* (L.) Sieb. var. *linaloolifera* from Vietnam. *J. Essential Oil Res.*, **5**, 451–453.

Dymock, W., Wardan, C.J.H. and Hooper, D. (1890) *Pharmacographia India* Part III, Thacker, Spink & Co., Calcutta (Reprint).

Fujita, Y. (1967) Classification and phylogeny of the genus *Cinnamomum* in relation to the constituent essential oils. *Bot. Mag. Tokyo*, **80**, 261–271.

Fujita, Y., Fujita, S. and Nishida, S. (1973) On the components of young and old shoot oils of *Cinnamomum camphora* Sieb. var. *linaloolifera*. *Nippon Nogei kagaku kaishi*, 47, 403–405.

Furia, T.E. and Bellanca, N. (1971) *Fenaroli's Handbook of Flavour ingredients*. The Chemical Rubber Co., Cleveland, Ohio.

Gildmeister and Hoffmann (1925) Quoted from Hiraizumi (1950).

Guenther, E. (1950) *The Essential Oils* Vol. 4, D. Var Hatrand Co., New York.

Gulati, B.C. (1982) Essential oils of *Cinnamomum* species. In C.K. Atal and B.M. Kapur (ed.), *Cultivation and Utilization of Aromatic Plants*. RRL, Jammu-Tabi, pp. 607–619.

Hiraizumi, T. (1950) Oil of camphor. In E. Guenther. *The Essential Oils*, Vol. 4. Robert E. Krieger Pub. Co., Florida, pp. 256–328.

Hirota, M. (1951) In: *Mem. Ehime Univ. Section-2, Sci.*, 1(2), 83–106 (cited from Kostermans, 1964).

Hirota, M. (1953) In: *Perf. & Essential Oil Rec.*, **44**, 4–10 (cited from Kostermans, 1964).

Hirota, N. (1956) Camphor and camphor oil production in Formosa before w.w.2. *Perf. Ess. Oil Record.*, 47(1), 17–22.

Hirota, N. and Hiroi, M. (1967) Later studies on the camphor tree, on the leaf oil of each practical form and its utilization. *Perf. Essential Oil Record.*, **58**(6), 364–367.

Huang, L.-C., Huang, B.-L. and Murashige, T. (1998) A micropropagation protocol for *Cinnamomum camphora. In vitro Cell. Dev. Biol-Plant.*, **34**, 141–146.

Kimmins, J.P. (1987) *Forest Ecology*. Mac Millan Publ. Co., New York.

Kostermans, A.J.G.H. (1964). *Bibliographia Lauracearum*. Min. of Natural Resources, Djakarta, Indonesia.

Manfred, L. (1947) 7000 *Recetas Botanicas, a base de 1300 Plantas Medicinales Americanas*. Editorial Kier, Buenos Ai res (quoted from Morton, 1977).

Morton, J.F. (1977) *Major Medicinal Plants: Botany Culture and Uses*, Charles C. Thomas, Illinois, USA, pp. 103–107.

Nirmal Babu, K., Ravindran, P.N. and Peter, K.V. (1997) *Protocols for micropropagation of Spices and Aromatic Crops*. Indian Institute of Spices Research, Calicut, Kerala, p. 35.

Nirmal Babu, K., Sajina, A., Minoo, D., John, C.Z., Mini, P.M., Tushar, K.V., Rema, J. and Ravindran, P.N. (2003) Micropropagation of camphor tree (*Cinnamomum camphora*). *Plant Cell Tiss. Org. Cult.*, **74**, 179–183.

Pelissier, Y., Marion, C., Prunac, S. and Bessiere, J.M. (1995) Volatile component of leaves, stems and barks of *Cinnamomum camphora* Nees et Ebermaier. *J. Essential Oil Res.*, 7, 313–315.

Pino, J.A. and Fuentes, V. (1998) Leaf oil of *Cinnamomum camphora* (L) J. Presl. from Cuba. *J. Essential Oil Res.*, **10**, 531–532.

Roig. Y. and Mesa, J.T. (1945) *Plantas, Medicinales, Aromaticas Venenosas de Cuba*. Cultural S.A., Havana (quoted from Morton, 1977).

Satoo, T. (1986) Materials for the studies of growth in stands. VII Primary production and distribution of produced dry matter in a plantation of *Cinnamomum camphora. Bull. Tokyo Univ. For.*, 64, 241–275.

Singh, R. and Negi, J.D.S. (1997) Biomass prediction and distribution of organic matter in natural *Cinnamomum camphora* stand. *Indian Forester*, **123**, 1161–1170.

238 *K. Nirmal Babu, P.N. Ravindran and M. Shylaja*

Tetenyi, P. (1974) *Infraspecific Chemical Taxa of Medicinal Plants*. Chemical Pub. Co, New York.

Van Khien, P., Chien, H.T., Dung, N.X., Leclercq, A.X. and Leclercq, P.A. (1998). Chemical segregation of progeny of camphor trees with high camphor c.q. linalool content. *J. Essent. Oil. Res.*, **10**, 607–612.

Wan-Yang, S., Wei, H., Guang-gu, W., Dexuam, G., Guang-yuan, L. and Yin-gou, L. (1989) Study on the chemical constituents of essential oil and classification of types from *C. camphora*. *Acta Bot. Sinica*, **31**(3), 209–214.

Watt, G. (1885) *Dictionary of the Economic Products of India* (1972 reprint), Periodical Experts, Delhi.

Watt, J.M. and Breyer-Brandwijk, M.G. (1962) *Medicinal and Poisonous plants of Southern and Eastern Africa*. 2nd ed. E&S Livingstone, Edinburgh.

Weiss, E.A. (1997) *Essential Oil Crops*. CAB International, U.K.

Windadri, I.F. and Rahayu, B.S.S. (1999) *Cinnamomum camphora* (L) J.S. Presl. In I.P.A. Oyen and N.X. Dung (ed.) *Plant Resources of South-East Asia, No. 19, Essential Oil Plants*. Backhuys Pub., Leiden, pp. 74–78.

Winters, H.I. (1950). Cited from Morton (1977).

Wren, R.W. (1970) *Potter's New Cyclopedia of Botanical Drugs and Preparations* 7th ed., Health Sci. Press, Rustington, England.

Zhang-Bin, Xu-Liyong, Zhang-B. and Xu-Ly (2001) Extracting camphor wood oil from camphor tree seeds. *J. Zhejiang-forestry coll.*, **18**, 57–59.

Zhu, I.F., Li, Y.H., Li, B.L., Lu, B.Y. and Xia, N.H. (1993). *Aromatic Plants and their Essential Constituents*. South China Inst. Bot., Chinese Acad. Sci., Peace Book Co., Hong Kong.

Zhu, I.F., Ding, D. S. and Lawrence, B.M. (1994). The *Cinnamomum* species in China: resources for the present and future. *Perfum. Flavour*, **17**, 17–22.

10 Pests and Diseases of Cinnamon and Cassia

M. Anandaraj and S. Devasahayam

Introduction

Cinnamon is reported to be infested by over 70 species of insect pests, especially in India and Sri Lanka. However, very little authentic information is available on the insect pests of cassia cinnamons (*C. burmannii*, *C. tamala* and *C. cassia*). Some information on the insect pests of cinnamon is offered by Singh *et al.* (1978), Rajapakse and Kulasekara (1982), Butani (1983), Kumaresan *et al.* (1988) and Premkumar *et al.* (1994). From Indonesia, Wikardi and Wahyono (1991) reviewed the insect pests and natural enemies of cinnamon and cassia. One of the most destructive of plant pathogens, *Phytophthora*, was first recorded on cinnamon (Rands, 1922). However information on diseases affecting this crop is few and scanty. The antimicrobial properties of essential oils present in cinnamon could be one of the reasons for fewer diseases recorded on this plant (Chandra *et al.*, 1982; Khanna *et al.*, 1988; Hili *et al.*, 1997; Baratta *et al.*, 1998). However, many of fungi have been recorded on dried and stored cinnamon (Balagopal *et al.*, 1973). The insect pests and diseases recorded on cinnamon and cassia are discussed in this chapter (see also Chapters 6 and 7 for the pests and diseases of Chinese and Indonesian cassias).

Insect Pests

Major insect pests of cinnamon include foliage feeders such as the cinnamon butterfly (*Chilasa clytia* L.), shoot and leaf webber (*Sorolopha archimedias* Meyr.), leaf miner (*Conopomorpha civica* Meyr.) and chafer beetle (*Popillia complanata* Newman) in India; and jumping plant louse (*Trioza cinnamomi* Boselli), cinnamon butterfly (*C. clytia*), cinnamon blue bottle (*Graphium sarpedon* Felder) and leaf webber (*Orthaga vitalis* Walk.) in Sri Lanka. Minor insect pests belong to diverse groups and are mostly foliage feeders; some are capable of causing severe damage to the crop in various locations. Major insect pests of cassia in India include the cinnamon butterfly (*C. clytia*) and leaf miner (*C. civica*). Information available on the distribution, nature of damage and life cycle of major insect pests of cinnamon and cassia are reviewed here. The minor insect pests of these crops (along with major insect pests) have been tabulated (Tables 10.1 and 10.2).

Major insect pests

Cinnamon butterfly (Chilasa clytia)

The cinnamon butterfly is the most serious pest of cinnamon and Chinese cassia, being widely prevalent in India and Sri Lanka. Larvae of the pest feed voraciously on tender and partly mature leaves, leaving only the midrib and portions of veins (Fig. 10.1). Both young plants in nurseries and plants in plantations are affected by the pest. Young plants are often completely defoliated by the pest. Heavy infestations of the pest

0-415-31755-X/04/$0.00 + $1.50
© 2004 by CRC Press LLC

Table 10.1 List of insects recorded on cinnamon (*Cinnamomum verum*)

Genus/species	Country	Reference
Order: Hemiptera		
Family: Psyllidae		
Paurapsylla depressa C.	India	Ayyar, 1940
Unidentified psyllid	India	Mani, 1973
Family: Triozidae		
Trioza cinnamomi Boselli	Sri Lanka	Rajapakse and Kulasekara, 1982
Family: Membracidae		
Gargara extrema Dist.	India	CPCRI, 1979
Family: Cicadellidae		
Bothrogonia sp.	India	Kumaresan *et al.*, 1988
Family: Aleyrodidae		
Bemicia tabaci Guen.	India	Koya *et al.*, 1983
Family: Aphididae		
Micromyzus nigrum Van der Goot	Sri Lanka	Van der Goot, 1918
Family: Margarodidae		
Icerya longirostris	Seychelles	Dupont, 1931
Family: Coccidae		
Ceroplastes rubens Mask.	Seychelles, India, Sri Lanka	Vesey-Fitzgerald, 1938; Singh *et al.*, 1978; Rajapakse and Kulasekara, 1982
Coccus mangiferae Green	Seychelles	Vesey-Fitzgerald, 1938
Eucalymnatus perforatus Newst.	Seychelles	Vesey-Fitzgerald, 1938
E. tesellatus (Sign.)	Seychelles, Samoa	Anon, 1913; Hopkins, 1927
Neolecanium cinnamomi	Sri Lanka	Rutherford, 1914
N. pseudoleae	Sri Lanka	Rutherford, 1914
Vinsonia stellifera Western	Seychelles	Hill, 1983
Family: Diaspididae		
Aulacaspis tubercularis Newst.	Java (Indonesia)	Williams, 1961
Chrysomphalus sp.	India	New record
C. dictyospermi (Morg.)	Seychelles	Vesey-Fitzgerald, 1938
C. ficus Ash.	Seychelles	Vesey-Fitzgerald, 1938
Pulvinaria pyriformis Ckll.	Seychelles	Vesey-Fitzgerald, 1938
Parasaissetia (*Saissetia*) *nigra* (Nietner)	India	Suresh and Mohanasundaram, 1996
Family: Miridae		
Helopeltis sp.	Java (Indonesia)	Roepke, 1916
Family: Plataspidae		
Coptosoma pygmaeum Mont.	Sri Lanka	Rajapakse and Kulasekara, 1982
Family: Membracidae		
Leptocentrus obliqus Walk.	Sri Lanka	Rajapakse and Kulasekara, 1982
Family: Scarabaeidae		
Apogonia proxima	India	Veenakumari and Mohanraj, 1993
Popillia complannta Newman	India	Singh *et al.*, 1978
Singhala helleri Ohs.	India	Singh *et al.*, 1978
Leucopholis pinguis Burm.	India	Beeson, 1921

Family: Dermestidae		
Evorinea hirtella Walk.	Sri Lanka	Rajapakse and Kulasekara, 1982
Order: Coleoptera		
Family: Chrysomelidae		
Coenobius lateralis Weise	Sri Lanka	Rajapakse and Kulasekara, 1982
Cryptocephalus snillus Suffr.	Sri Lanka	Rajapakse and Kulasekara, 1982
C. virgula Suffr.	Sri Lanka	Rajapakse and Kulasekara, 1982
Podagrica badia Harold	Sri Lanka	Rajapakse and Kulasekara, 1982
Family: Attelabidae		
Apoderus scitulus Walk.	India	Singh *et al.*, 1978
Centrocorynus dohrni Jekwl.	Sri Lanka	Rajapakse and Kulasekara, 1982
Family: Curculionidae		
Alcides morio Heller	India	Premkumar, 1988
Centhorrhynchus corbetti Marshall	Malaysia	Marshall, 1935
Family: Scolytidae		
Thamnurgides cinnamomi	Sri Lanka	Eggers, 1936
Order: Diptera		
Family: Cecidomyiidae	India	Mani, 1973
Unidentified cecidomyiid		
Order: Lepidoptera		
Family: Psychidae	India	Mathew and Mohandas, 1989
Unidentified psychid		
Family: Gracillariidae		
Acrocercops sp.	India	Singh *et al.*, 1978
Conopomorpha civica Meyr.	India	Devasahayam and Koya, 1993
Family: Phyllocnistidae		
Phyllocnistis citrella Straint	Sri Lanka	Rajapakse and Kulasekara, 1982
P. chrysoptohlama Meyr.	India	Meyrick, 1932
Family: Cossidae		
Zeuzera coffeae Nietn.	Sri Lanka	Rutherford, 1913
Family: Limacodidae		
Latoia lepida (Cram.)	India	Devasahayam (new record)
Family: Tortricidae		
Homona coffearia (Nietn.)	Malaysia	Hill, 1983
Lopharcha sp.	India	Devasahayam and Koya, 1993
Olethreutes (*Argyroploce*) *aprobola* Meyr.	Sri Lanka, Seychelles	Hutson, 1921; Dupont, 1921
O. semiculta Meyr.	Sri Lanka	Hutson, 1921
Sorolopha archimedias Meyr.	India, Sri Lanka	Singh *et al.*, 1978; Rajapakse and Kulasekara, 1982
Family: Pyralidae		
Anartula thurivora Meyr.	Sri Lanka	Meyrick, 1932
Orthaga vitalis Walk.	Sri Lanka	Rajapakse and Kulasekara, 1982
Family: Papilionidae		
Chilasa clytia L.	India, Sri Lanka	Singh *et al.*, 1978; Rajapakse and Kulasekara, 1982
Graphium doson L.	India, Sri Lanka	Singh *et al.*, 1978; Rajapakse and Kulasekara, 1982
G. saperodon L.	India, Sri Lanka	Singh *et al.*, 1978; Rajapakse and Kulasekara, 1982

Table 10.1 (Continued)

Genus/species	Country	Reference
Family: Geometridae		
Pingasa ruginaria Guen.	Malaysia	Corbett, 1929
Sauris sp.	India	Singh *et al.*, 1978
Semiothisa sp.	India	Singh *et al.*, 1978
Thalassodes sp.	Sri Lanka	Rajapakse and Kulasekara, 1982
Family: Lasiocampidae		
Bharetta cinnamomea Moore	India	Butani, 1983
Family: Saturniidae		
Attacus atlas L.	Sri Lanka	Hutson, 1921
Cricula trifenestrata Helf.	India, Burma	Ghosh, 1925; CPCRI, 1981
Family: Sphingidae		
Thevetra nessus Drury	Malaysia	Corbett, 1929
Family: Arctiidae		
Argina syringa Cram.	India	CPCRI, 1979
Diacrisia obliqua Walk.	India	CPCRI, 1981
Family: Noctuidae		
Selepa celtis Moore	India	Singh *et al.*, 1978
Family: Lymantriidae		
Dasychira horsfieldi Saunders	India	Butani, 1983
Dasychira mendosa Hab.	India, Sri Lanka	Butani, 1983; Rajapakse and Kulasekara, 1982
Euproctis fraterna Moore	India, Sri Lanka	Singh *et al.*, 1978; Rajapakse and Kulasekara, 1982
E. irrotata Moore	Sri Lanka	Rajapakse and Kulasekara, 1982
Hypsidera talaca Walk.	India	CPCRI, 1979
Leucoma submarginata Walk.	Malaysia	Corbett, 1929
Redoa submarginata Walk.	India	CPCRI, 1979
Order: Hymenoptera		
Family: Formicidae		
Anoplolepis longipes (Jerdon)	Seychelles	Haines and Haines, 1978
Oecophylla smaragdina (F.)	India, Sri Lanka	Singh *et al.*, 1978; Rajapakse and Kulasekara, 1982

Table 10.2 List of insects recorded on cassia (*Cinnamomum cassia*)*

Genus/species	Country	Reference
Order: Orthoptera		
Family: Acrididae		
Unidentified	India	Devasahayam (new record)
Order: Lepidoptera		
Family: Gracillaridae		
Conopomorpha civica Meyr.	India	Devasahayam (new record)
Family: Tortricidae		
Cophoprora sp.	China	Liu *et al.*, 1992
Unidentified	India	Devasahayam (new record)
Family: Papilionidae		
Chilasa clytia L.	India	Devasahayam (new record)

Note
* See also Chapter 6.

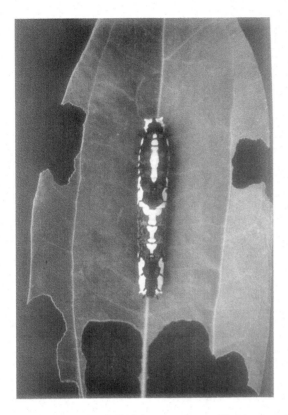

Figure 10.1 Young leaf of cinnamon damaged by cinnamon butterfly (*Chilasa clytia*) along with larva.

adversely affect plant growth. Infestation is more severe during December–June when numerous tender flushes are present on the plants.

The adults are large-sized butterflies with a wingspan of 90 mm and occur in two forms in the field, namely *clytia* and *dissimilis*. The form *clytia* mimics *Euploea* sp. where the wings are blackish brown with a series of marginal white arrowhead-shaped spots. The form *dissimilis* mimics *Danais* sp. and has wings that are black with elongated white spots and a series of marginal arrowhead-shaped spots.

Adults lay eggs singly on tender leaves and shoots. The incubation period lasts for three to four and three to five days in India (at Kasaragod) and Sri Lanka (at Motara), respectively. Newly hatched larvae are pale green with a pale yellow dorsal line and irregular white stripes. Fully grown larvae are dark brown and yellow with four rows of red spots and measure 25 mm in length. The larval period, comprising of five instars, is completed in 11–17 days in India and 12–18 days in Sri Lanka. The pre-pupal stage lasts for one day. The pupa is brownish black, elongated and is attached to the stem of the host plant with silken supports at the posterior end. The pupal period lasts for 11–13 days in India and 11–14 days in Sri Lanka. The total life cycle is completed in 24–36 days in India and 25–37 days in Sri Lanka (Singh *et al.*, 1978; Rajapakse and Kulasekara, 1982). The eggs of the cinnamon butterfly are parasitised by *Telenomus remus* Nixon (Hymenoptera: Scelionidae) in the field and over 50% of eggs collected at Kasaragod, India, during December–June were parasitised (Singh *et al.*, 1978).

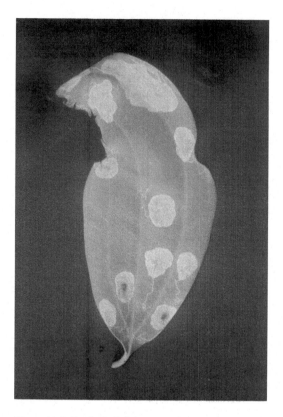

Figure 10.2 Tender leaf of cinnamon damaged by leaf miner (*Conopomorpha civica*), early stage of damage.

Leaf miner (Conopomorpha civica)

The leaf miner infests the tender leaves of cinnamon and cassia plants in the nursery and plantations in India, especially during the monsoon period. Larvae of the pest mine into tender leaves and feed on the tissues between the upper and lower epidermis making linear and tortuous mines, which end in patches (Figs 10.2 and 10.3). Infested portions appear as blisters with the larvae inside them. Infested leaves become crinkled and malformed, and the affected portions dry up resulting in large holes on the leaf lamina.

The adult is a minute silvery grey moth with narrow fringed wings and a wingspan of 5 mm. The larvae are pale creamy-white, becoming pinkish red when fully grown. Pupation occurs outside the larval mines on the leaf (Devasahayam and Koya, 1993). An unidentified species of *Conopomorpha* was also reported to cause serious damage to cinnamon in India (Singh *et al.*, 1978).

Shoot and leaf webber (Sorolopha archimedias)

Shoot and leaf webber is a serious pest of cinnamon in India although in Sri Lanka the pest is of minor concern. The larvae of the pest webs tender shoots and leaves and feeds from within. Infestation is generally serious in the field during the post-monsoon period.

The adult is a small grey moth with a wingspan of 15 mm and black spots on the forewings. Eggs are laid on newly emerged leaves and they hatch in three to four days.

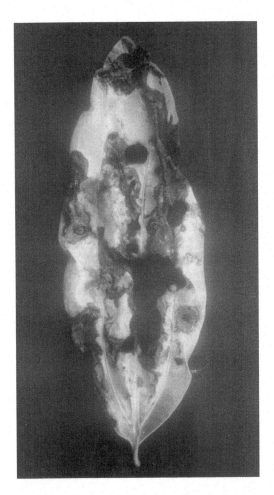

Figure 10.3 Tender leaf of cinnamon damaged by leaf miner (*Conopomorpha civica*), later stage of damage.

The larvae are pale green and the larval period lasts for ten days. Pupation occurs in a silken cocoon within the webbed leaves. The pupal period lasts for six to seven days (Singh *et al.*, 1978). Larvae are parasitised by *Goniozus* sp. (Braconidae) at south Andamans, India (Bhumannavar, 1991).

Chafer beetle (Popillia complanata)

The chafer beetle feeds on tender leaves of cinnamon in India (Fig. 10.4). They are abundant in the field during the monsoon season (July–August). Adults are brown with a metallic green head and thorax and measure 15 mm × 6 mm in size. The adults lay eggs near the root zone and the incubation period lasts for five days. Newly emerged grubs are creamy white and measure 2.5 mm in length. Grubs also feed on cinnamon roots and the grub period lasts for ten days. Pupation takes place in the soil in earthen cocoons and the pupal period lasts for 15 days (Singh *et al.*, 1978).

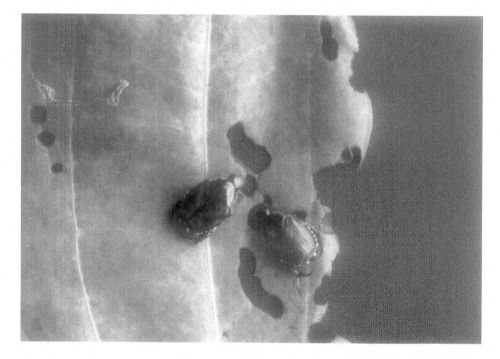

Figure 10.4 Tender leaf of cinnamon damaged by chafer beetle (*Popillia complanata*).

Jumping plant louse (Trioza cinnamomi)

The jumping plant louse is the most serious pest of cinnamon in Sri Lanka, causing major crops losses. The feeding activity of the nymphs of the pest result in the formation of galls on the leaf surface. These galls are mostly epiphyllous and are located on the leaf lamina and shoot buds. Galls dry up after the emergence of adults. Eggs are laid either singly or in groups on the newly emerged tender leaves. There are numerous generations of the pest in a year and each generation lasts for 35–45 days (Rajapakse and Kulasekara, 1982; Rajapakse and Ratnasekara, 1997).

Cinnamon blue bottle (Graphium sarpedon)

The cinnamon blue bottle is a major pest of cinnamon in Sri Lanka, whereas in India it is of minor concern as a pest. The larvae of this pest feed voraciously on tender leaves leaving behind only portions of midribs and veins.

The adults are large-sized butterflies with a wingspan of 80 mm with blackish brown wings and pale greenish-blue elongated spots arranged in a row in the mid fore and hind wings. Eggs are laid on the tender leaves and they hatch in five to six days. Newly hatched larvae are spiny and later as they grow, they become green, with a pale yellow line running longitudinally on the lateral and lower sides of the body. The larval period, comprising of five instars, is completed in 29–31 days. Pupation occurs on the undersides of leaves and lasts for 19–20 days (Rajapakse and Kulasekara, 1982).

Leaf webber (Orthaga vitalis)

The leaf webber is a major pest of cinnamon in Sri Lanka. The larvae of the pest web the tender leaves into clusters, which gradually dry up. A webbed-up cluster of leaves can harbour several larvae inside. Many such webbed clusters of leaves are seen in severely infested plants, which present a sickly appearance. Eggs are laid in small clusters on tender leaves and the incubation period lasts for four to five days. The larvae are initially gregarious and feed by scraping the plant tissues. Fully grown larvae are pale green with dark bands. The larval period, comprising of five to six instars, is completed in 28–30 days. Pupation takes place within a silken cocoon inside the webbed-up cluster of leaves and lasts for 11–14 days (Rajapakse and Kulasekara, 1982).

Minor insect pests

Various other insects such as leaf hoppers, psyllids, whiteflies, aphids, scale insects, mirid bugs, leaf-feeding caterpillars, leaf miners, leaf webbers, root grubs, leaf beetles and weevils have also been recorded on cinnamon in many regions. These species mostly infest foliage and can sometimes cause heavy damage in certain localities. A list of these species and their distribution in various parts of the world is provided in Table 10.1. Wikardi and Wahyono (1991), who reviewed the insect pests of cinnamon and their natural enemies from Indonesia, reported several insect pests and their natural enemies (eleven parasitoids, three predators and three microorganisms) from the Bogor region of Indonesia. The dominant pests are a kanani caterpillar (*Exicula triferestrata*), and a mite (*Eriophyes boiisi*). *C. burmannii* is never affected by these pests. Less important pests include a stem borer (*Coleoptera, Scolitidae*), which attacks seedlings in nurseries and a red caterpillar (Lepidoptera, Cossidae), which attacks young trees. *Mesocomys orientalis* (Hymenoptera, Eupelmidae), and a *Graphium* sp. are the dominant parasitoids. The predator population is high in cinnamon stands; however, their predatory potential is not known. Two unidentified microorganisms, one attacking a species of grasshopper and a rod-shaped bacterium attacking the kanani caterpillar have also been found.

Management

Very little published information is available on the management of the insect pests of cinnamon and cassia. The general recommendations suggested include spraying of quinalphos 0.05% for the management of leaf-feeding caterpillars and beetles. Monocrotophos 0.05% and quinalphos 0.05% is recommended for the management of leaf miner on cinnamon (Devasahayam, 2000).

Diseases

Though many diseases have been recorded on cinnamon (*C. verum*), only a few are serious, causing economic losses. These include stripe canker, leaf spot, dieback, brown root rot and grey blight diseases. Very few diseases have been recorded on cassia cinnamons (*C. cassia, C. burmanii* and *C. tamala*). Diseases of cultivated cinnamon are presented briefly here and have also been tabulated along with diseases that occur on other species of cinnamon (Table 10.3).

Table 10.3 Diseases recorded on cinnamon (*Cinnamomum* spp.)

Diseases	Species affected	Causal organism	Reference
Bark canker	*C. burmanni*	*Phytophthora cinnamomi*	Van Hall, 1921
			Deng and Guo, 1998
	C. camphora	*Botryosphaeria dothidea*	
Stripe canker	*C. verum*	*Phytophthora cinnamomi*	Rands, 1922
	C. camphora		
	C. culilawan		
	C. sintok		
Foot rot	*C. cassia*	*Fusarium oxysporum*	Dao (this volume)
Witches broom	*C. cassia*	Phytoplasma	Dao (this volume)
Decline	*C. osmophloeum*	*P. cinnamomi*	Chang, 1993a,b
	C. camphora		
Red root	*Cinnamomum* sp.	*Fomes pseudoferrus*	Van Overeem, 1925
Brown root rot	*C. camphora*	*Phellinus noxius*	Ann *et al.*, 1999
Canker	*C. camphora*	*Diplodia tubericola*	Martin, 1925
Canker	*C. dulce*	*D. cinnamomi*	Da Camara, 1933
Premature	*C. verum*	*Sphaerella*	Ciferri and
defoliation		*cinnamomicola*	Fragoso, 1927
		Pestalozzia funerea	
		Phyllosticta cinnamomi	Batista and Vital, 1952
		Cytosporella cinnamomi	Batista *et al.*, 1953
Foliar diseases	*C. verum*	*Exosporium cinnamomi*	Muthappa, 1965, 1967
	C. malabatrum	*Rosenscheldiella*	
		cinnamomi	
	C. verum	*Leptosphaeria almeidae*	Da Camera Edes, 1929
		L. cinnamomi	
		Physalospora cinnamomi	Cejp, 1930
		Fusarium allescherianum	
		Sclerotium cinnamomi	Sawada, 1933
		(Corticium) Hypnochnus	Ver Woerd and
		sasaki	Du Plessis, 1933
			Matsumoto and
			Hirane, 1933
	C. osmophloeum	*Colletotrichum*	Chang, 1993c
		gloeosporioides	
		(Glomerella cingulata)	
		Phoma multirostrata	
		Diaporthe sp.	
	C. verum	*Hansfordia cinnamomi*	Deighton, 1960
Wound parasites	*C. cecidodaphne*	*Polyporus gilvus*	Bagchee and Bakshi, 1950
		Ganoderma lucidum	
		G. applanatum	
		Fomes badius	
		F. rimosus	
Black mildews	*C. riparium*	*Amazonia cinnamomi*	Hosagoudar, 1984, 1988;
			Chang, 1993c
	C. verum	*Armatella cinnamomi*	Hansford and
			Thirumalachar, 1948
		A. balakrishnanii	Hosagoudar, 1988
		A. cinnamomicola	Hosagoudar and
			Balakrishnan, 1995
	C. malabatrum	*A. indica*	Hosagoudar, 1988

	C. malabatrum	*Meliola beilschmiediae* var. *cinnamomicola*	Hosagoudar and Goss, 1989 Hosagoudar and Goss, 1990
Exobasidium disease	*C. malabatrum*	*Exobasidium cinnamomi* *Exobasidium* spp.	Bilgrami *et al.*, 1979, 1981;
	C. japonicum		Zhang, 1998
	Cinnamomum sp.	*Clinoconidium farinosum*	Gomez and Kisimova, 1998
Rust	*C. tamala*	*Aecidium cinnamomi*	Goswami and Bhattacharjee, 1973
	C. malabatrum	*Caecoma keralense*	Hosagouder, 1984
	C. verum	*Uraecium nothopegiae*	Ramakrishnan, 1965
Dieback	*C. tamala*	*Colletotrichum gloeosporioides*	Roy *et al.*, 1976
	C. verum	*C. capsici*	Sharma, 1979; Karunakaran and Nair, 1980
	C. cassia	*Botryodiplodia theobromae*	Cen and Deng, 1994; Cen *et al.*, 1994
	C. cassia	*Diplodia* sp.	Anandaraj (new record)
	C. cassia	*Marasmius pulcherima*	Anandaraj (new record)
Thread blight	*C. tamala*	*M. pulcherima*	Anandaraj (new record)
Other diseases	*C. camphora*	*Corticium perenne* *Fomes pectinatus* *Glomerella cingulata* *Pestalotiopsis* sp.	Anonymous, 1950 Biharilal and Tandon, 1967 Purohit and Bilgrami, 1969
	C. cecidodaphne	*Ganoderma applanatum*	Anonymous, 1950
	C. obtusifolium	*Pestalotiopsis versicolor*	Purohit and Bilgrami, 1968
Shot hole	*C. cassia*	*Colletotrichum gloeosporioides*	Anandaraj (new record)
	C. tamala	*Phomopsis tezpatae* *Pestalotia cinnamomi* *Cercospora* sp.	Singh, 1978 Rahman, 1951
	C. verum	*Rosenscheldia orbis* *Pestalotia cinnamomi* *Zygosporium oscheoides*	Srivastava, 1979 Narendra and Rao, 1972 Patil and Thite, 1967 Rangaswamy *et al.*, 1970
	Cinnamomum sp.	*Colletotrichum* sp. *Colletotrichum cinnamomi*	Raju and Leelavathy, 1987 Seshadri *et al.*, 1972
	C. osmeophleum	*Calonectria theae* *Pythium* spp.	Chang, 1992a Chang, 1993b
	C. sulphuratum	*Stenella cinnamomi*	Hosagoudar and Braun, 1995
	C. kanehirai	*Androdia cinnamomea* *Colletotrichum gloeosporioides* *Phellinus noxius* *Pythium splendens* *Cylindrocladium clavatum*	Chang and Chou, 1995 Chang *et al.*, 1997 Chang, 1992b
Smut galls	*C. daphnoides*	*Sphaecelotheca cinnamomi*	Hirata, 1979
Withering *in vitro*	*C. verum*	*Clavibacter michiganensis* sub sp. *sepedonicus*	Trejo and Lopez, 1990

250   M. Anandaraj and S. Devasahayam

Major diseases

Stripe canker

Phytophthora cinnamomi, which causes severe damage to forest trees and avocado, was first recorded on *C. verum* causing stripe canker (Rands, 1922). Symptoms start as a vertical stripe on the stem with amber coloured exudate at the advancing margins, which hardens later. This disease is reported to be severe in ill-drained soils and causes up to 42% damage. This fungus also affects *C. camphora*, *C. culitlawan* and *C. sintok* (Rands, 1922; Ciferri and Fragoso, 1927). The fungus, *P. cinnamomi*, from pineapple was also pathogenic on cinnamon, but with reduced virulence. The fungus produces only chlamydospores characterised by non-papillate sporangia. Papillate sporangia can be obtained by incubating a non-sterile percolate of field soil (Mehrlich, 1934). Control measures include improving soil drainage conditions, phytosanitation and wound-dressing with tar. The disease is reported to spread along the west coast of Sumatera (Van Hall, 1921; 1924). This fungus has also been recorded on cloves in Malaysia (Lee, 1974; Purseglove *et al.*, 1981).

P. *cinnamomi* has also been recorded on *C. camphora* (Chang, 1993a) and *C. osmeophleum* from Taiwan, causing yellowing and wilting followed by death of plants from root rot. *P. cinnamomi* is heterothallic and both A1 and A2 mating types have been recorded in Hualing, whereas only the A1 mating type was recorded in Chia-yi. Isolates from *C. camphora* were of the A1 type (Chang, 1992a,b; 1993a). There is no report of this fungus on cinnamon in India. *P. cinnamomi* has been isolated from clove trees in India, which caused sudden death of the trees (Anandaraj, unpublished).

Foot rot

Foot rot is a major destructive disease of Chinese and Vietnamese cassia (*C. cassia*), caused by the fungus *Fusarium oxysporum*. This fungus attacks and destroys the vascular system by infecting through the roots. The disease is prevalent during the post-rainy season when the weather is humid and warm. Application of *Trichoderma* is effective in checking the disease incidence, especially in nurseries (Dao, this volume).

Witches broom

Witches broom disease affects more commonly cassia seedlings in nurseries and young trees in fields, though old trees are also infected. The Tra Giae hamlet of Vietnam (Tra Mi District, Quang Nam Province) is a hot spot area for this disease. Witches broom disease is caused by a phytoplasma. Preventive measures recommended for avoiding the disease include soaking the seeds in warm water (70 °C) containing an antibiotic before sowing. Not much is known about the mode of transmission, vector and other aspects of the disease. Phytosanitation is advocated to keep the incidence low (Dao, this volume).

Leaf spot and dieback

Leaf spot and dieback disease in cinnamon is caused by *Colletotrichum gloeosporioides*. Symptoms in young seedlings include small brown specks on leaf lamina which later coalesce forming irregular patches. In older leaves, these small specks coalesce and form necrotic blotches (Fig. 10.5). These later become papery with dark brown margins. In other cases the central papery portion is shed forming shot hole symptoms. In some seedlings, the infection spreads to the stem causing dieback (Karunakaran and Nair, 1980). A variant of this disease is reported from the lower Pulney Hills of Tamil Nadu, India, which produces reddish elongated spots arising from the margin and resulting in defoliation. This also causes shot hole symptoms in later stages. The fungus causing this disease has been identified as *Phytophthora capsici* (Prakasam, 1992). *C. gloeosporioides* has also been recorded on *Anacardium* sp., *Dalbergia* sp., *Polyalthia* sp. and *Strychnos nux-vomica* in Karnataka (Bhat *et al.*, 1998). The perfect stage of *Colletotrichum* and *Glomerella cingulata* has been recorded from Karnataka (Kumar, 1983). In Taiwan, anthracnose caused by *C. gloeosporioides* (*Glomerella cingulata*) was recorded on *C. verum* by Fu and Chang (1998). The symptom starts on leaves as brown to black spots which later coalesce and the infected leaves are then shed. The fungus has also been recorded on *C. kanehirae* causing anthracnose and dieback (Chang *et al.*, 1997; Cen *et al.*, 1994).

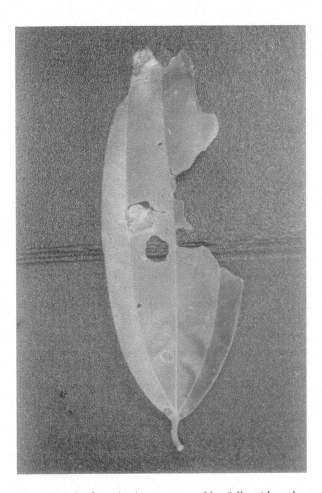

Figure 10.5 Leaf spot in cinnamon caused by *Colletotrichum gloeosporioides*.

Figure 10.6 Die back in *Cinnamomum cassia* caused by *Diplodia* spp.

A new dieback disease was noticed on *C. cassia* at Peruvannamuzhi, Kerala, India. This disease occurs on the lower branches of *C. cassia* and is caused by *Diplodia* sp (Fig. 10.6). Drying of branches leads to defoliation and the infection usually stops at the junction of the main stem (Anandaraj, unpublished).

Brown root rot

This disease has been recorded on *C. camphora* and is caused by the fungus *Phellinus noxius*. The affected roots turn brown and the disease results in wilting and death of aerial portions (Chang, 1992b). This fungus is also reported to affect adjacent shade and ornamental trees such as *Delonix regia* and other plants such as *Annona* sp. and *Prunus* sp. (Da Graca *et al.*, 1980).

Grey blight

The disease is caused by the fungus *Pestalotiopsis palmarum*. Small yellow brown spots which turn grey with a brown border are seen, which later spread to the lamina. In older lesions, dark acervuli are produced. Foliar damage up to 90% has been reported (Karunakaran *et al.*, 1993). A similar leaf spot caused by *P. furierea* was reported from the Dominican Republic (Ciferri, 1926; Ciferri and Fragosa, 1927) and later from Pakistan (Rahman, 1962).

Minor diseases

Many other fungal diseases have been recorded on cultivated cinnamon and their wild relatives (Table 10.3). These fungi occur both on the leaves and bark. Eight species of

Pythium were recorded on *C. osmeophleum* out of which *P. spendens* and *P. vexans* were widespread and caused 15–30% mortality. In addition, some new diseases have also been recorded on *C. cassia* from India, which are described below (Anandaraj, unpublished).

Thread blight

This disease is caused by the fungus *Marasmius pulcherima*. White mycelial strands traverse the ventral surface of leaves and branches of *C. cassia*. The affected leaves and branches dry up. The disease is controlled by removing the affected areas and spraying with Bordeaux mixture (1%) during the monsoon period.

Shot-hole

Shot-hole symptoms appear on leaves of *C. cassia* due to infection by the fungus *Colletotrichum gloeosporioides* (Fig. 10.7). Small brown lesions are formed on the leaf lamina. A yellow halo develops surrounding the lesion and the entire tissue along

Figure 10.7 Shot-hole in *Cinnamomum cassia* caused by *Colletotrichum gloeosporioides*.

254   M. Anandaraj and S. Devasahayam

with the lesions are shed, leaving shot-hole symptoms. On a single leaf many small holes could also be seen. (See Chapters 6 and 7 for pests and diseases of Chinese and Indonesian cassias.)

References

Ann, P.J., Wee, H.L. and Tsai, J.N. (1999) Survey of brown root diseases of fruit and ornamental trees caused by *Phellinus noxius* in Taiwan. *Plant Pathol. Bull.*, 8(2), 51–60.

Anonymous (1913) Insect notes from the Seychelles. *Annual Report on Agricultural and Crown Lands – 1912*, Seychelles, pp. 15–17.

Anonymous (1950) List of common names of Indian plant diseases. *Indian J. Agric. Sci.*, 20, 107–142.

Ayyar, T.V.R. (1940) *Hand Book of Economic Entomology for South India*. Government Press, Madras.

Bagchee, R. and Bakshi, B.K. (1950) Some fungi as wound parasites on Indian trees. *Indian For.*, 76, 244–253.

Balagopal, C., Bagyaraj, D.J., Varma, A.S. and Nair, E.V.G. (1973) Microflora of dried spices. *Agric. Res. J. Kerala*, 11, 88–89.

Baratta, M.T., Dorman, H.J.D., Deans, S.G., Figueiredo, A.C., Borroso, J.G. and Roberto, G. (1998) Antimicrobial and antioxidant properties of some commercial essential oils. *Flavour Fragrance J.*, 13, 235–244.

Batista, A.C. and Vital, A.F. (1952) Monograph of the species of *Phyllosticta* in Pernambuco. *Bot. Soc. Agric. Pernambuco*, 19(1–2), 1–80.

Batista, A.C., Vital, A.F. and Gayao Teresa de J. (1953) Some Leptostromataceae and other fungi recorded in Pernambuco. *An. Congr. Soc. Bot. Brasil*, pp. 134–143.

Beeson, C.F.C. (1921) The food plants of Indian Forest Insects. Part VI. *Indian For.*, 47, 247–252.

Bhat, N.M., Hedge, R.K. and Naik, K.S. (1998) Additional host records for *Colletotrichum gloeosporioides* from Karnataka. *Karnataka J. Agric. Sci.*, 11, 1090–1091.

Bhumannavar, B.S. (1991) New records of *Sorolopha archimedias* on cinnamon and *Mehteria hemidoxa* on betel vine in South Andamans. *J. Andaman Sci. Assoc.*, 7, 82–83.

Biharilal and Tandon (1967) Some new leaf spot diseases caused by *Colletotrichum* II. *Proc. Nat. Acad. Sci. India*, 37, 283–290.

Bilgrami, K.S., Jamaluddin and Rizwi, M.A. (1979) *Fungi of India, Part I*. Today and Tomorrow's Printers and Publishers, New Delhi.

Bilgrami, K.S., Jamaluddin and Rizwi, M.A. (1981) *Fungi of India, Part II*. Today and Tomorrow's Printers and Publishers, New Delhi.

Butani, D.K. (1983) Spices and pest problems – 2. Cinnamon. *Pesticides*, 17(9), 32–33.

Cejp, K. (1930) *Fusarium allescherianum* P. Henn, parasitic on certain glasshouse plants. *Ochrana Rostlin*, 10(3), 75–77.

Cen, B.Z. and Deng, R.L. (1994) Pathogenic identification of cinnamon die back. *J. South China Agric. Univ.*, 15(3), 28–34.

Cen, B.Z., Gan, W.Y. and Deng, R.L. (1994) Study of occurrence and control of cinnamon die back. *J. South China Agric. Univ.*, 15(4), 63–66.

Central Plantation Crops Research Institute (CPCRI) (1979) *Annual Report for 1977*. Central Plantation Crops Research Institute, Kasaragod.

Central Plantation Crops Research Institute (CPCRI) (1981) *Annual Report for 1978*. Central Plantation Crops Research Institute, Kasaragod.

Chandra, H., Asthana, A., Tripathi, R.D. and Dixit, S.N. (1982) Fungitoxicity of a volatile oil from the fruits of *Cinnamomum cecidodaphne* Meissn. *Phytopathologia Mediterranea*, 21, 35–36.

Chang, T.T. (1992a) A new disease of *Cinnamomum osmophloeum* caused by *Calonectria theae*. *Plant Pathol. Bull.*, 1, 153–155.

Chang, T.T. (1992b) Decline of some forest trees associated with brown root rot caused by *Phellinus noxius. Plant Pathol. Bull.*, 1(2), 90–95.

Chang, T.T. (1993a) Decline of two *Cinnamomum* species associated with *Phytophthora cinnamomi* in Taiwan. *Plant Pathol. Bull.*, 2, 1–6.

Chang, T.T. (1993b) Investigation and pathogenicity tests of *Pythium* species from rhizosphere of *Cinnamomum osmophloeum. Plant Pathol. Bull.*, 2, 66–70.

Chang, T.T., Chern, L.L. and Chiu, W.H. (1997) Anthracnose and brown root rot of *Cinnamomum kanehirae. Taiwan J. For. Sci.*, 12, 373–378.

Chang, T.T. and Chou-won (1995) *Androdia cinnamomea* sp. nov. in *Cinnamomum kanehirae*, in Taiwan. *Mycol. Res.*, 99, 756–758.

Ciferri, R. (1926) Report on Phytopathology principle of diseases of cultivated plants observed during 1926. *Segundo Informe Annual Estac. Nac. Agron. Moca, Republica Dominicana 1926*, pp. 36–44.

Ciferri, R. and Fragoso, G.R. (1927) Parasitic and saprophytic fungi of the Dominican Republic (11th series). *Bot. R. Soc. Espanola Hist. Nat.*, 27(6), 267–280.

Corbett, G.H. (1929) Division of Entomology. *Annual Report for 1928. Malayan Agric. J.*, 17, 261–276.

Da Camara Edes (1929) II Fungi associated with Phytopathological laboratory of the Lisbon Agricultural Institute. *Rev. Agron.*, pp. 1–16.

Da Camara Edes (1933) Contributions to the mycoflora of Portugal Century X. *Rev. Agron.*, 20, 5–63.

Da Graca, J.V., Vuuren, S.P., Van (1980) Transmission of avocado sunblotch disease to cinnamon. *Plant Disease*, 64, 475.

Deighton, F.C. (1960) African Fungi I. *Mycol. Papers*, 78, 43.

Deng, X.Q. and Guo, L.H. (1998) Research on camphor tree canker IV. Ways through which canker pathogen enters camphor tree and its prevention and cure. *J. Hunan Agric. Univ.*, 24, 300–304.

Devasahayam, S. (2000) Management of insect pests of spices. In: M.S. Madan and Jose Abraham (eds) *Spices Production Technology*. Indian Institute of Spices Research, Calicut. pp. 104–107.

Devasahayam, S. and Koya, K.M.A. (1993) Additions to the insect fauna associated with tree spices. *Entomon*, 18, 101–102.

Dupont, P.R. (1921) Entomological and mycological notes. *Annual Report on Agricultural and Crown Lands – 1920*, Seychelles, pp. 15–17.

Dupont, P.R. (1931) Entomological and mycological notes. *Annual Report, Department of Agriculture – 1930*, Seychelles, pp. 11–13.

Eggers, H. (1936) Neue Borkenkafer (Scolytidae, Col.) aus Indien. *Ann. Mag. Nat. Hist.*, 17, 626–636.

Fu and Chang, T.T. (1999) Anthracnose of cinnamon. *Taiwan J. For. Sci.*, 14, 513–515.

Ghosh, C.C. (1925) Report of the Entomologist, Mandalay and Sericultural Work for the year ended 30th June 1925, Rangoon, pp. 10–18.

Gomez, P.L.D. and Kisimova, H.L. (1998) Basidiomycetes of Costa Rica. New *Exobasidium* species (Exobasidiaceae) and records of cryptobasidiales. *Revista Biologia Tropical*, 46, 1081–1093.

Goswami, R.N. and Bhattacharjee, S. (1973) Rust, a new disease of tejpat. *Curr. Sci.*, 42, 257.

Green, E.E. (1916) On some animal pests of the Hevea rubber tree. *Trans. 3rd Intl. Cong. Trop. Agric.*, London. pp. 608–636.

Haines, I.H. and Haines, J.B. (1978) Free status of the crazy ant, *Anoplolepis longipes* (Jerdon) (Hymenoptera: Formicidae) in the Seychelles. *Bull. Ent. Res.*, 68, 627–638.

Hansford, C.G. and Thirumalachar, M.J. (1948) Fungi of South India. *Farlowia*, 3, 285–314.

Hili, P., Evans, C.S. and Veness, R.G. (1997) Antimicrobial action of essential oils; the effect of dimethyl sulphoxide on the activity of cinnamon oil. *Letters Appl. Microbiol.*, 24, 269–275.

Hill, D.S. (1983) *Agricultural Insect Pests of the Tropics and their Control* (2nd edn.), Cambridge University Press, Cambridge.

Hirata, S. (1979) A new species of the genus *Sphacelotheca* causing smut galls on *Cinnamomum daphnoides* Sieb & Zucc. *Bull. Faculty of Agric., Miyazaki University*, 26, 123–125.

Hopkins, G.H.E. (1927) Pests of economic plants in Samoa and other island groups. *Bull. Ent. Res.*, 18, 23–32.

Hosagoudar, V.B. (1984) Two interesting fungi on *Cinnamomum malabatrum* from Idukki, Kerala, India. *Indian J. Econ. Tax. Bot.*, 5, 209–211.

Hosagoudar, V.B. (1988) Meliolaceae of South India-V. *Nova Hedwigia*, 47, 535–542.

Hosagoudar, V.B. and Goss, R.D. (1989) Meliolaceous fungi from the state of Kerala, India – I *Mycotaxon*, 36, 221–247.

Hosagoudar, V.B. and Goss, R.D. (1990). Meleolaceous foungi from the state of Kerala, India – II. *Mycotaxon*, 37, 217–272.

Hosagoudar, V.B. and Balakrishnan, N.P. (1995) Diseases on *Cinnamomum* Schaefer (Lauraceae) in India. *J. Econ. Tax. Bot.*, 19, 361–370.

Hosagoudar, V.B. and Braun, U. (1995). Two Indian hypomycetes. *Indian Phytopath.*, 48, 260–262.

Hutson, J.C. (1921) Report of the Entomologist. *Report of the Ceylon Department of Agriculture – 1920*, pp. C 15–17.

Iijima, T., Kakishima, M. and Otani, Y. (1985) A new species of *Exobasidium* on *Cinnamomum japonicum* Sieb. *Trans. Mycol. Soc. Japan*, 26, 161–167.

Karunakaran, P. and Nair, M.C. (1980) Leaf spot and die back disease of *Cinnamomum zeylanicum* caused by *Colletotrichum gloeosporioides*. *Plant Disease*, 64, 220–221.

Karunakaran, P., Nair, M.C. and Das, L. (1993) Grey blight disease of cinnamon (*Cinnamomum verum* Bercht. & Presl.) leaves. *J. Spices Aromatic Plants*, 2(1–2), 66–67.

Khanna, R.K., Jouhari, J.K., Sharma, O.S. and Singh, A. (1988) Essential oil from the fruit rind of *Cinnamomum cecidodaphne*. *Indian Perfumer*, 32, 295–300.

Koya, K.M.A., Gautam, S.S.S. and Banerjee, S.K. (1983) *Bemisia tabaci* Guen. – A new pest of cinnamon. *Indian J. Ent.*, 45, 198.

Kumar, S.N. (1983) *Colletotrichum* blight of cinnamon. *Indian Cocoa Arecanut Spices J.*, 6, 106.

Kumaresan, D., Regupathy, A. and Baskaran, P. (1988) *Pests of Spices*, Rajalakshmi Publications, Nagercoil.

Lee, B.S. (1974) *Phytophthora cinnamomi*: a new pathogen on cloves in peninsular Malaysia. *MARD Research Bulletin*, 2(2), 26–30.

Liu, Z.C., Liu, J.F., Wang, C.X., Yang, W.H., Liang, Y.F., Peng, S.B., Jiang, Z.S., Li, J. and Liu, Z.Y. (1992) Control of the Chinese cinnamon tip moth *Cophoprora* sp. (Lep., Tortricidae) with *Trichogramma chilonis*. *Chinese J. Biol. Cont.*, 8(2), 61–63.

Mani, M.S. (1973) *Plant Galls of India*, Macmillan India.

Marshall, G.A.K. (1935) New injurious Curculionidae (Col.) from Malaya. *Bull. Ent. Res.*, 26, 565–569.

Martin, G.H. (1925) Diseases of forest and shade trees ornamental and miscellaneous plants in the United States in 1924. *Plant Disease Reporter* (Suppl.) 42, 313–380.

Matsumoto, T. and Hirane, S. (1933) Physiology and parasitism of the fungi generally referred to as *Hypochnus sasakii* Shrai. III. Histological studies in the infection by the fungus. *J. Soc. Trop. Agric. Formosa*, 5, 367–373.

Mathew, G. and Mohandas, K. (1989) Insects associated with some forest trees in two types of natural forests in the Western Ghats, Kerala (India). *Entomon*, 14, 325–333.

Mehrlich, F.P. (1934) Physiologic specialization in *Phytophthora* species. *Phytopathology*, 24, 1149–1150, 1139–1140.

Meyrick, E. (1932) Exotic Microlepideptera, iv pt. 10, Marborough, Wilts. pp. 289–320.

Muthappa, B.N. (1965) A new species of *Exosporium* on *Cinnamomum zeylanicum* L. from India. *Sydowia*, 19(1–6), 146–147.

Muthappa, B.N. (1967) A new species of *Rosenscheldiella* Theiss and Syd. from India. *Sydowia*, 21, 163–164.

Narendra, D.V. and Rao, V.H. (1972) Some additions to the fungi of India. *Sydowia*, 26, 282–284.

Patil, M.S. and Thite, A.N. (1967) Fungal flora of Rathnagiri (Kolhapur). *J. Shivaji University (Sci.)*, 17, 149–162.

Prakasam, V. (1992) Cinnamon red leaf spot disease in lower Pulney hills. *Spice India*, 2(8), 4.

Premkumar, T. (1988) *Alcides morio* Heller (Curculionidae: Coleoptera) cinnamon fruit borer. *Entomon* 13, 187.

Premkumar, T., Devasahayam, S. and Koya, K.M.A. (1994) Pests of spice crops. In K.L. Chadha and P. Rethinam (eds), *Advances in Horticulture Vol. 10 – Plantation and Spice Crops Part-2*, Malhotra Publishing House, New Delhi, pp. 787–823.

Purohit, D.K. and Bilgrami, K.S. (1968) Variation in fifteen different isolates of *Pestalotiopsis versicolor* (Speg) Stey. *Proc. Nat. Acad. Sci. India*, 38, 225–229.

Purohit, D.K. and Bilgrami, K.S. (1969) Variation in the conidial morphology and genus *Pestalotiopsis. Indian Phytopath*, 22, 275–279.

Purseglove, J.W., Brown, E.G., Green, C.C. and Robbins, S.R.J. (1981) *Spices Vol. 1* Longman Group Ltd, Harton, U.K. and New York, USA.

Rahman, S.M.A. (1951) Observations on the diseases of Tezpata. *Curr. Sci.*, 20, 135.

Rahman, S.M.A. (1962) Leaf spot and shot hole diseases of tezpata (*Cinnamomum tamala* Fr. Nees). *Biologia*, Lahore, 8(1), 51–54.

Rajapakse, R.H.S. and Kulasekara, V.K. (1982) Some observations on the insect pests of cinnamon in Sri Lanka. *Entomon*, 7, 221–223.

Rajapakse, R.H.S. and Ratnasekara, D. (1997) Studies on the distribution and control of leaf galls in cinnamon caused by *Trioza cinnamomi* Bosellis in Sri Lanka. *Intl. J. Tropical Agric.*, 15, 53–56.

Ramakrishna, T.S. (1965) Notes on some fungi from South India – IX: *Proc. Indian Acad. Sci.*, 62, 32–35.

Raju, A.R. and Leelavathy, K.M. (1987) Cross inoculation studies with *Colletotrichum gloeosporioides* causing leaf spot in cinnamon, nutmeg and cocoa. *J. Plantation Crops*, 15, 137–139.

Rands, R.D. (1922) Stripe canker of cinnamon caused by *Phytophthora cinnamomi* n. sp. *Meded. Inst. Voor Plantenziekten*, 54, pp. 53.

Rangaswamy, G., Seshadri, V.S. and Channamma, L.K.A. (1970) *Fungi of South India*, University of Agricultural Sciences, Bangalore.

Roepke, W. (1916) The *Helopeltis* question, especially in connection with Cacao. *J. Med. Proefsta*, No. 21, 40.

Roy, A.K., Jamaluddin and Prasad, M.M. (1976) Some new leaf spot diseases. *Curr. Sci.* 45, 604.

Rutherford, A. (1913) *Zeuzera coffeae* (Red Borer: Coffee Borer). *Trop. Agric.*, 41, 486–488.

Rutherford, A. (1914) Some Ceylon coccidae. *Bull. Ent. Res.*, 5, 259–268.

Sawada, K. (1933) Descriptive catalogue of the Formosan fungi Part VI. *Report Dept. Agric. Res. Inst., Formosa*, 61, 117.

Seshadri, V.S., Channamma, L.K.A. and Rangaswami, G. (1972) A few records of fungi from India. *Mycopath. Mycol. Appl.*, 25, 246–252.

Sharma, N.D. (1979) Some records of fungi from India. *Indian Phytopath.*, 32, 511–512.

Singh, S.M. (1978) Some Sphaeropsidales from Balaghat (M.P.). *Indian Phytopath.*, 31, 176–179.

Singh, V., Dubey, O.P., Nair, C.P.R. and Pillai, G.B. (1978) Biology and bionomics of insect pests of cinnamon. *J. Plantation Crops*, 6, 24–27.

Srivastava, A.K. (1979) Some additions to Indian Ascomycetes. *Indian Phytopath.*, 32, 445–446.

Suresh, S. and Mohanasundaram, M. (1996) Coccoid (Coccoidea; Homoptera) fauna of Tamil Nadu, India. *J. Ent. Res.*, 20, 233–274.

Trejo, G.E. and Lopez, F.M.C. (1990) Taxonomy of the plant bacteria involved in withering of cinnamon (*Cinnamomum zeylanicum*) cultivated *in vitro*. *Revista Chapingo*, 15, 67–68.

Van der Goot, P. (1918) Aphididae of Ceylon. *Spolia Zeylanica*, 61(40), 70–75.

Van Hall, C.J.J. (1921) Diseases and pests of cultivated plants in the Dutch East India during 1921. *Meded. Inst. Voor Planten Zicekten*, 53, 466.

Van Hall, C.J.J. (1924) Diseases and pests of economic crops in the Dutch East Indies in 1923. *Meded. Inst. Voor Plantenziekten*, 64, 47.

Van Overeem, C. (1925) Contribution to the fungus flora of the Dutch East Indies. II (No.10–13) 13. On the red rot fungus. *Bull. Jard. Bot. Buitenzorg ser. III, vii* 4 pp. 436–443.

Veenakumari, K. and Mohanraj, P. (1993) Insect pests of cinnamon (*Cinnamomum verum* Bercht. & Presl.) in the Andaman and Nicobar Islands. *J. Plantation Crops*, 21, 67–69.

Ver Woerd, L. and Du Plessis, S.J. (1933) Descriptions of some new species of South African fungi and of species not previously recorded from South Africa. *S. African J. Sci.*, 30, 222–233.

Vesey-Fitzgerad, D. (1938) Entomologists Report. *Report, Department of Agriculture, Seychelles – 1936*, pp. 17–18.

Wikardi, E.A. and Wahyono, T.E. (1991) Destructive insects of cinnamon (*Cinnamomum* spp.) and their natural enemies. Bull. *Penelitian- Tanaman- Rampah- dan- Obat (Indonesia)*, 6, 20–26.

Williams, D.J. (1961) Changes in nomenclature affecting some Coccoidea (Homoptera). *Ent. Mon. Mag.*, 97, 92–93.

Zhang, X.Y. (1998) A study on the taxonomy of *Exobasidium* spp. according to fuzzy analysis of cultural properties and the analysis of 28 Sr. DNA-PCR-RFLP. *Science Silvae Sinicae*, 34, 59–71.

11 Pharmacology and Toxicology of Cinnamon and Cassia

K.K. Vijayan and R.V. Ajithan Thampuran

Introduction

In the following text the term cinnamon is used to mean both Ceylon cinnamon and Chinese cinnamon or Cassia cinnamon.

The medicinal and aromatic properties of cinnamon are used in the traditional medicines of India and China. Cinnamon bark and cinnamon oil have also been used as food additives, condiments and flavouring agents due to their carminative, antioxidant and preservative actions. In the *Ayurveda* and *Sidha* medical systems, cinnamon bark, twigs, leaves and oil are used as ingredients of many multidrug preparations. In the pharmacopoeias of India, Britain, China, Australia, Belgium, Europe, France, Germany, Hungary, Japan, the Netherlands, Portugal and Switzerland both cinnamon bark and oil are included as official drugs (mainly as a carminative and flavouring compound, used for the treatment of dyspepsia and as a stomachic). *Ayurveda* texts describe cinnamon as *Katu-Mathiram* (pungent-sweet), *tiktarasm* (appetizer and produces dryness in mouth), *ushnaveeryam* (increases body temperature, improves blood circulation, stimulates appetite and digestion), *kaphavataharam* (subdues *vata** and *kapha**), *pittaharam* (subdues *pitta**), *laghu* (helps digestion), *ruksham* (produces dryness). These properties make cinnamon useful for the treatment of *aruchi* (anorexia), *hridrogam* (heart disease), diseases of the *vasthi* (bowel), *arsad* (piles), and *krimi* (helminthic infections). Since cinnamon bark contains many chemical constituents and oil, (the oil also contains several constituents), it is logical to assume that more than one might be the active principle and hence different modes of pharmacological activities are exhibited by cinnamon.

'The Indian Materia Medica' (Nadkarni, 1976) describes cinnamon bark as carminative, antispasmodic, aromatic, stimulant, astringent, antiseptic, stomachic and germicidal. Oil is a vascular and nervine stimulant. In large doses it is an irritant and a narcotic poison. In the book 'Materia Medica of India and their Therapeutics' (Khory and Katrak, 1994) many other valuable actions are attributed to cinnamon bark (powder) and oil.

---

* According to the *Ayurveda* system of medicine all the physiological functions of the human body are governed by three basic biological parameters – the *thridoshas* or the three basic qualities, namely *vata, pitta* and *kapha (kafa)*. *Vata* is responsible for all voluntary and involuntary movements in the human body; *pitta* is responsible for all digestive and metabolic activities; and *kapha (kafa)* provides the static energy (strength) for holding body tissues together, and also provides lubrication at the various points of friction. When these three *doshas* (qualities) are in the normal state of equilibrium the human body is healthy and sound, but when they lose equilibrium and get vitiated by various internal and external factors, they produce varied diseases. *Ayurveda* treatment of any disease is aimed at restoring the equilibrium of the three *doshas* or qualities.

0-415-31755-X/04/$0.00 + $1.50
© 2004 by CRC Press LLC

In medicinal doses it is a good remedy for flatulence, paralysis of the tongue, enteralgia (acute intestinal pain) and cramps of the stomach. As an antiseptic it is used as an injection in gonorrhea. As a germicide it destroys pathogenic bacilli and is used internally in typhoid fever. The bark is a homeostatic, has a specific action on the uterus and is given with other urine stimulants to promote parturition and to check haemorrhage. It is useful in bronchitis and pneumonia. The oil is used in the treatment of rheumatism and muscular pain. It is antitubercular and used as an injection in phthisis (tuberculosis of the lung). Many other books on medicinal plants ascribe similar activities and uses to cinnamon bark and oil (Kirtikar and Basu, 1975; Ambasto, 1986; Chatterjee and Prakash, 1994a,b; Agarwall, 1997; Bhatacharjee, 1998). Warriar *et al.* (1994) attribute many more medicinal properties and uses to cinnamon. According to them it is alexeteric (wards away infection), anthelminthic (expels worms), diuretic, stimulant, anti-inflammatory, expectorant and febrifuge (reduces fever). It is useful in hepatopathy and splenopathy (diseases of the liver and spleen), bronchitis (inflammation of bronchial tubes), cephalalgic (ache in the head and cephalic portion), odotalgia (toothache), cardiac disease, and halitosis (foul odour of breath). Cinnamon oil is a stomachic (increases appetite), carminative, emmanagogue (regulates the menstrual flow), stryptic (astringent) and is used in vitiated conditions of *Vata*. Cinnamon bark and oil are included in 'Martindale' as an official drug for its carminative and stimulant action and also for its use as a flavouring compound. Many multi-ingredient preparations containing cinnamon used in several countries in various therapeutic formulations are also listed therein. In Chinese traditional medicine cinnamon is used to induce perspiration, dispel pathogenic factors from the exterior of the body, as an analgesic and antipyretic against cold and associated fever and headache, myalgia (mascular pain), arthralgia (arthritic pain) and amenorrhea (failure of menstruation). In the Chinese Materia Medica many pharmacological activities are attributed to cinnamon. Cinnamon is considered to have antimalarial, antiallergic and immunostimulant, antiulcerogenic and hypotensive activities and cardiovascular effects. It is included in two forms, *Rou Gui* (dried bark) and *Gui Zhi* (dried twig). Many pharmacological investigations have been carried out mainly on *C. cassia*, *C. verum*, *C. tamala* and a few other species of *Cinnamomum*.

Pharmacological Studies

Analgesic, antipyretic and diaphoretic actions

A reduction of body temperature in mice was observed by the administration of a decoction of the dried twigs of cinnamon. The same result was obtained using cinnamaldehyde or sodium cinnamate (Chinese Materia Medica). Cinnamaldehyde provides a hypothermic and antipyretic action. Wang (1985) observed that cinnamaldehyde produced analgesic effects when tested in acetic acid-induced writhing in mice. Chatterjee and Prakash (1994b) also report the analgesic activity of cinnamon bark, but no experimental details were given. Atta and Alkofahi (1998) studied the anti-noiceptive activity (noiceptor – a receptor which transmits painful stimuli) of the ethanolic extract and reported that *C. verum* possessed an analgesic effect against both acetic acid-induced writhing and hot plate induced thermal stimulation. Newall *et al.* (1996) also reported analgesic and antipyretic activity. Cinnamon is a mild diaphoretic agent due to the vasodialatory effect it produces (Tanikawa *et al.*, 1999).

Anti-inflammatory action

Kirthikar and Basu (1975) and Ambasto (1986) have reported the anti-inflammatory action of cinnamon. It relieves pulmonary inflammation (Asolkar *et al.*, 1986; Agarwall, 1997; Warrier *et al.*, 1994). The Japanese plant *Cinnamomum seiboldii* has shown to possess anti-inflammatory action, which has been attributed to a series of tannins (Newall *et al.*, 1996). A study conducted on mice with 70% ethanolic extract of cinnamon gave promising results on acute inflammations (Kubo *et al.*, 1996). The extract inhibited the increase in vascular permeability induced by acetic acid. It inhibited the paw oedema induced by carragenan as well as seratonin, whereas it was ineffective against bradykinin and histamine-produced inflammations. Little effect was shown on secondary lesions in the development of adjuvant induced arthritis. It is also useful in pulmonary inflammations. A herbal ophthalmic medicament called *Ophthacare*, which contains 0.5% cinnamon, was tested for its anti-inflammatory activity on ocular inflammations in rabbit and found to be effective (Mitra *et al.*, 2000).

Action on the cardio-vascular system

The 'Indian Materia Medica' (Nadkarni, 1976) and the 'Indian Medicinal Plants – A Compendium of 500 species' (Warrier *et al.*, 1994) consider cinnamon as a herbal drug which has cardiovascular effects. A decoction of the herb increases coronary flow in guinea pig heart and antagonised pituitrin-induced reduction of flow. An oral dose of 1.2 g/kg of the decoction for six days improved pituitrin-induced acute myocardial ischemia. Intra-arterial injection of the decotion did not affect blood pressure but markedly decreased peripheral vascular resistance, suggesting a direct vasodilatation of peripheral vessels (Wang, 1983). In another study using anesthetised dogs and guinea pigs cinnamaldehyde produced a hypotensive effect mainly due to vasodilatation. An increase in cardiac contractile force and beating rate was exerted by cinnamaldehyde (Harada and Yano, 1975). Cinnamaldehyde induced the release of catacholamines from the adrenal glands of dogs. The positive inotropic and chronotropic effects (adrenergic impulses with an increase in rate and force of contraction) of cinnamaldehyde on perfused isolated guinea pig heart were presumably due to the release of endogenous catecholamines (Harada and Saito, 1978). Circulatory stimulant effects of cinnamon have been reported in several books on medicinal plants and also in *Ayurveda* (Bhattacharjee, 1998; Asolkar *et al.*, 1986). Tanikawa *et al.* (1999) observed that aqueous extract of the bark relaxed rat aorta ring (precontracted with $PGF_2\alpha$) preparation with intact epithelium. Tannins do not relax specimens without epithelium. They attributed the vasodialatory action to the tannins. It has also been observed that with increasing molecular weight of tannins the intensity of relaxation also increases. Inhibition of platelet aggregation and thrombosis has been reported by Matsuda *et al.* (1987). A 70% methanol extract of *C. cassia* bark has shown significant inhibition of intravascular coagulation in rats. Cinnamaldehyde exerts a similar action.

Hypoglycemic activity

Cinnamon is considered to be a herbal remedy for diabetes. *Ayurveda* texts and folk medicinal uses support this claim. Studies conducted on rabbit using an aqueous extract established the hypoglycemic effect (Asolkar *et al.*, 1986). It is reported in many

Ayurveda publications that cinnamon reduces the blood glucose level in non-insulin dependent diabetes. Polansky and Anderson (1992) in their study of insulin activity in isolated adipocytes observed that in the absence of added insulin, an extract of the herb potentiated the activity of insulin-stimulated utilisation of glucose. *In vivo* study in rats also gave encouraging results. Sharma *et al.* (1996a, b) investigated the hypoglycemic potential of *C. tamala* bark extract as a single dose of 0, 100, 200 and 500 mg/kg body weight in fasted rats. They obtained a good hypoglycemic effect, which was more pronounced at lower doses (100–200 mg/kg). The same group extended the study with a 50% alcoholic extract of the leaves and obtained a significant reduction in plasma glucose levels in normoglycemic rats at a dose of 250 mg/kg body weight (Sharma *et al.*, 1996b). In another *in vitro* study using pancreatic cell culture, Patole and Agte (1998) observed that a 10% alcoholic extract of cinnamon enhanced insulin secretion (P < 0.01). Broadhurst *et al.* (2000) investigated several medicinal herbs for their activity in reducing blood glucose levels, and found that aqueous extract of cinnamon gave good results. Using rat epidydimal adipocytes they studied the insulin dependent utilisation of glucose and found that aqueous extract of cinnamon was very effective and concluded that dietary cinnamon played a major role in improving glucose and insulin metabolism. Cinnamon containing multi-herbal formulations used in traditional therapeutic practices has also given well-defined hypoglycemic activity in *in vitro* studies. Using a Chinese drug formulation *Sangbackpitang* (SBPT), Lee *et al.* (1999) obtained a significant reduction of blood and serum glucose levels. The mechanism proposed was that of inhibition of glycosidase catalysed reaction and upregulation of mascular GLU-T4 m-RNA expression. Inhibition of aldose reductase enzyme activity was obtained for cinnamon in a study conducted by Aida *et al.* (1987). In this investigation carried out on traditional oriental herb prescriptions used for the treatment of diabetes and its implications, the investigators found that an extract of cinnamon at a concentration of 0.1 mg/ml completely inhibited the activity of aldose reductase enzyme. Aldose reductase enzyme has been implicated for the pathogenesis of some chronic complications of diabetes like retinopathy, neuropathy and nephropathy. Aldose reductase inhibitors in animal experiments, and recently in clinical trials, improved cataract and peripheral neuropathy conditions. Even though very promising results were reported on the high potential of using cinnamon as a therapeutic agent for diabetes, no clinical study has been reported. The potential of this herb in reducing blood glucose level and inhibiting complications due to diabetes definitely is a very promising activity which requires further research to establish its utility.

Hypocholesterolemic action

Studies on the anti-cholesterolemic potential of cinnamon have given encouraging results. Sambaiah and Sreenivasan (1991) observed that dietary cinnamon increased bilary secretion of cholesterol and phospholipids without affecting the bile content. Sharma *et al.* (1996b) studied the effect of a 50% alcoholic extract of cinnamon on rats and reported a significant anti-hypercholesterolemic action and reduced serum triglyceride at a single dose of 250 mg/kg body weight. In another investigation using triton WR-1339-induced hyperlipidaemic rats, Kim *et al.* (1999, 2000) obtained suppression of total serum cholesterol, triglycerides, phospholipids and low density lipoprotein levels. The same group extended the study with 80% methanolic extract and its chloroform fraction of different species of *Cinnamomum* and found that the extracts suppressed the elevated serum total cholesterol and triglyceride levels in corn oil-induced hyperlipidaemic rats. The chloroform fraction exhibited remarkable inhibitory effects on HMG-CoA reductase, an enzyme that catalyses cholesterol biosynthesis.

Antioxidant action

The use of spices and their constituent essential oils as additives for the prevention of food deterioration is an ancient practice. The basis of action involves antioxidant and/or antimicrobial activity. Antioxidants scavenge free radicals and control lipid peroxidation in mammalian systems. Lipid peroxidation is a chain reaction, providing a continuous supply of free radicals that initiate further peroxidation. This cycle causes the deterioration of food. *In vivo* it causes tissue damage, which can lead to inflammatory diseases, aging, atherosclerosis, cancer, etc. Within the cell free radicals can cause peroxidation of polyunsaturated fatty acids in the phospholipid membranes, the formation of cytotoxic peroxides, the oxidation of proteins and the denaturation of DNA. These phenomena contribute to death, true death and aging.

Investigations on the antioxidative potential were carried out by several groups of workers and their results indicated that cinnamon has a very high potential in countering the phenomena listed above. In a study conducted to ascertain the antioxidant effect of spices, Reddy and Lokesh (1992) observed that cinnamon at high doses (600 μg/ml) inhibited lipid peroxidation of rat microsomes. The antiperoxidative ability of cinnamon extract was studied by Yokozawa *et al.* (1977) in a screening test carried out on traditional Chinese prescriptions and they obtained promising results. In yet another study to explore the ability of crude drugs on free radical scavenging activity on 1,1-diphenyl-2-picrylhydrazyl radical (DPPH), Yokozawa *et al.* (1998) reported strong free radical scavenging activity for cinnamon cortex. Trombetta *et al.* (1998) correlated the antioxidant effect of cinnamon with that of the phenolic concentration (hydroxycinnamaldehyde, hydroxycinnamic acid, etc.) of the extract. The antioxidant activity of cinnamon is attributed to its ability to activate the hepatic and cardiac antioxidation enzymes. Dhuleep (1999) studied the hepatic and cardiac antioxidant enzyme (GSH) content and lipid conjugated dienes in rats fed a high fat diet along with spices. The result of the study showed that cinnamon and cardamom counteracted increases in lipid conjugates, the primary product of lipid peroxidation. Hence they concluded that the spice significantly activates the production of antioxidant enzymes. DPPH free radical trapping activity has also been reported by the methanol extract of *C. cassia* (Kim *et al.*, 1999). Stoltz (1998) indicated that the cinnamon powder containing cream epicontrol A5 (cinnamon powder ligated to a lipoaminoacid, such as lipoglycine) has high free radical scavenging action, which is attributed to the cinnamon component. Its activity is much greater than known free radical scavengers such as phenoxypropanol and zinc gluconate, but less effective than vitamin C.

Recently Hsiao *et al.* (2001) reported a new antiperoxidative cytoprotectant and free radical scavenger, cinnamophilin, from *Cinnamomum philippinense*. The antioxidant properties of cinnamophilin (a natural compound isolated from *Cinnamomum philippinense*) were evaluated for its ability to react with relevant reactive oxygen species. Its protective effect on cultured cells and biomacromolecules under oxiditive stress was also studied. Cinnamophilin suppressed non-enzymatic iron-induced lipid peroxidation in rat brain homogenates, with an IC_{50} value of $8.0 \pm 0.7\,\mu M$ and an iron ion/ADP/ascorbate-initiated rat liver mitochondrial lipid peroxidation, with an IC_{50} value of $17.7 \pm 0.2\,\mu M$. It also exerted an inhibitory activity on NADPH-dependant microsomal lipid peroxidation with IC_{50} value of $3.4 \pm 0.1\,\mu M$ without affecting the microsomal electron transport of NADPH-cytochrome P-450 reductase. Both 1,1-diphenyl-2-picrilhydrazyl and $2,2^1$-azo-bis (-amidinopropane) dihydrochloride-derived peroxyl radical tests demonstrated that cinnamophilin possessed a marked free radical scavenging capacity. Cinnamophilin significantly protected cultured rat aortic smooth muscle cells (A7r5) against alloxan/iron ion/H_2O_2-induced damage that would result in cytoplasmic membranous disturbance and mitochondrial decay. Cinnamophilin inhibited copper-catalysed oxidation of human low-density lipoprotein, as measured by fluorescence intensity and thiobarbituric acid-reactive substance formation in a concentration-dependant manner. On the other hand it was reactive towards superoxide anions generated by the xanthine/xanthine oxidase system and the aortic segment from aged spontaneously hypertensive rat.

Antiulcerogenic effects

In *Ayurveda* texts and other books on medicinal plants the main activity and use attributed to cinnamon (and its related taxa) is as a curative agent for gastrointestinal problems. In the pharmacopoeias of different countries and in 'Martindale' cinnamon is included as an official drug for carminative action and for gastrointestinal problems. The various preparations of cinnamon described in 'Martindale', which are employed in the therapeutic practices of various countries, are used mainly for gastrointestinal disorders. Cinnamon is used in traditional medicine as a cure for diarrhoea, dyspepsia and for its carminative action. In modern medicine it is regarded as a stomachic and carminative. In the 'Materia Medica of India', cinnamon is characterised as a drug active against diarrhoea and dysentery. The effect of cinnamon oil on digestive enzymes was studied by Yoshiki *et al.* (1981). An aqueous extract of Chinese cinnamon suppressed the serotonin-induced gastric mucosal lesions in mice and rats and the antiulcerogenic activity was attributed to the inhibition of both the gastric secretion and pepsin output. It potentiated gastric mucosal flow (Akina, 1986). Later Tanaka *et al.* (1989) isolated and identified the active compounds which prevented the ulcerogenesis. These were 3-(2-hydroxyphenyl) propanoic acid and its o-glycoside. The former compound administered orally or parenterally to rats at a remarkably low dose of $40\,\mu g/kg$ body weight produced the effect. This compound also inhibited gastric ulcers induced by other ulcerogens such as phenylbutazone, aspirin and water immersion stress. However, it failed in the case of indomethacin induced ulcers. The result suggests that the antiulcerogenic effect of this compound is attributable to the potentiation of defensive factors through the improvement of gastric blood flow and gastric cytoprotection. Cassioside, cinnamoside and 3,4,5-trimethoxyphenol-β-D-apiofuranosyl $(1 \rightarrow 6)$-β-D-glucopyranoside were isolated as antiulcerative factors (Shiraga *et al.*, 1988).

Sedative and anticonvulsant effects

Oral administration of cinnamaldehyde (250 mg/kg B.W.) reduced spontaneous motor activity in mice. This also antagonised the drug-induced locomotor activity and prolonged the hexobarbital induced sleeping time. Similar effects were produced by an intraperitonial dose of 125–250 mg/kg (Harada and Ozaki, 1972). A Chinese herbal preparation *Saiko-keishi-to* (SK), which contains cinnamon, clearly inhibited the calcium binding state change near the cell membrane during bursting activity, which is characteristic of seizure discharge (Sugaya *et al.*, 1985).

Immunological effects

Several *Ayurveda* preparations and aquous herbal extracts exert immunomodulatory activity. It is hypothesised that many multidrug preparations produce their therapeutic action through the modulation of immune systems. The immune system is a highly sophisticated defence mechanism against external biological invaders through the inter-connected network between the brain, endocrine and immune system. It also serves to regulate the internal environment by eliminating aberrant cells or misplaced tissues within the body. The immune system is composed of two major branches, the humoral immune system, primarily the domain of the B-lymphocytes, and the cell mediated immune system, the domain of the T-lymphocytes. The humoral immune system produces the antibodies that react specifically with the antigens (the invaders), and the cell-mediated immune system mobilises phagocytic leukocytes for the ingestion and subsequent destruction of invading organisms. The exposure of antigens to the cells of the immune system will elicit an immune response, often associated with clonal prolif-eration, an expression process by which a few antigen responsible cells can generate a large number of immune competent cells specifically to cope with the foreign invaders. Complement is a powerful effector system involved in the body's immunological defence. They are a group of self-reacting proteins in blood serum and body fluids that play an important role as a mediator of immune and allergic reactions. Complement represents the humoral arm of natural host defence mechanisms. Complement acti-vation forms a major part of natural defence, with a range of mediators possessing immuno-inflammatory potency. However, complement activation may also damage host tissue as in autoimmune diseases, immunodeficiency diseases, etc.

In vitro inhibitory activity against complement formation has been documented for cinnamon cortex and cinnamon oil (Norhara *et al.*, 1980; Toshihiro *et al.*, 1981; Yoshiki *et al.*, 1981). These workers have studied the extract of cinnamon bark and reported anticomplementery activity. An aqueous extract of the herb showed an inhibitory effect on the complement activity in the heterophile antibody (antibodies that react with multiva-lent antigens) reaction (Wang, 1983). Tang and Eisenbrand's (1992) study ascribes immunosuppressive action to the extract. They have found that oral administration of the extract prevented an increase in protein levels in the urine of rats with nephritis. Nephritis is an autoimmune disease caused by activation of the complement system. Several diterpenoids have been isolated and characterised from the stem and root bark extracts of cinnamon and allied species. Cinncassiol C_1 and its glycosides, cinncassiol C_2, C_3, and cinncassiol D_1 and its glycoside were identified as the active constituents responsible for the anticomplementery activity. Koh *et al.* (1999) isolated two cinnamaldehyde derivatives, 2-Hydroxy cinnamaldehyde (HCA) and 2-Benzoxycinnamaldehyde (BCA), from the stem bark of cinnamon and studied their immunomodulatory effects. These compounds suppressed lymphoproliferation when treated with Con A stimulated mouse splenocytes cultures in a dose dependent manner. Both the compounds at micromolar dose inhibited Con A stimulated proliferation by 60–69%. Decreased levels of antibody production fol-lowing HCA or BCA treatment were observed. The inhibitory effect of cinnamaldehyde on lymphoproliferation was specific to the early phase of cell activation. It was suggested that HCA and BCA inhibit lymphoproliferation and T-cell differentiation. However, Shan *et al.* (1999) attributed an immunostimulant effect on human lymphocytes for the extract. They have studied the action of eight medicinal herbs and reported that the extract of cinnamon stimulated lymphocyte proliferation, enhanced cytotoxic T-lymphocyte activity, stimulated immunoglobulin production by B-cells and interleukin (IL-1α) production by monocytes. These activities are attributed to the glycoprotein of MW 100 KDa from the extract. An *et al.* (1987) studied the hot water extract of tender stems of Chinese cassia for its immunolomodulatory action and found that a water soluble polysaccharide isolated from the extract showed a significant complement system activating effect.

The different investigations reveal that cinnamon shows both immune system potentiating and inhibiting effects. The macromolecules such as the glycoproteins and water-soluble polysaccharides are responsible for the potentiating activity, whereas the diterpenoid moieties and the phenolic molecules give inhibitory properties. *Kaishi-ni-eppi-ichi-to* (TJS-664), a Chinese herbal preparation containing cinnamon as its main constituent, has been shown to exhibit antiviral action against the influenza A$_2$ virus. It has been suggested that the antiviral effects of TJS-064 are expressed through the host's immune system. This inference is based on the observation that a 100% survival rate of the virus infected mice was obtained on a treatment with 70 mg/kg p.o. of the drug, whereas the control mice had a mean survival of 11.2 days and all the mice died in 25 days. *In vitro* studies did not give any viricidal or viristatic effects (Ball *et al.*, 1994). In another study Kobayashi *et al.* (1994) investigated the effect of TJS-038, another herbal preparation (cinnamon with a combination of four herbs), which is used as a traditional medicine to treat influenza following rheumatoid arthritis. TJS-038 at a dose of 50 mg/kg p.o. stimulated the generation of contrasuppressor T-cells in mice (selective elimination or inactivation of Suppressor T-cells enhances immune reactions).

Antiallergenic activity

Allergy is a hypersensitivity reaction due to an excessive or inappropriate immunological response to an antigen (IgE). This occurs when an antigen reacts with antibody bound to mast cells or basophilic blood cells, leading to disruption of these cells with the release of vasoactive mediators such as histamine. The antigens that initiate this response are called the allergens. Compounds, which can inhibit or antagonise this excessive immunological response and intervene in the multitude of complex reaction sequences of the immune system, may function as antiallergenic. Cinncassiol C$_1$, cinncassiol D$_4$ and its glycosides, which are constituents of cinnamon bark, exhibit anti-allergic activity (Torhara, 1980). In another investigation, Tanaka *et al.* (1991) found that cinnamaldehyde inhibited allergic reactions caused by the release of chemical mediators from basophils and mast cells. The antigen (allergen) and antibody (IgE) reaction caused a cell degranulation function. Shichinohe *et al.* (1996) studied the anti-allergic effect of a Chinese herbal formulation *Moku-boi-to* (M-711), which contained cinnamon as the main constituent. A model dermatitis was created in rats, which when administered with 20 mg/kg B.W of M-711 significantly suppressed the capillary permeability induced by histamine. The antihistamine effect of 40 mg/kg was equivalent to the optimum dose of other anti-inflammatory drugs, azelastine and diphenhydramine. Ikwati *et al.* (2001) found that the hexane extract of *C. massoia* bark from Indonesia had a strong inhibitory effect on IgE-dependent hisstamine release from RBL 2 H_3-cells. The result indicates that the extract contains active compounds that inhibit mast cell granulation.

Anticancer activity

Investigations have also been carried out on the anticancer activity of cinnamon. Several groups of workers studied the problem in *in vivo* and *in vitro* models. The results obtained were encouraging. Konoshima *et al.* (1994) and others studied the effect of cinnamon bark extract on skin and pulmonary tumours *in vivo* in animal models. They investigated EBV-EA early antigen activation, a tumour-initiating factor, and found that the extract strongly inhibited the activation of the antigen. The study on polynuclear carbon induced skin tumour on mouse also has shown inhibition of skin change with that of the control group. In another study Jin Won *et al.* (1994) using gastric and colon cancer cell lines, reported that the extract inhibited growth. 2'-Hydroxycinnamaldehyde extracted from the stem bark of cinnamon was studied for its activity on Farnesyl-protein-transferase, an enzyme involved in the initiation of tumour formation, and Byoung Mog *et al.* (1996) reported that the enzyme activity was completely inhibited at a concentration of 22 µg/ml. A cinnamon containing multiherbal formulation shows antimetastatic potential to experimental liver and lung metastasis (Onishi *et al.*, 1998). *Juzen-taiho-to*, a Kampo Japanese herbal medicine, was administered orally seven days prior to tumour inoculation, and it significantly decreased the number of liver metastatic colonies of colon 26-L5 carcinoma cells and attenuated the increase of liver weight at a dose range 4–40 mg/day. Oral administration of the medicine also inhibited lung metastasis of B16-BL6 melanoma cells. 2'-Hydroxycinnamaldehyde and 2'-benzoyloxycinnamaldehyde, constituents of cinnamon which showed other useful pharmacological activities, are also active as anticancer agents. These compounds inhibited the growth of 29 human cancer cell lines *in vitro*. Both compounds are active *in vivo* also. They inhibited the growth of SW-20 human tumour xenograft without loss in body weight in nude mice (Chang Woo *et al.*, 1999). Hyun *et al.* (1994) studied the effect of the extracts of several medicinal herbs and found that Chinese cassia extracts possessed an antineoplastic effect against human gastric and colon carcinoma cell lines. The effect was produced at a dose of 250 µg/ml concentration. Mukherjee *et al.* (1994) isolated two monoacylglycerols and 2-flavanols from the methanol extract of *C. camphora* and tested these compounds *in vitro* against L1210 cell lines and found cytotoxic activity at an ED_{50} value of 209 µg/ml.

Metabolic studies

Samuelson *et al.* (1986) studied the metabolism of o-methoxycinnamaldehyde in rats. A major pathway was found to be the oxidation to the corresponding cinnamic acid and phenylpropanoic acid, which were largely excreted as glycine conjugate. No evidence of conjugation to glutathione was obtained. Peters and Cadwell (1994) investigated the metabolism of *trans*-cinnamaldehyde in rats and mice. They reported that over 90% of the cinnamaldehyde administered was recovered from the excreta 72 hours after administration, most of which (75–81%) was present in the urine produced over the first 24 hours. Less than 2% of the dose was found in the carcasses 72 h after administration. The major urinary metabolite was hippuric acid. The metabolic profiles of cinnamaldehyde in rats and mice were not systematically affected by sex, dose size or route of administration.

Antimicrobial Activity

Antibacterial activity

The antimicrobial effect of cinnamon is one area of activity that has been extensively investigated. Reports are available from 1944 onwards. These studies include antibacterial and antifungal effects. An antiviral effect has also been reported. Kobayashi *et al.* (2000) reported the inhibition of HIV-1 by the methanolic extract of cinnamon due to the inhibition of reverse transcriptase; the IC_{50} value being $4.3\ \mu g/ml$. Premanathan *et al.* (2000) carried out a survey of Indian medicinal plants for anti-HIV activity and reported that cassia extract is most effective in inhibiting HIV-1 and HIV-2. Cinnamon was found to be active against most of the foodborne pathogenic organisms (see Table 11.1). Many other types of human pathogens have also been found to be susceptible to the activities of cinnamon and its derived compounds. Scientific evidence on the preservation potential of spices emerged in the early nineteenth century. Chamberland first reported the antimicrobial activity of cinnamon oil against spores and anthrax bacilli (Webb and Tanner, 1944). Later studies indicated the usefulness of aqueous and alcoholic extracts of ground cinnamon in preserving tomato sauce. Following a study conducted on different strains of *Clostridium botulinum* (an anaerobic spore forming bacteria producing neurotoxins causing fatal food poisoning), Hall and Maurer (1986) reported that cinnamon was active in inhibiting the growth of three different strains of *C. botulinum* at 150–200 ppm level. Ismaiel and Pierson (1990) observed in their *in vitro* study that an extract of cinnamon at 150–200 ppm inhibited the growth and germination of *C. botulinum*. Spice oils have an advantage over other antimicrobial agents in that they prevent the germination of *C. botulinum*. Zhang *et al.* (1990) compared the bacteriostatic ability of cinnamon with that of two standard bacteriostatic substances, nipagin-A and benzoic acid, and found that the same amount of activity was associated with cinnamon. Iyangar *et al.* (1994) attributed the bactericidal activity of the herb to benzylbenzoate, an ester constituent of the oil. A 40% ethanolic extract of the bark had a good growth inhibitory effect on *Enterobacter aerogenes*, a bacterium that causes urinary tract infections. Wendakoon and Sakagudi (1995) attributed the antibacterial effect to cinnamaldehyde and eugenol. From a study of the antimicrobial effect of spices Bara and Venetti (1995) reported that cinnamon bark powder exhibited a bacteriostatic effect. The effect was obtained after 12 hours at a concentration of 4.1–4.4% w/v of cinnamon powder against *Yersinia enterocolitica*, a bacteria causing food borne infections and food poisoning. In an *in vitro* screening study of 17 essential oils for antibacterial activity against the growth of *Trichomonas vaginalis*, a protozoan parasite on humans that causes infections of the urethra and vagina, Viollon *et al.* (1996) found that cinnamon oil had the maximum growth inhibiting effect with an LD_{100} value of 50 mg/ml.

Quale *et al.* (1996) studied the effect of cinnamon and its constituents against *Candida* species and extended a clinical study to patients with candidiasis. They studied the activity against fluconazole-resistant *Candida* species isolates from HIV-infected patients *in vitro* and found that cinnamon extract completely inhibited growth at a MIC (minimum inhibitory concentration) value of 0.5–30 mg/ml. *trans*-cinnamaldehyde and *o*-methoxycinnamaldehyde also exhibited activity at a MIC value of 0.03–0.5 mg/ml. A pilot study on five patients with oral candidiasis was made using commercially available cinnamon preparation. The patients were treated with the formulation for one week. Three patients exhibited improvement. However, to establish the

Table 11.1 Antibacterial effects of cinnamon and derived compounds on toxigenic foodborne bacteria

Microorganism	Toxic effect	Sources	Nature of activity	Dose	Reference
Clostridium bolulinum	Botulism intoxication caused by toxin – paralysis and death	Meat, sausage, canned foods etc.	Growth and germination inhibition/bactericidal	150–200 ppm of extract/oil	Hall and Maurer (1986) Ismail and Pierson (1990) Iyangar et al. (1994)
Clostridium perfringens	Clostridial myonecrosis/ gasgangrene	Meat and meat products	Bacteriostatic	0.5 mg/disk (paper disk method)	Lee and Ahn (1998)
Bacillus cereas	Food poisoning	Grains, vegetable, dairy products	Bactericidal	Not mentioned	De et al. (1999)
Bacillus subtilis	Allergic reactions	Cereal foods	Bactericidal	Not mentioned	De et al. (1999)
Campylobacter jejuni	Gastroentritis, bloody diarrhea with abdominal pain and fever	Chicken and milk products	Bacteriostatic	0.075% oil	Smith-Palmer et al. (1998)
Enterobacter cloacae	Opportunistic pathogen causes intestinal disease & diarrhea. Also causes neonatal meningitis, urinary tract infections	Dairy products Soil, water sewage Intestinal tract	Bacteriostatic & Bactericidal Bactericidal	Qualitative	Chao et al. (2000)

Table 11.1 (Continued)

Microorganism	Toxic effect	Sources	Nature of activity	Dose	Reference
Escherichia coli	Gastroenteritis, enterocolitis	Several strains, intestinal bacteria contaminated food fish, meat, poultry	Bacteriostatic & Bactericidal	0.075% oil qualitative	Smith-Palmer *et al.* (1998) De *et al.* (1999)
Listeria monocytotogens	Listerosis, affects nervous system	Dairy products, sausage, cabbage, raw vegetables	Bacteriostatic	0.075% oil	Smith-Palmer *et al.* (1998)
Salmonella enteritids	Gastroenteritis meningitis, salmonellosis	Dairy products, meat, fish, canned foods, poultry, etc.	Bacteriostatic	0.075% oil	Smith-Palmer *et al.* (1998)
Yersinia enterocolitica	Gastroenteritis	Meat, dairy products	Bacteriostatic	40% ethanol extract	Bara and Venetti (1995)
Staphylococcus aureus	Food toxicity, intestinal inflammation; also a causative agent of abscesses, skin lesion, pneumonia	Meat, poultry, fish, dairy products, raw vegetables, eggs, cream	Bacteriostatic	0.075% G.I.	Smith-Palmer *et al.* (1998)

Pharmacology and Toxicology of Cinnamon and Cassia 271

therapeutic use of cinnamon or its constituents, a clinical trial must be conducted. Nevertheless, the results of this study are very significant because a resistant variety of organism isolated from HIV patients was used for the investigation.

Another significant antibacterial activity exhibited by cinnamon is against *Helicobacter pylori*, a human pathogen producing cytokines which cause peptic ulcers. Chronic infection can lead to stomach cancer. A ethanol and methylene chloride extract of powdered cinnamon was studied by Tabak *et al.* (1996). The methylene chloride extract completely inhibited the growth at a concentration of 50 μg/ml in solid medium and by 15 μg/ml in liquid medium. It inhibited the urease activity also. Smith-Palmer *et al.* (1998) studied the effect of essential oils against five food borne pathogens and obtained very significant results. They found that Gram positive bacteria was more sensitive than Gram negative. A concentration of 0.075% cinnamon oil was found to be bacteriostatic against *Campylobacter jejuni* (causes food poison in infants and bacterial gastroenteritis, and usually grows in chicken, milk products); *Salmonella enteritidis* (causes salmonellosis associated with chicken, meat, milk and egg products); *Escherichia coli* (several pathogenic strains, produces toxins, which causes diarrhea, gastroenteritis, enterocolitis, etc.); *Staphylococcus aureus* (causes food poisoning, intestinal inflammation, gastroenteritis); *Listeria monocytogens* (a pathogenic organism, which causes meningitis, encephalitis, septicemia, endocolitis, abscesses in animals and man). Another observation in this study was the temperature effect. The bactericidal and bacteriostatic concentration at 35 °C was less than 0.05% and 0.01%, respectively, whereas when the temperature was reduced to 4 °C an increase in concentration was required for both activities, i.e. 0.5% and more than 1% for bacterisotatic and bactericidal activity. Mikamo *et al.* (1998) carried out an *in vivo* study using rat models and an *in vitro* study in intrauterine infections of *E. coli*. In the *in vitro* experiments the minimum inhibitory concentration was more than 100 μg/ml. In the *in vivo* experiments, extract was administered (125 mg/kg) to rats 16 hours after inoculating them with 8.1×10^6 colony forming units, three times a day p.o. for seven days. Cinnamon extracts significantly reduced the viable *E. coli* counts in rat utrine infections.

Lee and Ahn (1998) investigated the effect of cinnamon bark derived materials against human intestinal bacteria using an impregnated paper disc method. They compared the effect of cinnamaldehyde, *trans*-cinnamic acid, cinnamyl alcohol and eugenol with that of tetracycline and chloramphenecol as standards. The growth percentage varied with each bacterial strain. With 1 and 0.5 mg/disk, cinnamaldehyde has shown a potent inhibitory action against *Clostridium perfringens* (a food infecting bacteria usually growing in cooked, cooled and reheated foods; it causes gasgangrene, toxic shock, etc.) and *Bacteroides fragilis*, major inhabitants of the human intestinal tract (causative agents of several pathogenic conditions like interabdominal infections, gastrointestinal abscesses and ulcers). At 1 and 0.5 mg/disk, the growth of *Bifidobacterium bifidum* (pathogenicity doubtful) was inhibited and no inhibition was obtained against *B. longum* and *Lactobacillus acidophilus* (useful intestinal bacteria). Cinnamaldehyde was found to be the most potent inhibitor, and little inhibitory activity was obtained for *trans*-cinnamic acid, cinnamyl alcohol and eugenol. Tetracycline and chloramphenicol inhibited growth of all test bacteria at a low dose of 0.01 mg/disk. The result of this study indicates the high potential of cinnamom for the development of specific antibacterial agents for pathogenic bacteria. Sasidharan *et al.* (1998) also studied the effect of cinnamon extract on human pathogenic bacteria and reported the significant antibacterial effects.

Azumi *et al.* (1997) found that cinnamon bark contained an inhibitor of bacterial endotoxin. The inhibition of the activity of the bacterial endotoxins (LPS) was caused by the direct interaction of the LPS and inhibitory molecule. This finding was significant in the sense that it was the first report of the presence of a plant-derived bacterial toxin inhibitor. In yet another antimicrobial screening study De *et al.* (1999) tested the activity of 35 spices against *Bacillus subtilis, E. coli* and *Saccharomyces cerevisiae* and found that cinnamon had potent activity against both the bacteria and the fungus. The growth inhibiting potential of spice oils were studied by Chao *et al.* (2000), who observed that cinnamon oil was active against four Gram positive bacteria (*Bacillus cereas, Micrococcus luteus, Staphylococcus aures, Enterococcus faecalis*) and four Gram negative bacteria (*Alcaligens faecalis, Enterobacter cloacae, E. coli, Pseudomonas aeruginosa*). The results show that cinnamon oil had a broad spectrum of activity and a high degree of inhibition. The inhibitory effects were shown towards two fungi and on *Candida albicans* also. The results of the above investigations show the high potential of cinnamon as an antibacterial agent. It should be specifically noted that on some beneficial intestinal flora it does not have any activity. This increases the potential of developing cinnamon as a therapeutic agent against many of the pathogenic bacteria. The antibacterial studies described above provide a scientific basis for the age-old practice of using cinnamon as a food preservative. The spoilage of cooked/processed food occurs because of bacterial action. Moreover, the use of cinnamon protects food from the effect of harmful bacteria by inhibiting the growth in storage. Also, the therapeutic use of this herb in the Indian and Chinese systems of medical practice is scientifically supported.

ShangTzen *et al.* (2001) recently investigated the antibacterial action of essential oil of *C. osmeophleum*. The oil displayed a very good inhibitory effect on bacterial growth. The minimum inhibitory concentration (MIC) of the native variety of *C. osmeophleum* leaf oil is 500 μg/ml when tested against seven bacterial species (*E. coli, Pseudomonas aeruginosa, Enterococcus faecalis, Staphylococcus aureus, Staphylococcus epidermidis*, Methylene blue resistant *Staphylococcus aureus, Vibrio parahaemolyticus*).

Yadav and Dubey (1994) studied the effect of the oil on organisms causing ringworm infection in animals and humans. They found that the minimum inhibitory concentration (MIC value) for the two fungi *Trichophyton mentagrophytes* (Tinea infections) and *Microsporum audounil* (Tinea ring worm infections) was 500 ppm and was more effective than the synthetic antifungal agents clotrimazole, griseofulvin or nystatin. Yadav *et al.* (1999) extended the study on ringworm infections caused by the two fungi using ointment prepared from the oil at 500 ppm concentration. They treated the dermatomycosis of guinea pigs by twice daily applications of 2 ml of ointment in PEG (0.5 ml oil/50 ml PEG) and obtained good results. The oral toxicity of the oil in mice was recorded as an LD_{50} of 5.36 ml/kg body weight.

Cinnamon-containing creams are found to be useful in fighting 'acne' producing bacteria (such as *Propionibacterium acnes*). A commercial product, Sepicontrol A5, consisting of cinnamon bark powder and a lipo-aminoacid (lypoglycine) in a cream formulation is reported to be very effective in improving the state of oily acne-prone skin situations.

Antifungal activity

Cinnamon is a potent antifungal substance. It is fungistatic or fungicidal against many species of pathogenic fungi (Table 11.2). The preservative action of cinnamon in food and food products is partially due to the fungitoxic effect. The growth of all *Aspergillus*

Table 11.2 Antifungal activity of cinnamon against toxic fungi

Organism	Toxic effect	Source	Nature of activity	Dose	Reference
Aspergillus flavus	Hepatotoxic, carcinogenic, mutagenic, teratogenic	Cereals, corn, peanuts, tree nuts, figs, sorghum	Growth inhibition, inhibition of myclelial growth	Oil 1 mg/ml/, oil at 3–8%/3000 ppm level	Masimango (1978) Misra *et al.* (1987) Mishra *et al.* (1991)
Aspergillus parasiticus	Teratogenic	Sorghum	Growth inhibition, inhibition of myclelial growth	Not mentioned	Valearcel *et al.* (1986)
Aspergillus species			Growth inhibition, inhibition of myclelial growth	Not mentioned	Montes-Balmont and Carvagal (1998)
Alternaria alternata	Mutagenic, hemorrhagic	Cereals, grains, tomato, animal feeds	Growth inhibition	100 ppm in agar plats	Kim Young Ho (1996)
Fusarium graminearm	Nurotoxic, mutagenic	Wheat, corn	Growth inhibition	Not mentioned	Baruah *et al.* (1996)
Candida species	Candidosis	Airborn	Growth inhibition	0.5–30 mg/ml	Quale *et al.* (1996)

species of fungi and aflatoxin production by them are inhibited by cinnamon bark or cinnamon oil. Studies carried out in recent years have proved the advantages of using cinnamon as a fungistatic agent in stored cereals, pulses and other type of food articles including processed foods (see 13.2). It can be used as a preservative in fruit juice and fruit products at a wide range of pH conditions. Masimango *et al.* (1978) reported that cinnamon inhibited *Aspergillus flavus* growth and aflatoxin production. *Aspergillus* species of fungi are usually grown on storage cereals and pulses and produces toxins like aflatoxin, which are highly poisonous and carcinogenic. Valearcel *et al.* (1986) in an *in vitro* study observed that cinnamon inhibited the growth of *A. parasiticus* at 1 mg/ml concentration. Misra *et al.* (1987) observed that cinnamon leaf oil exerted the fungitoxic effect on *A. flavus* and *A. parasiticus* at 3000 ppm and 1000 ppm, respectively, and was not affected by temperature, autoclaving or storage. They attributed this activity to euginol, a minor constituent of cinnamon. Mishra *et al.* (1991) reported that the mycostatic effect was produced at a concentration of 4000 ppm and was as effective as many synthetic antifungal agents commonly employed, such as dithane M-45, thiovic, etc. Tiwari and Dixit (1994) studied the effect of cinnamon bark oil on the storage fungi *A. flavus* and *A. niger* and found that it was effective in inhibiting the mycelial growth at a MIC of 400 ppm. Tiwari *et al.* (1994) found that at the concentration of 400 ppm, cinnamon oil completely inhibited the mycelial growth of *Alternaria* spp., *Aspergillus* spp., *Bipolaris oryzae*, *Chaetomium hispanicum*, *Cladosporium* spp., *Curvularia* spp., *Fusarium* spp., *Mucor* spp., *Penicillium* spp., *Phoma* spp., *Rhizopus arrhizus.*, *Thelavia terricola*, and *Trichoderma* spp. Mukherjee and Nandi (1994) investigated the fungistatic effect of cinnamon oil on poultry feed. Poultry feed starter mash was treated with 0.1 and 0.2% w/v of cinnamon oil and stored for 30 days. In treated feeds the fungal population decreased with the length of storage compared to untreated feeds. The fungal contamination was due mainly to *A. flavus*, *A. niger*, *A. candidus*, *A. fumigatus* and *Rhizopus nigricans*. The optimal quality of cinnamon oil to protect stored maize from fungal growth was assessed by Montes-Belmont and Carvajal (1998) who found that 3 to 8% of the oil gave protection. No phytotoxic effect on germination and corn growth was observed.

Studies on the wood-destroying fungi *Gloeophyllum trabeum*, *Coriolus versicolor*, and *Botryodiploidea theobromeae* by agar diffusion technique by Jantan *et al.* (1994) indicated that cinnamon was very effective, exhibiting an ED_{50} of 60.3 μg/ml for *C. versicolor*, 58.8 μg/ml for *G. trabeum* and 48.0 μg/ml for *B. theobromeae*. Baruah *et al.* (1996) investigated the antifungal action against *Fusarium moniliforme*, a postharvest fungal pathogen of cereal crops, and observed that it completely inhibited the fungal growth. Wilson *et al.* (1997) tested cinnamon on *Botrytis cinerea* and observed that cinnamon oil was very effective against this plant pathogen. The bark extract of *C. loureirii* was investigated for antifungal activity against *Alternaria alternata*, a plant pathogen. The ether fraction of the extract was found to be highly active. A visible inhibition zone was caused by a 1000-fold dilution of the extract and the activity was comparable to polyoxin. Further investigation revealed that cinnamaldehyde was the most active component (Ho *et al.*, 1996). Cinnamon bark oil is a potent fungitoxicant against fungi causing respiratory tract mycoses. Singh *et al.* (1995) studied *in vitro* the effect of the oil on *A. niger*, *A. fumigatus*, *A. nidulans*, *A. flavus*, *Candida albicans*, *C. tropicalis*, *C. pseudotropicalis* and *Histoplasma capsulatum* and determined the minimal inhibiting concentration (MIC),

Pharmacology and Toxicology of Cinnamon and Cassia   275

minimal lethal concentration (MLC), and exposure duration for fungicidal action at MIC and higher doses. *Aspergillus fumigatus* was the most susceptible and the *H. capsulatum* and *Candida* species were most resistant. The MIC varied from 16–40 ppm. They have suggested that vapour inhalation of cinnamon oil could be useful in the treatment of respiratory tract mycosis. Inoneye *et al.* (2000) studied the inhibitory effect of essential oil on apical growth of *A. fumigatus* by vapour contact. The results showed that the growth of the fungus was retarded at 63 µg oil/ml of air concentration.

Insecticidal activity

The use of plants or extracts of plants to control, repell or to eradicate pests and harmful insects stretches back over the agricultural history of mankind. Many of the plant-derived insecticides, insect repellants and antifeedants have the advantage that while being toxic to the pests, they are nontoxic to humans. Studies conducted in different laboratories have shown promising results with cinnamon and cinnamon oil in pest control (see Table 11.3). Gracia (1990) studied the effect of the extracts of cinnamon and cinnamon oil toxicity, repellency, growth inhibition and ovicidal actions on the bean weevil *Callosobruchus chinensis*. The petroleum-ether extract and oil were found to be toxic to the bean weevil with an LD_{50} of less than 200 mg/ml. Cinnamon oil showed fumigation action, which caused 100% mortality at 50 mg of 100% oil in 40 cc of space. Both petroleum and alcoholic extracts exhibited concentration-dependent repellent action. Cinnamon oil could be used for the management of honeybee diseases. The results of the study by Calderon *et al.* (1994) on *Bacillus laevis*, the causative agent of American foulbrood diseases, have shown that the oil completely inhibited the growth at 10 ppm level for 72 hours, *Ascosphaera apis* at 100 ppm for 168 hours and *Bacillus alvei* at 10 ppm for 72 hours. Floris *et al.* (1996) also investigated the same problem and extended the study to field trials and obtained beneficial results. The oil was tested in nutrient broth and found that the minimal bactericide concentration was 50 mg/kg and the sporicide concentration was 100 mg/kg. The oil at 400 mg/kg concentration was found to be nontoxic to adult bees. In field trials, they administered the oil in solid food and it was effective at a concentration of 400 mg/kg. Another beneficial activity of cinnamon oil obtained from the studies by Hauhong and Shan Huan (1994) was the complete inhibition of the reproduction of *Sitophilus zeamais*, *Rhyzopertha dominica* (lesser grain borer) and *Tribolium castaneum* (flour beetle) at a concentration of 0.1–0.2% mixed with wheat or wheat flour. Topical application of oil at a concentration of 0.1–0.5 µl/larva effectively inhibited the growth of larvae of *Tenebrio molite*. The active constituent producing the inhibitory action was found to be cinnamaldehyde. The same authors studied the effect of the oil from another species of cinnamon (*C. micranthum*) and found that it inhibited the growth of *T. castaneum* effectively in stored crops and that the oil was safe for mammals even at high concentrations (Hauhong and Shan Huan, 1996). The antifeedant effect of cinnamaldehyde against the grain storage insects *Tribolium castaneum* and *Sitophilus zeamais* was studied by Huang and Ho (1998). A methylene chloride extract was insecticidal to both the insects. (The contact fumigant and antifeedant effects were also tested and found to have an LC_{50} value of 0.7 mg cm^{-2} and LC_{95} of 0.9 mg cm^{-2}.) However, *T. castaneum* had a higher fumigant toxicity than *S. zeamais* (LC_{50} 0.28 mg cm^{-2} and 0.54 mg cm^{-1} respectively). The adults

Table 11.3 Insecticidal activity of cinnamon extract/oil

Organism	Action	Dose	Reference
Callosobruchus chinensis (been weevil, pulse beetle)	Growth inhibition, ovicidal action	Pet. ext. 200 mg/ml LD50	Gracia (1990)
	Fumigation, repellant action	500 mg of oil in 40 cc of space	
Sitophilus zeamais (rice weevil)	Complete inhibition	0.1–0.2% oil	Xu-Hauhong and Zhao-Shauhuan (1994)
Rhyzopertha dominica (lesser grain borer)	Complete inhibition	0.1–0.2% oil	Xu-Hauhong and Zhao-Shauhuan (1994)
Tribolium castaneum (flour beetle)	Complete inhibition	0.1–0.2% oil	Xu-Hauhong and Zhao-Shauhuan (1994)
Tenebrio molitor (meal worm)	Contact fumigants	0.9 mg/cm^2 LC$_{95}$ – value	Xu Hang, Hong (1996)
Ceratitis capitata (amphibian pest)	Complete mortality	5% emulsion	Moretti *et al.* (1998)
Acanthoscelides oblectus (storage pest)	Fumigant toxicity ovicidal & larvicidal action	5–25% ext.	Lee Hoiseon *et al.* (1999)

were more susceptible than larvae. Cinnamaldehyde also exhibited good antifeedant activity. The toxicity of the oil on *Ceratitis capitata* (an amphibian pest which causes serious damage to fruit crops) was investigated by Moretti *et al.* (1998). Toxicity produced by oral administration of emulsions containing the active constituent was assessed and it was found that over 90% mortality was obtained with 5% emulsion. The pesticidal effect was caused by the irreversible damage caused by the active constituent to the gut of the organism.

Cinnamon wood exhibits considerable termiticidal activity. The wood meal was used for a study and compared with other pine wood meals. It was found that cinnamon wood meal exhibited a significantly higher effect than other wood meal studied (Hashimoto *et al.*, 1997). Morallo Rejeus and Punzalan (1997) reported on the molluscicidal action of cinnamon oil. The LC_{100} value was assessed to be less than 200 ppm. Another study by Roger and Hamraoui (1994) on *Acanthoscelides oblectus*, a storage pest that damages its host plant *Phaseolus vulgaris* (kidney bean), showed that cinnamon oil exhibited fumigant toxicity to adults and inhibited reproduction through ovicidal and larvicidal action. These actions can be combined to improve the management of this bruchid. The antignawing activity of cinnamon bark was studied using laboratory reared mice and it was found to be very effective (Hoiseon *et al.*, 1999). Both cinnamaldehyde and cinnamylalcohol were the active constituents which produced the action. They exhibited potent activity at a concentration of 5 and 2.5%.

Toxicology

Cinnamon bark and cinnamon oil are non-toxic and generally considered as safe (GRAS) substances. The ADI (allowable daily intake) recommended by WHO/FAO is 700 μg/kg body weight when used as food additives (FAO/WHO Tech. Report, 1984). Allergic reactions, mainly contact sensitivity, to cassia oil and bark have been reported. Cinnamaldehyde in toothpastes and perfumes has also been reported to cause contact sensitivity. In a double blind placebo-controlled personal study Minimaki (1995) reported that out of 29 patients studied six positive cases were obtained. Allen and Blozis (1988) reported the formation of oral lesions induced by contact allergy with cinnamon flavoured chewing gum. The symptoms subsided two days after discontinuing the product containing cinnamon. Sensitivity reactions due to cinnamaldehyde in toothpaste have also been reported (Lamey *et al.*, 1990). Cassia oil is stated to cause dermal and mucous membrane irritation. The irritant and sensitising properties of cassia oil have been attributed to cinnamaldehyde. The LD_{50} value of cinnamaldehyde was assessed to be 132, 6110 and 2225 mg/kg body weight in mice by intravenous, intraperitonial and oral administration, respectively. The intravenous LD_{50} of the decoction of the herb was 18.48 ± 1.80 g/kg (Wang, 1983). The LD_{50} of the oil in an oral toxicity study was assessed to be 5.36 ml/kg in mice (Yadav *et al.*, 1999). Cinnamaldehyde at low dosage caused inhibition of the activity of animals and at high dosage caused strong spasms, motor disturbances, shortness of breath and lastly death from respiratory paralysis (Wang, 1983). However, no human study has been reported which assessed the acute and chronic toxicity at high doses. Because of the pungent taste of the bark and oil, ingestion of high doses is highly unlikely. The recommended dose as a food additive is 10–20 ppm.

Therapeutic Uses

In the allopathic system of medicine neither cinnamon nor cinnamon oil is used. However, in multidrug formulations it forms an important ingredient. Tincture of cinnamon, cinnamon oil, etc. has been used in the pharmacy when medicaments are dispensed in various forms. In the pharmacopoeias of different countries cinnamon and cinnamon oil are included as official drugs mainly for the treatment of gastrointestinal disorders. In 'Martindale' (Reynold, 1996) and the Extrapharmacopoeia, several multidrug preparations containing cinnamon are described which are used therapeutically in various countries for the treatment of gastrointestinal disorders. The main use of cinnamon is as a food additive. In the 'Indian Systems of Medicine', cinnamon is used in the treatment of rheumatism, colic and diarrhoea, dyspepsia, flatulence, nausea and vomiting. It is used in many *Ayurvedic* formulations such as *Sudarsan choorna*, *Yograj gulgul*, *Aswagandharista*, *Dasamoolarishta*, *Vyaghri*, *Harithaki*. Bark forms a constituent of the Unani composition *Jawarish jalinoos* used for the treatment of gastrointestinal complaints. In folk medicine the seeds are bruised and mixed with honey and used for the treatment of coughs and dysentry in children. Doses usually used for bark powder are 6.0 to 1200 mg as an infusion three times daily; for liquid extract 0.5–1.0 ml three times daily; and, for tincture cinnamon 2–4 ml.

Chinese cassia is important in Chinese medicine. The 'Chinese Materia Medica' recognises the bark and twigs of the cassia tree as two separate drugs having different therapeutic effects (see Chapter 6 for details).

Conclusion

The broad spectrum of pharmacological actions of cinnamon cortex and cinnamon oil, together with its use as a spice, makes it a wonder plant. In the Indian and Chinese systems of traditional medicine cinnamon is used in many therapeutic formulations as a stimulant, a tonic, and in the relief of rheumatic disorders, of stomach disorders, of diabetes, etc. But in these systems an active principle-based pharmacological activity is not considered as the basis of therapeutics. Therefore, investigations employing modern tools of pharmacological parameters have not been undertaken to ascertain the wide range of actions attributed to cinnamon. However, the past two decades have seen a spurt in research activities in *Ayurveda* medicines and medicinal plants due to the revival of popularity of this system of traditional medicine. Many studies have been reported from China and Japan on their traditional medicines and these investigations provide pharmacological supports for many of the traditional uses of medicinal plants including cinnamon and cassia.

Most studies are confined to *in vitro* and *in vivo* in small laboratory animals. The results are all encouraging. Cinnamon is mainly used as a food additive and flavouring agent. The antimicrobial and antioxidant actions protect food from oxidative spoilage and also from bacterial growth. Cinnamon is a potent antioxidant and free radical scavenger. Free radical scavenging inhibits tissue damage of the host cell. Another significant activity is the hypoglycemic effect of this herb. It reduces the blood glucose level and also modulates the enzyme aldose reductase, thereby preventing complications of diabetes such as retinopathy, neuropathy, etc. The hypocholesterolemic potential of cinnamon is also high. It reduces not only serum cholesterol but also triglycerides, low density lipoproteins and phospholipids. These findings are very significant because these factors are decisive in cardiac disorders. The results of studies on the antiulcerogenic effects have also yielded valuable information. Cinnamon inhibits gastric secretions and potentiates mucosal flow. The hypoglycemic, antihyperlipedemic and antiulcerogenic effects are exerted at low doses. These beneficial effects could be obtained even at low concentrations through daily food intake.

Another important activity is the immunomodulatory effects exerted by cinnamon. Immunomodulation (immunostimulants and suppressants) plays an important role in human health. An interesting factor regarding cinnamon is that it can act both as immunostimulant and suppressant. Macromolecules isolated from cinnamon, such as glycoproteins and water soluble polysaccharides, were found to stimulate the immunological system, whereas smaller molecules such as cincassiol – C_1, C_2, C_3, C_4 and their glycosides suppress the system. The antiallergenic potential of this plant is a complementary reaction to the immunosuppressant action. Anticancer studies also yielded encouraging results. The modulative effect of cinnamon on certain enzymes such as Farnesyl-prolein-transferase shows that fruitful results may be obtained by further research in this area. Cytotoxicity effects were also obtained with certain molecules isolated from cinnamon.

The antimicrobial study unambiguously proved the high beneficial effect of cinnamon. It exerts selective action against many human pathogenic bacteria but spares some beneficial ones. The bactericidal and bactereostatic effects are the aspects which have been studied very widely. Cinnamon completely inhibits the growth of five foodborne pathogenic bacteria at low doses. Further studies may result in developing cinnamon as a selective antibacterial agent specific for human pathogenic organisms. Cinnamon is bactericidal not only to foodborne infective organisms but to other pathogenic microbes also. Similarly, it is a broad spectrum antifungal substance. Both cinnamon bark and oil inhibit growth of many storage fungi on cereals and pulses. In the area of insecticidal activity it has also yielded beneficial effects and is nontoxic to humans. This wide spectrum of beneficial effects makes cinnamon a wonder plant with a very high potential for future development. However, further research is required to establish and validate the beneficial activities documented in the traditional systems of medicines.

References

Agarwall, V.S. (1997) *Drug Plants of India*, Vol.I, pp. 274, Kalyan Publishers, New Delhi.

Aida, K., Shindo, H., Tawata, M. and Onaya, T. (1987) Inhibition of aldose reductase activities by Kampo Medicines. *Pl. Medica*, 53, 131–134.

Akina, T., Tanaka, S. and Tabata, M. (1986) Pharmacological studies on antiulcerogenic activity of Chinese cinnamon. *Pl. Medica*, 6, 440–443.

Allen, C.M. and Blozis, G.G. (1988) Oral mucosal reactions to cinnamon-flavoured chewing gum. *J. American Dental Assoc.*, 116, 664–667.

Ambasto, S.P. (1986) *The Useful Plants of India*, Publication and Information Directorate, New Delhi, p. 125.

An, H.J., Yang, H.C., Kweon, M.H., Shin, K.S. and Sung, H. (1987) Purification of a complement system – activity of polysaccharide from hot water extract of young stems of *C. cassia*. *Korean J. of Food Sci. and Technol.*, 29, 1–8.

Asolkar, L.V., Kakkar, K.K. and Chakae, O.J. (1986) *IInd Supplement to Glossary of Indian Medicinal Plants Part – I* (1965) Publication and Information Directorate, CSIR, New Delhi, pp. 203–204.

Atta, A.H. and Alkofahi, A. (1998) Antinociceptive and anti-inflammatory effects of some Jordanian medicinal plant extracts. *J. Ethanopharm.*, **60**, 117–124.

Azumi, S., Tanimura, A. and Tanamoto, K. (1997) A novel inhibitor of bacterial endotoxin derived from Cinnamon bark. *Biochemical and Biophysical Research Comm.*, **234**, 506–510.

Ball, M.A., Utsunomica, T., Ikemoto, K., Kobayashi, M., Pollard, R.B. and Suzuki, F. (1994) The antiviral effects of TJS-064, a traditional herbal medicine on Influenza A_2 virus infection in mice. *Experimentia*, **50**, 774–779.

Bara, M.T.F. and Vanetti, M.C.D. (1995) Antimicrobial effect of spices on the growth of *Yersinia euterocolitica. J. Herbs, Spices and Medicinal Pl.*, **3**, 51–58.

Baruah, P., Sharma, R.K., Singh, R.S. and Gosh, A.C. (1996) Fungicidal activity of some naturally occurring essential oils against *Fusarium moniliforme. J. Ess. Oil Res.*, **8**, 411–412.

Bhatacharjee (1998) *Hand Book of Indian Medicinal Plants*, Pointer Publishers, Jaipur, pp. 97–98.

Broadhurst, C.L., Polansky, M.M. and Anderson, R.A. (2000) Insulin like biological activity of culinary and medicinal plant aqueous extract *in vitro. J. Agri. Food Chem.*, **48**, 847–852.

Byoung Mog, K., Youngkwon, C. and Seuygtto, L. (1996) 2′-Hydroxy cinnamaldehyde from stem bark of *C. cassia. Pl. Medica*, **62**, 183–184.

Calderon, N.W., Shimanuki, H. and Allen-Wardell, G. (1994) An *in vitro* evaluation of botanical compounds for the control of honeybee pathogen *Bacillus larvae, Ascosphaera apis* and secondary invaders *Bacillus alvei. J. Ess. Oil Res.*, **6**, 279–287.

Chang Woo, L., Dong Ho, H. and Sang Bae, H. (1999) Inhibition of human tumor growth by 2′-hydroxy- and 2′-benzoyloxycinnamaldehydes. *Pl. Medica*, **65**, 263–266.

Chao, S.C., Young, D.G. and Oberg, C.J. (2000) Screening for inhibitory activity of essential oils on selected bacteria, fungi and viruses. *J. Ess. Oil Res.*, **12**, 639–649.

Chatterjee, A. and Prakash, S.C. (1994a) *Glossary of Indian Medicinal Plants*, Publication and Information Directorate, New Delhi, pp. 104–105.

Chatterjee, A. and Prakash, S.C. (1994b) *The Treatise on Indian Medicinal Plants*, Publication and Information Directorate, New Delhi, pp. 104–105.

Chinese Materia Medica (Eng. Trans.), p. 51 (1996).

De, M., De, A.K. and Banerjee, A.B. (1999) Screening of spices for antimicrobial activity. *J. Spices and Aromatic Crops*, **8**, 135–144.

Dhuleep, T.N. (1999) Antioxidant effects of cinnamon bark and cardamom seeds in rats fed high fat diet. *Indian J. of Exp. Biol.*, **37**, 238–242.

FAO/WHO: Food Additive Series – Toxicological Evaluation of Certain Food Additives and Contaminants (1984), 28th Report, *WHO Technical Report Series 19*, 710, World Health Organization, 1986.

Floris, I., Carta, C. and Morethi, M.D.L. (1996) Activity of various essential oils against *Bacillus* larvae with *in vitro* and *Apis* trials. *Apidologia*, **27**, 111–119.

Gracia, J.R. Jr (1990) Bioassay of five botanical materials against the been weevil, *Callosobruchus chinensis*. M.S. Thesis, Laguna (Philippines).

Hall, M.A. and Maurer, A.J. (1986) Spice extracts, lauricidine and propyleneglycol as inhibitors of *C. botulinum* in Turkey Frankfurter. *Poultry Sci.*, **65**, 1167–1171.

Harada, M. and Ozaki, Y. (1972) Pharmacological studies on Chinese cinnamon – Central effect of cinnamaldehyde, *Yakuzaku Zasshi*, **92**, 135–140; *Chinese Materia Medica*, Eng. Tran. pp. 51–54, 1996.

Harada, M. and Yano, S. (1975) Pharmacological studies on Chinese cinnamon-II Effects of cinnamaldehyde on the cardiovascular and digestive systems. *Chem. Pharmaceu. Bull.*, **23**, 941–947.

Harada, M. and Saito, A. (1978) Effects of cinnamaldehyde on the isolated heart guinea pigs and its catacholamine releasing effect from the adrenal glands of dogs. *J. Pharmacobiodynamics*, **1**, 89–97.

Hashimoto, K., Ohtani, Y. and Sameshima, K. (1997) The termiticidal activity and its transverse distribution in *Cinnamomum camphora* wood. *J. Japan Wood Res. Soc.*, **43**, 566–573.

Hauhong, X. and Shan Huan, Z. (1994) Studies on insecticidal activity of Cassia oil and its toxic constituent analysis. *J. South China Agricultural Uni.*, **15**, 27–33.

Hauhong, X. and Shan Huan, Z. (1996) Studies on insecticidal activity of the essential oil from *Cinnamomum micranthem* and its bioactive component. *J. South China Agricultural Uni.*, **17**, 10–17.

Ho, K.Y., Hynn, Y.Y., Seungttwan, O. and Kim, H.Y. (1996). Screening of antagonistic natural materials against *Alternaria alternata*. *Korean J. Pl. Path.*, **12**, 66–71.

Hoiseon, L., Young Joon, A., Lee, H.K., Lee, H.S. and Ahn, Y.J. (1999) Antignawing factor derived from *Cinnamomum cassia* bark against mice. *J. Chem. Eco.*, **25**, 1131–1139.

Hsiao, G., Ming, T.-C., Rong, S.J., Cheng, Y., Keung, L.-K., Mei, L.-Y., Shung, W.-T., Hsiung, Y.-M., Teng, C.M., Sheu, J.R., Chen, Y.-W., Lam, K.K., Lee, Y.M., Wu, T.S. and Yen, M.H. (2001) Cinnamophilin as a novel antiperoxidative cytoprotectent and free radical scavenger. *Bioch. Biophy. Acta*, **152**(1–2), 72–78.

Huang, Y. and Ho, S.H. (1998) Toxicity and antifeedant activities of cinnamaldehyde against the grain storage insects, *Tribolium castaneum* (Herbst) and *Sitophilus zeamais* (Motsch) *J. Stored Products Res.*, **34**, 11–17.

Hussain, R.A., Kim, J., Hu, T.W., Soejarto, D.D. and Kinghorn, A.D. (1986) Isolation of a highly sweet constituent from *Cinnamomum osmophloem* leaves. *Planta Medica*, **5**, 403–404.

Hyun, J.W., Lim, K.H. and Shin, J.E. (1994) Antineoplastic effect of extracts from traditional medicinal plants. *Korean J. Pharmacognosy*, **25**, 171–177.

Ikawati Z., Wahyuono, S. and Maeyama, K. (2001) Screening of several effect on histamine release from RBL-2H3 cells *J. Ethnopharmacology*, **75**, 249–256.

Inoneye, S., Tsuruoka, T., Watanabe, M., Takeo, K., Akao, M., Nishiyama, Y. and Yamaguchi, H. (2000). Inhibitory effect of essential oils on apical growth of *Aspergillus fumigatus* by vapour contact. *Mycoses*, **43**, 17–23.

Ismaiel, A. and Pierson, M.D. (1990) Inhibition of growth and germination of *C. botulinum*, 33A, 40B and 1623E by essential oils of spices. *J. Food. Sci.*, **55**, 1676–1698.

Iyangar, M.A., Gosh, T.K. and Nayak, S.G.K. (1994) Evaluation of a locally growing *Cinnamomum* species. *Indian Drugs*, **31**, 87–89.

Jantan, I., Ali, R.M. and Goh, S.H. (1994) Toxic and antifungal properties of the essential oils of cinnamon species. *J. Tropical Forest Sci.*, **6**, 286–292.

Jin Won, H., Kyoungttwa, L. and Jin, E.S. (1994) Antineoplastic effects of extracts from traditional medicinal plants. *Korean J. Pharmacognosy*, **25**, 171–177.

Khory, R.N. and Katrak, N.N. (1994) *Materia Medica of India and their Therapeutics*, Komal Prakashan, Delhi (reprinted 1999), pp. 527–528.

Kim, N.J., Jung, E.A., Kim, D.H. and Lee, S. (1999; 2000) Studies on the development of antihyperlipidemic drugs from Oriental herbal medicine. *Korean J. Pharmacognosy*, **30**, 368–374, **31**, 190–195.

Kirtikar, K.R. and Basu, B.D. (1975) *Indian Medicinal Plants*, Vol.III, Bishen Singh Mahendrapal Singh, Dehra Dun, pp. 2145–2150.

Kobayashi, M., Utsunomiya, T., Herndon, D.N., Pollard, R.B. and Suzuki, F. (1994) Effect of traditional Chinese herbal medicine on the production of interleukin-4 from a clone of burn-associated suppressor T cells. *Immunology Let.*, **40**, 13–20.

Kobayashi, Y., Watanabe, M., Ogihara, J., Kato, J. and Oshi, K. (2000) Inhibition of HIV-1 reverse transcriptase by methanol extracts of commercial herbs and spices. *J. Jap. Soc. Food Sci. & Technol.*, **47**, 642–645.

Koh, W.S., Yoon, S.Y., Kwon, B.M., Jeong, T.C., Nam, K.S. and Han, A.U. (1999) Cinnamaldehyde inhibits lymphocyte proliferation and modulates T-cell differentiation. *International J. Immunopharm.*, **20**, 643–660.

Konoshima, T., Takashi, M., Kozuka, M. and Takuda, H. (1994) Antitumor promoting activities of Kampo prescriptions II. Inhibitory effect of mouse skin tumors and pulmonary tumors. *J. Pharmaceutical Soc. Japan*, **114**, 248–256.

Kubo, M., ShiPing, M., JianXin, W., Matsuda, M., Ha, S.P. and Wu, J.X. (1996) Anti-inflammatory activities of 70% methanolic extract of Cinnamomi cortex. *Biol. Pharmaceu. Bull.*, **19**, 1041–1045.

Lamey, P.J., Lewis, M.A., Rees, T.D., Fowler, C., Binnie, W.H. and Forsyth, A. (1990) Sensitivity reactions to cinnamaldehyde content of toothpaste, *British Dental J.*, **168**, 115–118.

Lee, H.S. and Ahn, Y.J. (1998) Growth inhibiting effects of *Cinnamomum cassia* bark derived materials on human intestinal bacteria. *J. Agri. and Food Che.*, **46**, 8–12.

Lee, S.H., An, S.Y., Du, H.K. and Chung, S.K. (1999) Blood glucose lowering activity and mechanism of Sangbackpitang in mouse. *Yakhak Hoyi*, **43**, 818–826.

Masimango, N., Ramut, J.L. and Remade, J. (1978) Study on the role of chemical additives in combating aflatoxins. *Rev. Fermont. Ind. Aliment*, **33**, 116–23; *Chem. Abstr.* 90, 1119835z.

Matsuda, M., Matsuda, R., Fukuda, S., Shimoto, H. and Kubo, M. (1987) Antithrombotic actions of 70% methanolic extract and cinnamic aldehyde from cinnamomi cortex. *Chem. and Pharmaceu. Bull.*, **35**, 1275–1280.

Minimaki, A. (1995) Double blind placebo-controlled peroral challenges in patients with delayed-type allergy. *Contact Dermatitis*, **33**, 78–83.

Mikamo, H., Kawazoe, K., Izumi, K., Sato, Y. and Tamaya, T. (1998) Effects of crude herbal ingredients on intrauterine infection in a rat model. *Therapeutic Res.*, **59**, 122–127.

Mishra, A.K., Dwivedi, S.K., Kishore, N. and Dubey, N.K. (1991) Fungistatic properties of essential oil of *C. camphora*. *Intern. J. Pharmacog.*, **29**, 259–262.

Misra, N., Batra, S. and Batra, S. (1987) Efficiency of essential oil of *Cinnamomum tamala* against *Aspergillums flavus* and *A. parasiticus* producing mycotoxins in stored seeds. *Indian Perfumer*, **31**, 332–334.

Mitra, S.K., Sundaram, R. and Venkataranganna, M.V. (2000) Antiinflammatory, antioxidant and antimicrobial activity of 'Ophthacre' brand, a herbal eye drop. *Phytomedicine*, 7, 123–127.

Montes-Belmont, R. and Carvajal, M. (1998) Control of *Aspergillus flavus* in maize with plant essential oils and their components. *J. Food Protection*, **61**, 616–619.

Morallo-Rejeus, B. and Punzalan, E.G. (1997) Molluscicidal activity of some Phillippine plants on golden snail. *Phillippine Entom.*, 11, 65–79.

Moretti, M.D.L., Bazzoni, E., Passino, G.S. and Prota, R. (1998) Antifeedant effects of some essential oils on *Ceratitis capitata*. *J. Essential Oil Res.*, **104**, 405–412.

Mukherjee, P.S. and Nandi, B. (1994) Poultry feed preservation from fungal infection by Cinnamon oil. *J. Mycopath. Res.*, **32**, 1–5.

Mukherjee, R.K., Fujimoto, Y. and Kakimema, K. (1994) 1-(ω-Hydroxy-fattyacyl)glycerols and two flavonols from *Cinnamomum camphora*. *Phytochemistry*, **37**, 1641–1643.

Nadkarni, K.M. (1976) *Indian Materia Indica*, Popular Prakshan, Bombay, pp. 329. (Originally published in 1899), Scientific Publishers (India) Jodhpur.

Newall, C.A., Anderson, L.A. and Philipson, J.D. (1996) *Herbal Medicines*, The Pharmaceutical Press, London.

Nohara, T., Kashiwada, Y., Tomimatsu, T. and Nishikawa, I. (1980) Two novel diterpenes from the Bark of Cinnamomi cortex. *Chem. Pharm. Bull.*, **28**, 1969–1970.

Onishi, Y., Yamamura, T. and Tauchi, K. (1998) Expression of antimetastatic effect induced by Juzeu-Taiho-to is based on the content of Shimosu-to constituents. *Biol. Pharmaceu. Bull.*, **21**, 761–765.

Patole, A.P. and Agte, V.V. (1998) Effect of various dilatory constituents on insulin secretion on pancreas culture. *J. Medicine and Aromatic Plant Sci.*, **20**, 413–416.

Peters, M.M.C.G. and Cadwell, J. (1994) Studies on *trans*-cinnamaldehyde: Influence of dose, sex and size in rats and mice. *Food and Chem. Toxico.*, **32**, 869–876.

Polansky, M.M. and Anderson, R.A. (1992) Stimulatory effects of cinnamon and brewers yeast as influenced by insulin. *Hormone Res.*, **37**, 225–229.

Premanathan, M., Rajendran, S., Ramanathan, T., Kathiresan, K., Nakashimo, H. and Yamamoto, N. (2000) A survey of some Indian medicinal plants for anti-human immuno deficiency virus (HIV) activity. *Indian J. Med. Res.*, **112**, 73–77.

Pharmacology and Toxicology of Cinnamon and Cassia 283

Quale, J.M., Landman, D., Zaman, M.M., Burney, S. and Sathe, S.S. (1996) *In vitro* activity of *Cinnamomum zeylanicum* against azole resistant and sensitive *Candida* species and a pilot study of cinnamon for oral candidiosis. *American J. Chinese Medicine*, 24, 103–109.

Reddy, A.C. and Lokesh, R.R. (1992) Studies on spice principles as antioxidants in the inhibition of lipid peroxidation of liver microsomes. *Molecular Cell Bioch.*, 111, 117–124.

Reynold, J.E.F. (ed.) (1996) *Martindale – The Extra Pharmacopoeia* Royal Pharmaceutical Society, London, pp. 1686–3.

Roger, R.C. and Hamraoui, A. (1994) Inhibition of reproduction of *Acanthoscilides oblectus* Say (Coleoptera), a kidney bean bruchid by aromatic essential oils. *Crop Protection*, 13, 624–628.

Sambaiah, K. and Sreenivasan, K. (1991) Secretion and composition of bile in rats fed diets containing spices. *J. Food Sci. and Technol.*, 28, 35–38.

Samuelson, D.B., Brenna, J., Solheim, E. and Solheim, R.R. (1986) Metabolism of the cinnamon constituent *o*-methoxy cinnamaldehyde in rats. *Xenobiotica*, 9, 845–47.

Sasidharan, V.K., Krishnakumar, T. and Manjula, C.B. (1998) Antimicrobial activity of nine common plants in Kerala, India. *Phillippine J. Sci.*, 127, 65–72.

Shan, B.E., Yoshida, Y., Sugimura, T. and Yamashita, U. (1999) Stimulating activity of Chinese medicinal herbs on human lymphocytes *in vitro. International J. of Immunopharm.*, 21, 149–159.

Sharma, S.R., Dwivedi, S.K. and Swarup, D. (1996a) Hypoglycemic and hypolipidemic effects of *C. tamala* leaves. *Indian J. Experimental Biol.*, 34, 372–374.

Sharma, S.R., Dwivedi, S.K. and Swarup, D. (1996b) Hypoglycemic effect of some indegeneous medicinal plants in normoglycemic rats. *Indian J. Animal Sci.*, 66, 1017–1020.

ShangTzen, C., Paifun, C., Shachwen, C., Chang, S.T., Chen, P.F. and Chang, S.C. (2001) Antibacterial activity of leaf essential oils and their constituents from *Cinnamomum osmophloeum. J. Ethnopharmacology*, 77, 123–127.

Shichinohe, K., Shimizu, H. and Kurokawa, K. (1976) Effect of M-711 on experimental skin reactions induced by chemical mediators in rats. *J. of Veterinary Medical Sci.*, 58, 419–443.

Shiraga, Y., Okano, K., Akira, T., Fukaya, C., Yokoyama, K. and Tanaka, S. (1988) Structures of potent antiulcerogenic compounds from *Cinnamomum cassia. Tetrahedron*, 44, 4703–4711.

Singh, H.B., Srivastava, M., Singh, A.B. and Srivastava, A.K. (1995) Cinnamon bark oil, a potent fungitoxicant against fungi causing respiratory tract mycosis. *Allergy*, 50, 995–999.

Smith-Palmer, A., Stewart, J. and Fyfe, L. (1998) Antimicrobial properties of plant essential oils and essences against five important food-borne pathogens. *Lett. Appl. Microbio.*, 26, 118–122.

Stoltz, C. (1998) Sepicontrol A5. *SEPIC product Newsletter*, pp. 152–156.

Sugaya, A., Tsuda, T., Yasuda, K., Sugaya, E. and Onozuka, M. (1985) Effect of Chinese Herbal Medicine "Saiko-keishi-to" on intracellular calcium and protein behaviour during PT 2-induced bursting activity in snail neurons. *Pl. Medica*, 1, 2–6.

Tabak, M., Arimon, R., Potasman, J. and Neeman, I. (1996) *In vitro* inhibition of *Helicobacter pylori* by extracts of thyme and cinnamom. *J. Appl. Bacteriol.*, 80, 667–672.

Tanaka, S., Yoon, Y.H., Fukui, H., Tabata, M., Okira, T., Okano, K., Iwai, M., Iga, Y. and Yokoyame, K. (1989) Antiulcerogenic compound isolated from Chinese cinnamon. *Pl. Medica*, 55, 245–248.

Tanaka, Y., Takagaki, Y. and Nishimura, T. (1991) Effect of food additives on β-hexosamidase release from rat basophil leukemia cells. *EiseiKagaker*, 37, 370–378.

Tang, W. and Eisenbrand, G. (1992) *Chinese Drugs of Plant Origin – Chemistry, Pharmacology and Use in Traditional and Modern Medicine*, Springer-Verlag, Berlin, pp. 319–330.

Tanikawa, K., Goto, H., Nakamura, N., Tanaka, N., Hattori, M., Itoh, T. and Terasawa, T. (1999). Endothelium-dependent vasodilatation effect of tannin extract from Cinnamoni cortex in isolated rat aorta. *J. Traditional Medicine*, 16, 45–50.

Tiwari, R. and Dixit, V. (1994) Fungitoxic activity of vapours of some higher plants against predominant storage fungi. *National Academy of Sci. Lett.*, 17, 55–57.

Tiwari, R., Dixit, V. and Dixit, S.N. (1994) Studies on fungitoxic properties of essential oil of *Cinnamomum zeylanicum* Breyn. *Indian Perfumer*, **38**, 98–104.

Torhara, T., Kashiwada, Y., Tonimastu, T. and Nishikoa, I. (1980) Two novel diterpenes from bark of *Cinnamomum cassia. Chem. Pharm. Bull.*, **28**, 1969.

Toshihiro, N., Yoshiki, K., Kotaro, M., Toshiaki, T. and Masuru, K. (1981) *Chem. Pharm. Bull.*, **29**, 2451–2459.

Trombetta, D., LoCasero, R., Pellegrino, M.L. and Tomarino, A. (1998) Antioxidant properties and phenolic content of essential oils from Mediterranean plants. *Fitoterapia*, **69**, 42.

Valearcel, R., Bennett, J.W. and Vitanza, J. (1986) Effect of selected inhibitors on growth, pigmentation and aflatoxin production by *Aspergillus parasiticus. Mycopathologia*, **94**, 7–10.

Viollon, C., Mandin, D. and Chaumont, J.P. (1996) Antagonistic activity, *in vitro*, of several essential oils and constituent volatile compounds against the growth of *Trichomonas vaginalis. Fitoterapia*, **67**, 279–281.

Wang, J.H. (1985) *Chinese Herbal Pharmacology*. Shanghai Science & Technology Press, Chinese Materia Medica, pp. 27–28.

Wang, Y.S. (1983) *Pharmacology and Applications of Chinese Materia Medica*. Peoples Health Publishers, Beijing, pp. 442–446.

Warriar, P.K., Nambiar, V.P.K. and Ramankutty, C. (1994) *Indian Medicinal Plants – a Compendium of 500 Species*, Vol.II. Vaidyaratnam P.S. Varier's Arya Vaidyasala, Kottakkal, Kerala, pp. 80–83.

Webb, A.H. and Tanner, F.W. (1944) Effect of spices and flavouring materials on growth of yeast. *Food Resis.*, **10**, 273–282.

Wendakoon, C.N. and Sakagudi, M. (1995) Inhibition of amino acid decarboxylase activity of *Enterobacter aerogenes* by active components in spices. *J. Food Protection*, **58**, 280–283.

Wilson, C.L., Solar, J.M., El-Ghaouth, A. and Wisnieweski, M.E. (1997) Rapid evaluation of plant extracts and essential oils for antifungal activity against *Botrytis cinerea. Pl. Diseases*, **81**, 204–210.

Yadav, P. and Dubey, P.N.K. (1994) Screening of some essential oils against ringworm fungi. *Indian J. Pharmaceutical Sci.*, **56**, 227–230.

Yadav, P., Dubey, N.K., Joshi, V.K., Chansouria, J.P.N. and Yadav, P. (1999) Antidermatophytic activity of essential oil of *Cinnamomum* as herbal ointment for cure of dermatomycosis. *J. Medicinal and Aromatic Plant Sci.*, **21**, 347–351.

Yokozawa, T., Dong, E., Liu, W. and Oura, H. (1977) Antiperoxidation activity of traditional Chinese prescriptions and their main crude drugs in vitro. *Natural Medicine*, **51**, 92–97.

Yokozawa, T., CuiPing, C. and ZhongWu, L. (1998) Effect of traditional Chinese prescriptions and their main crude drugs on 1,1-diphenyl-2-picrylhydrazylradical. *Phytotherapy Res.*, **12**, 94–97.

Yoshiki, K., Toshihiro, N., Toshiaki, T. and Itsuo, N. (1981) *Chemical Pharmaceutical Bull.*, **29**, 2686–2688.

Zhang, T., Su, C. and Chen, D. (1990) Comparison of bacteriostatic ability of oleum *Perilla frutescens* and *Cinnamomum cassia. Chung Kuo Chung Yao Chich*, **15**, 95–97, 126–127.

12 Economics and Marketing of Cinnamon and Cassia – A Global View

M.S. Madan and S. Kannan

Introduction

Cinnamon (or true cinnamon or Sri Lankan cinnamon) is produced and exported from Sri Lanka, Madagascar and Seychelles, while cassia cinnamons are produced and exported from China and Vietnam (Chinese cassia), Indonesia (Indonesian cassia) and India (Indian cassia) (Anon, 1995; Madan, 2001). Together cinnamon and cassia constitute important spices that are traded in the international market and are used by people of many countries. Official statistics on area, production and productivity of cinnamon from producing countries are conflicting and are of doubtful reliability. For many countries only the production figures are available. Further more, as in the case of FAOSTAT, the available statistics do not distinguish between cinnamon and cassia, and provide data on area, production and yield under a single crop head of cinnamon (*canella*). However, the trade related figures are comparatively complete and make a distinction between cassia and cinnamon. The U.N. Comtrade Database provides country-wise data the on import of whole cinnamon and ground cinnamon listed separately (ITC, 2000). USDA/FAO data sources provide export-import data for ground and whole cinnamon and cassia separately. In both databases cinnamon and cassia are not differentiated. However, cassia leaf oil and cinnamon bark oil imports are listed separately by USDA/FAO.

Despite certain limitations in the availability of data, this chapter makes use of the time series data obtained from FAO, to analyse the trend in country-wise area, production and export and import. The aim of this effort is to get some broad indications on the possible changes that have taken place in the crop economy during the last three decades since 1970–71, and future prospects based on observed trends.

As per the official statistics available from FAO (FAOSTAT) the world production of cinnamon (*canella*) during 2000 was 90,213 from an area of approximately 132,970 ha (FAO, 2000). Table 12.1 provides decade-wise (period) average production and their percentage share in total world production. As can be seen from Table 12.1, during the past three decades from 1971 to 2000, both area and production have increased slowly but steadily.

Country-wise contribution to the world production over the period as presented in Table 12.1 indicates that Sri Lanka, the traditional producer of cinnamon, contributed about 39.3% of the total world production in the 1970s (period I). Its share came down to 19.7% during period II (1980–90). Countries like Indonesia and China improved their contribution from 30.7% and 21.7%, respectively, in period I to 40.3% and 32.9% in the period II. Vietnam is the other country to contribute considerably to world production of cassia cinnamon with an annual production of more than 4000 t in the recent past.

0-415-31755-X/04/$0.00 + $1.50
© 2004 by CRC Press LLC

Table 12.1 World production of cinnamon during 1971–2000

Countries	Period I (1971–1980)		Period II (1981–1990)		Period III (1991–2000)	
	Area (ha)	*Prodn.* (t)	*Area* (ha)	*Prodn.* (t)	*Area* (ha)	*Prodn.* (t)
Sri Lanka	21,628 (43.3)	10,324 (39.3)	21,088 (23.1)	10,318 (19.7)	23,292 (18.5)	12,010 (15.0)
Indonesia	18,790 (37.6)	8076 (30.7)	41,873 (45.8)	21,100 (40.3)	56,603 (44.2)	35,742 (44.5)
China	–	5700 (21.7)	35,000 (38.2)	17,200 (32.9)	35,000 (27.7)	26,350 (32.8)
Madagascar	467 (0.9)	282 (1.1)	1660 (1.8)	760 (1.5)	1800 (1.4)	900 (1.1)
Seychelles	6421 (12.9)	1138 (4.3)	3285 (3.6)	661 (1.3)	2800 (2.8)	467 (0.6)
Vietnam	1500 (3.0)	425 (1.6)	3295 (3.6)	1700 (3.3)	7640 (6.1)	4210 (5.2)
World	49,937	26,281	91,421	52,337	126,207	80,361

Source: FAOSTAT.

Notes
1 Production figures are decadal average.
2 Figures in brackets indicate percentage in totals.

Emerging Trends in Area and Production

An analysis of time series data on area and production for the period 1971 to 1999–2000 revealed much information on the crop. For the sake of comparison the three decades were treated independently (I, II and III). As can be seen from Tables 12.2 and 12.3, during the first ten year period (1971–1980) there was a slow improvement in area expansion and production. By the end of the period (i.e. in 1980) the area increased by 19.4% and the total world production by 33.4% over the base year (1971). In the following two decades, there was a steady increase in area and production. Growth in production was around 160% by the end of the second period (i.e. 1990), and 276% by 2000. The area expansion of 112.38% and 171.24%, respectively, during the same periods was comparatively less. In countries like India, where the crop is cultivated mostly, under homestead conditions, reliable statistics are lacking and in Seychelles the bark is harvested from natural forests, for which statistics are not available.

Keeping 1970–71 as the base year, country-wise growth indices for subsequent years were worked out for area and production, which indicated that, China and Indonesia achieved an increase of 600% and 595%, respectively, by the end of 2000. The highest (952%) increase in production level was achieved by Vietnam. The estimated growth index is negative in most years for the traditional cinnamon producing countries like Seychelles and Madagascar. Sri Lanka, the major producer of true cinnamon in the world, lost its monopoly to other competing countries. There was a continuous decline in area as well as in production. However, Sri Lanka

Table 12.2 Estimated growth index for area under cinnamon in major producing countries (1971–2000)

Period	Year	Indonesia	Madagascar	Seychelles	Sri Lanka	Vietnam	China	World
	1971	100	100	100	100	100	100	100
	1972	3.03	−28.79	59.72	−27.00	–	–	−3.26
	1973	27.27	−32.20	30.56	−7.17	–	–	9.99
Period I	1974	20.61	−10.50	18.06	−4.97	–	–	7.18
(1971 to	1975	−16.64	−82.67	−11.11	−6.65	–	–	−11.20
1980)	1976	−5.50	−81.99	−8.33	−6.11	–	–	−6.78
	1977	−15.15	−78.85	−34.72	−5.78	–	–	−13.69
	1978	45.45	−76.13	−51.39	−5.71	–	–	4.30
	1979	65.81	33.02	−58.89	−1.10	–	–	12.72
	1980	87.62	−5.18	−52.08	−4.67	–	–	19.39
	1981	172.95	6.41	−63.19	−5.87	33.33	–	46.92
	1982	205.60	10.50	−33.33	−4.91	33.33	–	62.77
	1983	193.94	13.23	−27.78	−4.54	33.33	–	59.83
Period II	1984	187.88	20.05	−72.22	−9.27	33.33	–	49.03
(1981 to	1985	139.39	36.43	−27.78	−10.05	66.67	677.78	111.50
1990)	1986	125.41	63.71	−18.06	−10.85	100.00	677.78	109.07
	1987	109.03	131.92	−63.89	−9.36	133.33	677.78	99.06
	1988	120.00	350.20	−75.00	−12.37	176.67	677.78	102.45
	1989	137.58	459.35	−77.78	−12.12	280.00	677.78	111.85
	1990	145.95	172.85	−84.72	−13.07	306.67	677.78	112.38
	1991	164.55	241.06	−62.50	−12.58	306.67	677.78	122.34
	1992	164.75	186.49	−77.78	−13.00	313.33	677.78	119.99
	1993	218.18	186.49	−79.17	4.07	326.67	677.78	146.29
	1994	245.45	159.21	−62.50	4.22	373.33	677.78	159.42
Period III	1995	263.64	131.92	−52.78	3.29	400.00	677.78	167.15
(1991 to	1996	284.85	104.64	−51.39	3.29	400.00	677.78	174.29
2000)	1997	280.00	104.64	−54.17	3.29	493.33	677.78	171.24
	1998	269.70	118.28	−56.94	3.29	493.33	677.78	171.24
	1999	269.70	118.28	−56.94	3.29	493.33	677.78	171.24
	2000	269.70	104.64	−56.94	3.29	493.33	677.78	171.24

Source: Based on FAOSTAT.

is still the leader in the production of true cinnamon. As far as India is concerned, there was no improvement in overall production, though import substitution was the main thrust in spice development during the eigth and ninth five-year plan periods (Spices Board, 1992). Against the world trend, area and production were declining in India year after year. Research programmes to improve production and productivity of cinnamon and cassia in the country are in operation at the Indian Institute of Spices Research, Calicut (Rema and Krishnamoorthy, 1999).

Growth estimates

In order to get the long-term trends in area, production and productivity globally and in major producing countries, semi-logarithmic growth equations are estimated and the results are presented in Table 12.4. The overall trend in area under cinnamon

288   M.S. Madan and S. Kannan

Table 12.3 Estimated growth index for production of cinnamon in major producing countries (1971–2000)

Period	Year	Indonesia	Madagascar	Seychelles	Sri Lanka	Vietnam	China	World
	1971	100	100	100	100	100	100	100
	1972	7.46	−28.79	61.54	−27.03	0.00	11.11	−5.79
	1973	27.61	−32.20	27.69	−7.21	0.00	11.11	6.23
	1974	25.51	−10.50	15.38	−4.95	−50.00	11.11	5.47
Period I	1975	12.75	−82.67	−13.85	−6.76	−50.00	11.11	−2.85
(1971 to	1976	2.69	−81.99	−9.85	−6.31	0.00	33.33	0.16
1980)	1977	25.32	−78.85	−37.62	−5.86	12.50	33.33	4.60
	1978	124.48	−76.13	−52.31	−1.31	25.00	37.78	28.35
	1979	85.39	33.02	−61.54	−4.67	50.00	40.00	25.12
	1980	98.97	−5.18	−53.85	−5.83	75.00	77.78	33.43
	1981	137.94	6.41	−63.85	−4.92	100.00	122.22	50.62
	1982	126.94	10.50	−33.08	−4.57	125.00	166.67	58.96
	1983	197.56	13.23	−26.92	−12.41	150.00	211.11	85.16
	1984	273.55	20.05	−26.92	−15.59	175.00	255.56	106.46
Period II	1985	279.69	36.43	−28.46	−13.00	200.00	300.00	119.26
(1981 to	1986	266.04	63.71	−7.69	−7.79	275.00	344.44	125.94
1990)	1987	372.03	131.92	−64.62	−5.83	350.00	344.44	151.75
	1988	343.32	350.20	−75.38	−11.11	525.00	388.89	163.43
	1989	324.39	459.35	−78.46	−5.10	650.00	344.44	158.37
	1990	362.84	172.85	−86.15	9.91	700.00	344.44	160.12
	1991	372.31	241.06	−64.46	7.41	700.00	366.67	167.41
	1992	412.73	186.49	−79.77	8.11	750.00	388.89	182.36
	1993	465.13	186.49	−81.38	7.84	775.00	411.11	206.33
	1994	518.75	159.21	−63.92	9.64	900.00	344.44	207.68
Period III	1995	551.89	131.92	−52.31	5.77	1000.00	455.56	238.21
(1991 to	1996	588.76	104.64	−47.69	9.64	1000.00	522.22	258.73
2000)	1997	544.79	104.64	−57.54	9.64	1100.00	566.67	258.56
	1998	595.53	118.28	−63.23	5.77	1100.00	600.00	275.23
	1999	595.53	118.28	−65.38	10.09	1100.00	600.00	277.11
	2000	595.53	104.64	−65.38	8.11	1100.00	600.00	275.78

Source: Based on FAOSTAT.

registered an average annual growth rate of 5.2% for the period from 1971 to 2000. The increase in world production was at the rate of 5.6% during the same period, indicating a slight improvement in productivity. In order to examine the cyclical fluctuation if any, growth analysis was done for each decade separately. The estimated growth rate was 1.1 and 4.5% in period I and period II, while there was a negative growth rate of − 7.1% in period III. Production had a positive growth rate of 3.4, 6.7 and 4.1% in all the three periods, respectively. The estimated negative growth rate in area and positive growth rate in production during the 1990s indicate the improvement in productivity, i.e. with less area under the crop more quantity is produced.

Country-wise estimates

As regards country-wise performance the major cassia producing countries (Indonesia, China and Vietnam) registered a positive growth rate both in area expansion and production during all three periods (decades), i.e. their overall performance is more

Economics and Marketing of Cinnamon and Cassia 289

Table 12.4 Growth estimates for area and production of cinnamon in major producing countries

Countries	Estimated compound growth rate (%)			
	1971–80	1981–90	1991–2000	Overall
Area				
Indonesia	5.43	−3.18	3.98	5.10
China	5.30	18.80	27.70	–
Sri Lanka	1.07	−1.00	1.60	0.39
Seychelles	−12.91	−11.41	6.33	−4.17
Madagascar	−3.40	19.13	−5.13	7.68
Vietnam	–	14.87	5.13	8.21
World	1.10	4.46	−7.10	5.22
Production				
Indonesia	8.02	8.22	4.23	7.78
China	5.35	8.10	4.62	7.65
Sri Lanka	0.85	1.05	0.08	0.77
Seychelles	−14.5	−14.39	4.66	−4.61
Madagascar	−3.5	19.1	−5.2	7.68
Vietnam	7.6	17.91	4.34	12.11
World	3.41	6.70	4.09	5.63

than the world average. Vietnam outperformed all other cinnamon producing countries in area expansion (8.2%) and production (12.1%). Sri Lanka, the major producer of true cinnamon, had a marginal increase of 0.7% per annum for the entire period from 1971 to 2000. Seychelles is the only country that had a negative growth rate in area expansion and production.

Productivity

The average annual productivity is around 350 kg of 'Sri Lankan Cinnamon' per hectare. In Madagascar, the average annual yield from a cinnamon plantation is around 180–220 kg/ha of cinnamon sticks in quills, and 60–65 kg/ha of cinnamon chips. Productivity of true cinnamon in Sri Lanka was around 4700 kg/ha only, and 1999–2000 registered only a marginal increase over the average (Ratwatte, 1991). In Vietnam the mean productivity is around 6500 kg/ha of dry bark and 280 kg of essential oil from leaf distillation for a 20-year cycle (Anon, 2001). The highest recorded yield of above 8000 kg/ha is from China in Chinese cassia and the lowest productivity is registered from Seychelles in true cinnamon. In China the productivity of cassia cinnamon is around 3–5/ha per year. Cinnamon and cassia are perennial trees and are cultivated under natural forestry conditions. Not much effort on productivity improvement has been made in these crops.

Trade and Commerce

World trade in Sri Lankan cinnamon is centered around London and the Dutch ports of Amsterdam and Rotterdam, which are the main transshipment points for the leading buyers such as Mexico, US, UK, Germany, Holland, Colombia, etc. The traded quantity of cinnamon was between 7500 t and 10,000 t annually. Sri Lanka contributes about 80–90%, the remaining comes from Seychelles and Madagascar. The world trade in cassia is between 20,000 and 25,000 annually, of which Indonesia accounts for two-thirds and China most of the remainder. Other producers include Vietnam and India. About 2000–3000 t of cassia bark are exported from Vietnam annually. Taking Sri Lanka, Madagascar and Seychelles as the cinnamon exporting countries, during 1975 the ratio of cinnamon to cassia in the total world export was 53:47. The ratio came down to an average of 19:81 in the 1980s and in the 1990s the ratio further came down to 13:87 of the average world export of 75,763 t of cinnamon and cassia.

The market for cinnamon and cassia

Most countries import some quantity of cinnamon or cassia or both. In many of the importing countries cinnamon is the most important spice commodity in use after black pepper. The available statistics do not differentiate between cinnamon and cassia, as both are treated under the common name of cinnamon. However, whole and ground cinnamon, bark and leaf oils are tabulated separately in the export-import statistics. Imports of total cinnamon (including cassia) by major-importing countries are given in Table 12.5. The USA is the largest importer of cinnamon (mostly cassia), averaging 18,173 t during 1997–99, which is about 20% of the world exports. During 1999 India was the second largest importer (12,467 t). In both cases, based on the country of origin, it can be concluded that the maximum imported commodity was cassia. The Item-wise import of cinnamon and cassia by India during 1995–96 to 1999–2000 is given in Table 12.6. The main items of import are cinnamon, cassia and *tejpat*. The total imports went up many-fold between the reference period. Cassia and cassia leaf (*tejpat*) account for more than 90% of the total imports.

United States of America

Imports of cassia and cinnamon (whole) are given in Table 12.7, while imports of ground cinnamon and cassia are given in Table 12.8. Indonesia is the major supplier (>90%) to the USA, followed by Sri Lanka and Vietnam. India also supplies a sizable quantity to the US market. The annual imports of cassia and cinnamon averaged 9262 t in 1976–80 and has almost doubled (18,175 t) in recent years (FAOSTAT, 2000). The ratio of cassia to cinnamon in the total imports by the USA was previously 2:1, but recently the import of cassia has increased. The cinnamon market was supplied almost solely by Sri Lanka, though a small quantity was shipped from Seychelles. These countries supplied 7% of the unground cinnamon imported into the USA during 2000. However, in terms of value cinnamon accounts for about 28% of the total, indicating the higher price of cinnamon in comparison to cassia in the USA market. Indonesia was the leading supplier of cassia from 1996–2000, followed by Vietnam (USDA/FAS, 1998, 1999).

European Union

The combined cinnamon and cassia imports into European Union countries is given in Table 12.9. Although cinnamon imports into the European Union are comparatively small, it is an important market from the point of view of end use. European countries like Germany, Netherlands, France and the United Kingdom are the major re-exporting

Economics and Marketing of Cinnamon and Cassia 291

Table 12.5 Import of cinnamon and cassia by major importing countries (1997–99)

Country & year	Total spice imports		Cinnamon & cassia		Cinnamon & cassia ground		Total import of C&C		Share in total spice import (%)	
	Qty	Value	Qty	Value	Qty	Value	Qty	Value	Qty	Value
Mexico										
1997	25,488	50,554	5449	25,982	744	406	6193	26,388	24.3	52.2
1998	29,550	65,314	4880	29,601	396	430	5276	30,031	17.9	46.0
1999	31,372	60,567	5402	28,341	443	428	5845	28,769	18.6	47.5
US										
1997	242,309	496,354	16,431	31,228	1065	2045	17,496	33,273	7.2	6.7
1998	266,691	553,103	18,134	27,204	1325	2053	19,459	29,257	7.3	5.3
1999	286,855	598,846	16,346	22,059	1220	1968	17,566	24,027	6.1	4.0
Netherlands, the										
1997	94,370	156,262	3211	4942	71	244	3282	5186	3.5	3.3
1998	83,967	155,406	5333	6181	103	152	5436	6333	6.5	4.1
1999	82,626	175,667	7148	7010	140	157	7288	7167	8.8	4.1
Germany										
1997	111,799	225,766	1961	4011	860	2237	2821	6248	2.5	2.8
1998	115,625	242,105	1878	4040	553	1392	2431	5432	2.1	2.2
1999	121,066	242,366	1732	2984	582	1425	2314	4409	1.9	1.8
Korea										
1997	33,225	41,980	2482	3136	146	290	2628	3426	7.9	8.2
1998	40,670	41,875	2180	2413	175	204	2355	2617	5.8	6.2
1999	32,100	40,443	3546	3471	65	121	3611	3592	11.2	8.9
Japan										
1997	162,747	268,092	1462	3286	347	671	1809	3957	1.1	1.5
1998	160,825	217,379	1310	3001	489	908	1799	3909	1.1	1.8
1999	162,881	225,886	971	2188	509	967	1480	3155	0.9	1.4
India										
1997	35,225.51	32,957.1	–	–	–	–	3048	4389	8.3	12.0
1998	67,436.97	69,900.6	–	–	–	–	6973	8307	9.9	11.4
1999	63,488.39	66,539.9	–	–	–	–	12,467	13,711	19.6	20.6
Singapore										
1997	130,917	209,356	4265	5562	1248	1593	5513	7155	4.2	3.4
1998	85,960	159,055	2165	2479	976	1100	3141	3579	3.7	2.3
1999	106,084	212,898	2397	2782	1018	1079	3415	3861	3.2	1.8

Source: UN Comtrade Database.

Note
Qty in tonnes, values in thousand US$.

countries for cinnamon and cassia. As it can be seen from the above Table, the combined percentage for Sri Lanka, Madagascar and the Seychelles, the true cinnamon producing countries, is around 28% of the total imports. China, Vietnam, and Indonesia, the cassia producing countries, are the other suppliers to this market. The main sources for European Union countries in 1991 were Indonesia (36%), Malagasy Republic (21%), Sri Lanka (11%), China (6%), Seychelles (1%) and others, (15%), indicating the dominance of cassia type in the total imports. Nearly 36% of the ground cinnamon and cassia for the EU market is sourced from the Intra-EU countries. Indonesia, with a 23.7% share, tops the list of suppliers of ground cinnamon and cassia to this market. Thus, among the cassia exporting countries, Indonesia has taken the

Table 12.6 Item-wise import of cinnamon and cassia into India (1995–96 to 1999–2000)

Years	Items of import											
	Cinnamon				*Cassia*				*Tejpat cassia leaf*		*Total*	
	Whole		*Powder*		*Whole*		*Powder*					
	Qty	*Value*	*Qty*	*Value*	*Qty*	*Value*	*Qty*	*Value*	*Qty*	*Value*	*Qty*	*Value*
1995–96	–	–	–	–	1346.76	554.75	–	–	0.89	0.06	1347.7	554.81
1996–97	–	–	25.69	9.15	2243.46	1017.1	–	–	–	–	2269.2	1026.25
1997–98	–	–	–	–	2918.01	1578.53	–	–	9.8	0.61	2927.8	1579.14
1998–99	33.81	20.96	14.54	6.86	6581.8	3297.9	–	–	24.46	8.17	6654.6	3333.89
1999–2000	1.0	0.76	35.62	15.79	12,294.9	5829.99	14.5	6.42	121.04	42.77	12,467.1	5895.73

Source: Directorate of Arecanut and Spices, Calicut.

Note
Qty in tonnes, value in Rs Lakhs.

Table 12.7 US imports of whole cassia and cinnamon – country-wise sources

Country	1996		1998		2000	
	Qty	*Value*	*Qty*	*Value*	*Qty*	*Value*
Canada	0	0	5252	18,531	0	0
China	423,964	755,276	83,027	92,845	27,137	81,147
Germany	0	0	42,000	182,220	0	0
Hong Kong	726	1980	0	0	0	0
India	170,906	321,033	0	0	50,310	78,962
Indonesia	1,459,584	2,764,444	1,159,383	1,491,452	1,319,944	1,007,645
Mexico	0	0	1500	5319	1046	3188
Netherlands, the	2250	13,868	1500	8850	2334	5965
Singapore	0	0	0	0	0	0
Sri Lanka	628,816	2,569,850	0	0	985,071	4,518,960
Thailand	11,123	35,489	6564	35,347	8000	12,659
Turkey	0	0	0	0	0	0
United Kingdom	996	3386	3409	6375	0	0
Vietnam	341,920	877,939	0	0	431,690	1,076,549
Other	2868	5421	21,343	35,370	59,369	95,030
Total	1,617,941	3,222,869	1,323,978	1,876,309	1,476,439	1,594,891

Source: USDA/FAS, Tropical Products – World Markets and Trade, various issues of Circular Series.

Note
Qty in kg, value in US$.

Table 12.8 US imports of ground cassia and cinnamon – country-wise sources

Country	1996		1997		1998		1999	
	kg	*Dollars*	*kg*	*Dollars*	*kg*	*Dollars*	*kg*	*Dollars*
Canada	2085	8455	18,901	56,438	5252	18,531	66,608	162,049
China	118,281	166,224	133,607	164,021	83,027	92,845	67,306	70,751
Germany	50,642	210,773	55,467	240,536	42,000	182,220	90	6628
Hong Kong	21,358	70,642	23,671	50,089	0	0	301	2007
India	33,821	62,525	0	0	0	0	7467	24,978
Indonesia	580,787	117,653	804,097	128,436	115,938	149,145	104,788	137,808
Italy	0	0	0	0	1500	5319	1720	6710
Netherlands, the	1497	8834	1000	5781	1500	8850	1497	8850
Russia	0	0	0	0	0	0	0	0
Singapore	8000	18,354	0	0	0	0	0	0
Sri Lanka	78,504	313,957	28,035	73,825	6564	35,347	11,390	71,312
Switzerland	0	0	0	0	0	0	0	0
Thailand	20,526	70,759	0	0	3409	6375	0	0
Vietnam	0	0	0	0	0	0	15,000	12,300
Other	3580	19,949	0	0	21,343	35,370	653	5052
Total	919,081	212,701	106,477	187,505	132,397	187,630	121,991	174,872

Source: USDA/FAS, Tropical Products – World Markets and Trade, various issues of Circular Series.

294   M.S. Madan and S. Kannan

Table 12.9 Cassia and cinnamon imports by European Union with country of source

| Country | 1992 | | 1994 | | 1996 | | (1992–96) | Average |
	Qty	Value	Qty	Value	Qty	Value	Qty	value
Cinnamon, whole								
Inra-EU	305	673	661	1393	643	1646	757.8	1582.4
Extra-EU	5807	12,676	6645	14,108	7497	15,904	6763.2	14426.0
Sri Lanka	795	3905	883	3353	919	3381	858.6	3466.8
Vietnam	226	434	312	523	684	914	394.0	610.4
Seychelles	321	410	241	350	586	874	359.2	509.8
China	431	745	510	906	576	853	541.0	885.8
Madagascar	1037	1262	1099	1432	347	492	910.0	1140.8
Others	2997	5920	3600	7544	4385	9390	3700.4	7812.4
Total	6112	13,349	7306	15,501	8140	17,550	7521.0	16,008.4
Cinnamon, ground								
Inra-EU	540	1910	480	1686	807	2770	592.2	2068.0
Extra-EU	1003	2077	963	1956	1466	3066	1056.6	2263.6
Indonesia	357	820	327	847	530	1231	390.6	931.4
China	115	320	86	118	378	519	158.6	259.6
US	–	12	37	112	82	306	–	117.0
Sri Lanka	93	412	63	267	80	298	83.4	369.4
Seychelles	0	0	–	–	94	206	–	–
Others	438	513	450	612	302	506	368.6	534.0
Total	1543	3987	1443	3642	2273	5836	1648.8	4331.6
Total	10,465	24,092	11,794	25,707	13,525	29,900	12,232.6	26,953.6

lion's share in the European market for cinnamon. Recently, Vietnam is emerging as a major supplier of cassia and cassia products, including oils, to the EU market. The EU market was importing around 226 t of cinnamon from Vietnam in 1992; by 1996 that figure rose to 684 (a 391% increase). Statistics on cinnamon and cassia oil and oleoresin imports from Vietnam are not included here. During 1995 Vietnam exported nearly 480 t of cassia cinnamon oil. After its re-entry into world trade Vietnam's export is increasing steadily both to US and to EU markets (Anon, 2001).

Mexico

Cinnamon is the most important spice imported into Mexico, accounting for an average of two-thirds of the country's total imports of spices in the 1970s. It is estimated that more than 50% of the world trade in true cinnamon is absorbed by this market. Further, over the five-year period of 1976–80, imports of cinnamon into Mexico averaged 2224 t. In recent years (between 1997 and 1999) the average import has been around 5771 t valued US$28.4 million. As can be seen from Table 12.10, Sri Lanka provided around 94% of the total imports during 1998. The USA is the major non-producing supplier of both ground and unground cinnamon to the Mexican market. The Indian export to Mexico has come down from 80 t in 1981 to 10 t in 1998, while imports from the Netherlands have risen dramatically.

Economics and Marketing of Cinnamon and Cassia 295

Table 12.10 Mexico's imports of ground cassia and cinnamon – country-wise sources

Country of origin	1995		1996		1997		1998	
	Qty	*Value*	*Qty*	*Value*	*Qty*	*Value*	*Qty*	*Value*
Cinnamon, whole								
Sri Lanka	4390	17,902	5128	20,442	5052	25,522	4577	28,552
US	33	201	170	248	338	268	184	615
Singapore	–	–	–	–	–	–	41	248
Indonesia	12	52	22	76	20	51	66	126
India	14	52	1	4	8	31	10	48
China	–	3	9	7	7	31	2	12
Cinnamon, ground								
US	30	146	17	87	625	170	311	164
Sri Lanka	16	35	4	12	11	11	24	126
Indonesia	6	25	27	120	26	136	15	106
India	2	13	5	22	19	48	42	31
China	–	–	1	9	8	13	1	1
Spain	–	–	1	19	–	–	–	–
Honduras	–	–	3	12	–	–	–	–
Grenada	–	–	5	24	–	–	–	–
Brazil	–	5	–	3	20	15	–	–
Total	4510	18,458	5416	21,175	6193	26,389	5275	30,031

Note
Qty in tonnes, value in thousand US$.

Japan

Cinnamon and cassia are the most important spices imported into Japan. In the import statistics both spices are treated under cinnamon, although the bulk is, in fact, cassia. China is by far the largest supplier of cassia, accounting for over 90% of total imports. In 1996 Japan imported 2361 t of cinnamon and cassia worth US$5.35 m. The cinnamon import was mainly from Sri Lanka. During the same year Japan imported cinnamon worth 12.3 million Sri Lankan rupees. Cinnamon quills accounted for about 21% of the total imports of cinnamon from Sri Lanka. In 1998 Japan's total import was 1797 t worth US$3.9 m. China was the leading supplier of both ground (333MT) and unground (1085MT) cinnamon and cassia to Japan in 1998 (Table 12.11).

Australia

Cinnamon and cassia are imported into Australia in the forms of cinnamon bark and quills, ground cinnamon and cassia and cassia bark.

Sri Lanka has maintained its position as the major supplier of cinnamon to the Australian market. Indonesia is the other major source. The US, Singapore and Germany are the non-producing suppliers, and are constant sources of the ground product, albeit at a fluctuating level. The imports of ground cinnamon accounted for more than 50% in the beginning of 1990s and the share came down to around 33% in 1998. The total import of cinnamon (both whole and ground) has remained around 350 t during the past ten years.

296 M.S. Madan and S. Kannan

Table 12.11 Japan's imports of ground cassia and cinnamon – country-wise sources

Country of origin	1995		1996		1997		1998	
	Qty	*Value*	*Qty*	*Value*	*Qty*	*Value*	*Qty*	*Value*
Cinnamon, whole								
China	1657	3552	1548	3288	1132	2223	1085	2117
Vietnam	242	701	243	767	291	806	163	441
Sri Lanka	14	191	17	156	19	197	23	354
Indonesia	28	90	49	144	18	36	27	45
US	–	–	–	–	–	9	4	13
Hong Kong	13	23	1	11	–	2	5	13
Malaysia	–	–	–	–	2	11	3	11
Cinnamon, ground								
China	241	452	328	533	231	369	333	512
Vietnam	148	424	155	347	94	183	110	240
Malaysia	–	–	–	–	1	7	27	92
Indonesia	13	76	17	87	20	89	10	41
US	–	3	2	19	1	20	1	12
Hong Kong	–	–	–	–	–	–	8	11
Total	2356	5519	2360	5359	1809	3957	1797	3908

Note
Qty in tonnes, value in thousand US$.

Cinnamon is used in the food manufacturing sector and the bakery industry. As in the US, cassia is preferred in the Australian bakery sector as well, because of its stronger flavour.

Middle East countries

The Middle East market is a growing market for cinnamon and cassia products. Next to pepper, cassia and cinnamon are the most popular spices in Egypt, Saudi Arabia, Iran, Iraq and the UAE. Their import mainly consists of cassia, though a small quantity of cinnamon is also imported from Sri Lanka. China is the major supplier for this vast and vibrant market.

The period-wise growth rate estimated for individual country's imports indicates that there is a surge in demand for the commodity both in traditional and in new markets. The estimated demand of growth rate is more than that of the population growth rate, indicating enormous market potential for the crop.

Export

About 94% of the total production in the world is exported, indicating the commercial nature of the crop. Untill the end of 1970s (period I), Sri Lanka was the leader in production and export of this commodity controlling the markets of North and South America. The other major importers, like France, Germany and the UK, re-export value added cinnamon products. The share of Sri Lanka's export has been reduced from one third of the global demand to one/tenth over the past two decades.

Table 12.12 Period-wise cinnamon export by the major producing countries (1971–2000)

Year	Indonesia	Madagascar	Seychelles	Sri Lanka	Vietnam	China	India	World
Period I (1971–80)								
Average	6100.2	333.3	1111.3	6164.5	108.5	1269.8	700.0	17,983.8
% share	33.9	1.9	6.2	34.3	0.6	7.1	3.9	100.0
Period II (1981–90)								
Average	16226.3	1235.2	557.6	7366.6	1894.2	17200	750.8	49,255.6
% share	32.9	2.5	1.1	15.0	3.8	34.9	1.5	100.0
Period III (1991–99)								
Average	21137.1	913.7	465.9	8271.8	3087.8	22374.6	689.2	75,763.6
% share	27.9	1.2	0.6	10.9	4.1	39.7	0.9	100.0

Source: Based on FAOSTAT.

Note
Qty in tonnes.

Indonesia and China, the major producers of the cassia type of cinnamon increased, their production from the beginning of 1980s and together captured almost 70% of the world market. Consequently, export from traditional cinnamon producers like Sri Lanka, Madagascar and Seychelles declined drastically from 34.3, 1.9 and 6.8 per cent, respectively, in the 1970s to only about 10.9%, 1.2% and 0.6%, respectively, in the 1990s. China and Indonesia substantially increased their share in world trade during this period to about 39.7% and 27.9%, respectively (Table 12.12). There has been an increasing trend in export of cassia products from Indonesia, China and Vietnam, while the true cinnamon export could not keep up the pace. The export performance by non-producing (re-exporting) countries is given in Table 12.13. Among the re-exporting countries, US, the Netherlands and

Table 12.13 Period-wise exports of cinnamon by non-producing countries (1975–2000)

Year	USA	The Netherlands	Singapore	United Kingdom	Germany	France	World
Period I (1971–80)							
Average	812.3	267.0	231.4	90.4	97.5	13.3	17,984
% share	4.52	1.48	1.29	0.50	0.54	0.07	100.0
Period II (1981–90)							
Average	583.1	379.1	1374.7	55.8	177.0	741.7	49,256
% share	1.18	0.77	2.79	0.11	0.36	1.49	100.0
Period III (1991–2000)							
Average	844.0	1443.1	6201.0	43.0	362.3	136.4	75,764
% share	1.11	1.90	8.18	0.06	0.48	0.18	100.0

Source: Based on FAOSTAT.

Note
Qty in MTs.

Singapore account for more than 90% of the total exports. Singapore alone accounts for more than 50% of the re-exports in the world market. Produce from Indonesia and Vietnam is routed through this important entry port for spices. The US, the largest importer in the world, re-exports a sizeable quantity of both cinnamon and cassia. Of the total, imports, the US re-exported around 674 t of cinnamon and cassia, mainly to Mexico, Germany and Canada, accounting for more than 70% of total exports (Table 12.14).

India exports and re-exports a greater amount of cassia type cinnamon than true type cinnamon. Item-wise and year-wise total exports of both cinnamon and cassia from India are given in Table 12.15, however, there is no definite pattern in export of either commodities. The export of cassia and cassia products has remained greater than that of cinnamon, and there has been a steady decline in export of cinnamon products during the period 1995–2000. India's export is more than that of its production, indicating that the country re-exports a sizeable quantity of imported cinnamon and cassia.

Vietnam's export earnings from spices indicated an annual average growth rate of 13.1% during 1994–2000, which declined by 3.6% in 2000 mainly due to a negative growth rate (−9.2%) registered by cinnamon export. Earnings from cinnamon export, which accounted for 59% of the total spice exports from Vietnam, declined mainly due to low international prices.

Competitiveness

The opening up of the Vietnamese economy has introduced competition into this commodity market. Saigon cassia, a much sought after variety prior to the Vietnam war, is now available on the market. During 1995, among the individual products exported from Vietnam to the European Union, the highest market shares were for cinnamon (6.8%) and essential oils (6.4%), mainly from cinnamon. However, the unit price received by Vietnamese cinnamon is less than the market average. An exceptionally higher price deviation was recorded in cinnamon (price ratio 0.61). Cinnamon and its essential oil have similar price pattern, and are influenced by the same market trends (Anon, 2001).

In order to understand the position and competitiveness of individual exporters in the world trade of cinnamon, market shares and unit value ratios were calculated and presented in Table 12.16. In the absence of time series data on prices, the unit price was worked out from the value of export and quantity exported. While calculating the unit price, individual items of export were not taken into account. So, there is bound to be a slight variation depending upon the share of value added products in the export basket of individual countries. However, the estimated unit value ratios help in comparing the prices of each exporting country with one another and with the average of total imports. The ratio is computed by dividing the country's unit value by the unit value of total imports. The data shows that:

- Sri Lanka's share of total exports has declined substantially since 1970, despite the higher unit value ratios when compared to other exporting countries. The decline is only in terms of a percentage share, not in terms of actual quantity.
- Two major suppliers of cassia cinnamon taken together dominate the market. Both Indonesia and China have improved their share in the import market by many-fold since 1970, not at the cost of other exporting countries, but by the steady growth in their export to meet the world demand. China with its faster growth rate has overtaken Indonesia in export performance. In future one country's success depends on the failure of the other in light of the emergence of new competitors like Vietnam.

Table 12.14 US exports of cassia and cinnamon by country of destination

Country	1996		1997		1998		1999		Avg. qty (1996–99)	% in total imports
	Qty	Value	Qty	Value	Qty	Value	Qty	Value		
Australia	816	5058	0	0	904	5319	13,430	61,418	4778	0.7
Canada	253,199	814,638	217,623	676,996	168,331	4,696,640	120,557	325,407	126,628	18.8
Colombia	1880	7952	0	0	0	0	0	0	1880	0.3
Costa Rica	0	0	0	0	660	3758	0	0	660	0.1
France	–	–	10,096	35,573	0		0	0	10,096	1.5
Germany	–	–	293,236	567,963	91,714	184,035	3147	3757	129,366	19.2
Haiti	–	–	0	0	0	–	16,244	41,385	16,244	2.4
Honduras	–	–	0	0	0	–	22,161	22,881	22,161	3.3
Jamaica	–	–	0	0	0	–	13,545	31,723	13,545	2.0
Korea, Republic of	1526	7970	0	0	24,715	68,107	0	0	12,358	1.8
Mexico	225,172	903,243	265,423	981,003	302,988	1,342,268	272,880	872,330	210,323	31.2
Netherlands, the	0	0	0	0	0	0	846	12,852	846	0.1
Panama	1814	5240	5991	17,113	3755	10,018	4535	9000	3570	0.5
Singapore	0	0	0	0	0	–	33,834	42,867	33,834	5.0
Sweden	1713	4719	0	0	0	–	14,479	21,336	7240	1.1
Switzerland	5292	47,784	0	0	0	–	0	0	47,784	7.1
Thailand	0	0	17,005	29,622	0	–	0	0	4251	0.6
Trinidad and Tobago	0	0	1000	2750	0	–	0	0	1000	0.1
United Kingdom	8957	3,515,210	27,626	84,225	15,909	45,483	10,274	24,530	13,452	2.0
Venezuela	7347	34,571	10,808	42,056	748	2739	0	0	3852	0.6
Other	24,534	133,282	8731	31,213	15,872	37,663	16,809	42,050	10,353	1.5
Total	532,250	1,999,609	857,639	2,468,514	625,596	2,196,030	542,741	1,511,536	674,220	100.0

Source: USDA/FAS, Tropical Products – World Markets and Trade, various issues of Circular Series.

Note
Qty in kilograms, value in US$.

300 M.S. Madan and S. Kannan

Table 12.15 Item-wise export of cinnamon and cassia products from India (1989–90 to 1998–99)

Year	Whole		Powder		Oil		Oleoresin	
	Qty	Value	Qty	Value	Qty	Value	Qty	Value
Cinnamon								
1989–90	70.26	23.43	–	–	–	–	0.05	0.31
1990–91	4.66	1.85	–	–	0.03	16.15	0.06	0.42
1991–92	8.4	4.50	15.5	5.26	0.1	15.80	0.3	3.16
1992–93	0.0	0.00	0.0	0.00	0	0.00	1.5	18.35
1993–94	0.4	0.22	5.0	0.38	neg	2.20	0.1	0.88
1994–95	0.61	1.09	0.4	1.20	neg	17.00	0.1	0.98
1995–96	2.48	0.58	0.0	0.00	0.5	3.13	0.52	3.99
1996–97	8.13	7.06	1.8	1.12	0.08	0.60	1.1	35.73
1997–98	1.85	3.30	1.93	3.18	0.09	1.03	0.11	3.46
1998–99	1.41	2.02	2.2	1.30	1.14	12.55	0.54	13.63
Cassia								
1989–90	512.7	37.68	67.0	1.47	–	–	–	–
1990–91	544.73	47.26	142.06	5.93	–	–	–	–
1991–92	957.7	146.59	805.6	485.38	0.0	0.00	0.0	–
1992–93	20.5	10.35	1549.8	145.77	0.0	0.00	5.3	169.24
1993–94	1442.7	61.68	805.4	34.96	2.0	77.62	12.4	45.60
1994–95	617.5	25.95	265.0	19.42	0.6	14.27	neg	0.03
1995–96	3733.95	246.11	65.74	10.44	1.38	9.47	4.06	28.84
1996–97	3159.09	277.02	289.98	48.65	0.74	4.00	1.93	48.13
1997–98	932.9	63.11	186.6	98.73	1.63	49.67	9.65	321.32
1998–99	60.04	16.15	1114.0	781.78	0.02	0.65	15.55	604.14

Note
Qty in tonnes, value in Rs Lakhs.

- Sri Lanka's prices tend to be above the level of the average prices for total imports because of the higher price of true cinnamon exported by that country. China's prices are generally below the average, while Indonesian prices are slightly above the Chinese prices. In recent years the Indonesian product has earned an above average price in the world market. This may be because of the increased share of value added products in the total export.

Price analysis, price formation

An analysis of the structure and behaviour of farm prices is of considerable interest in the context of finding ways and means for increasing production and productivity. Prices often act as a guide to indicate the change in production decisions. Cinnamon is a moderately storable export commodity. Long-term storage is not possible as in the case of black pepper. In the absence of year to year violent fluctuation in production, the domestic price of the commodity is totally dependent on international price.

For marketing, cinnamon and cassia are graded according to quality. Though the producer does not have any idea of the oil content of his produce, the general trend is the higher the essential oil content, the better the price. Grade A cinnamon has the highest oil content – at least 2.75% and up to 4%, with a better flavour and aroma. The length of the quills is another factor that decides the quality and price. In Sri Lanka rolled and

Table 12.16 World exports of cinnamom – market shares (quantity) and unit price ratios

Year	Indonesia		Madagascar		Seychelles		Sri Lanka		Vietnam		China		India		World	
	Share (%)	Unit value ratio	Share (%)	Unit value ratio	Share (%)	Unit value ratio	Share (%)	Unit value ratio	Share (%)	Unit value ratio	Share (%)	Unit value ratio	Share (%)	Unit value ratio	Export (MT)	Unit price (US$/t)
1971–75	25.78	0.92	3.62	0.67	10.57	0.58	40.59	1.20	–	–	3.96	–	5.51	0.61	13,176	1106.8
1972–76	26.20	0.86	2.55	0.68	9.92	0.56	40.90	1.21	–	–	6.46	0.34	6.24	0.60	14,239	1139.8
1973–77	26.84	0.86	1.90	0.62	7.60	0.53	40.96	1.21	–	–	8.85	0.26	6.27	0.56	15,020	1208.0
1974–78	29.72	0.78	1.31	0.56	5.54	0.45	37.46	1.27	–	–	11.69	0.33	5.02	0.53	16,433	1270.8
1975–79	34.84	0.65	0.96	0.96	3.61	0.44	33.93	1.29	–	–	14.20	0.52	3.40	0.54	18,551	1189.7
1976–80	38.63	0.62	0.83	0.86	2.73	0.44	30.63	1.32	–	4.77	16.43	0.65	2.96	0.54	22,792	1218.1
1977–81	40.20	0.60	0.82	0.75	1.87	0.43	28.40	1.36	–	4.57	18.46	0.67	2.40	0.59	26,223	1346.5
1978–82	42.44	0.59	0.87	0.69	1.65	0.40	24.33	1.37	–	4.21	20.81	0.71	2.07	0.64	30,288	1460.8
1979–83	43.18	0.64	1.03	0.97	1.68	0.39	20.86	1.29	–	4.40	23.32	0.64	2.29	0.66	34,498	1535.0
1980–84	41.64	0.69	1.11	0.97	1.49	0.37	19.69	1.21	–	–	26.82	0.60	2.77	0.63	38,636	1565.8
1981–85	39.93	0.71	1.24	1.42	1.47	0.33	18.15	1.16	–	–	29.04	0.59	2.50	0.57	41,538	1600.6
1982–86	39.07	0.76	1.31	1.46	1.69	0.30	15.55	1.07	–	2.80	29.26	0.65	2.27	0.50	46,479	1572.4
1983–87	36.46	0.84	1.50	1.43	1.47	0.31	15.29	1.02	–	2.58	30.43	0.68	1.96	0.43	48,926	1600.1
1984–88	33.28	0.88	2.38	1.06	1.12	0.32	15.18	1.14	–	2.34	30.05	0.69	1.47	0.44	50,807	1670.6
1985–89	29.40	0.96	3.37	0.67	1.05	0.35	13.53	1.31	–	2.01	27.89	0.63	0.89	0.41	54,778	1847.8
1986–90	27.85	0.96	3.43	0.27	0.73	0.39	12.63	1.52	3.32	1.76	27.91	0.57	0.82	0.40	56,973	1998.6
1987–91	26.07	0.95	3.74	0.26	0.38	0.43	12.24	1.74	3.71	1.40	28.35	0.52	0.89	0.38	59,255	2071.1
1988–92	22.80	1.05	3.49	0.28	0.35	0.47	11.56	1.96	3.51	1.24	26.55	0.49	1.04	0.34	63,959	2061.1
1989–93	23.47	1.07	2.78	0.33	0.40	0.51	10.47	2.11	3.60	1.13	27.42	0.46	1.37	0.23	68,738	1994.8
1990–94	25.45	1.04	2.03	0.39	0.43	0.54	9.64	2.17	3.54	1.07	29.48	0.46	1.40	0.19	68,963	1912.9
1991–95	24.48	1.09	1.81	0.45	0.58	0.55	9.56	2.12	4.49	1.04	29.65	0.47	1.56	0.16	73,321	1860.6
1992–96	23.87	1.11	1.36	0.50	0.72	0.60	9.77	2.04	4.46	1.09	30.99	0.49	1.38	0.14	73,718	1841.4
1993–97	27.33	0.95	1.14	0.51	0.75	0.63	10.32	1.57	4.77	1.08	35.03	0.52	1.10	0.24	74,448	1845.8
1994–98	28.89	1.09	0.87	0.64	0.69	0.80	11.31	1.40	4.66	1.31	36.08	0.67	0.72	0.51	74,223	1511.6
1995–99	30.05	1.24	0.60	0.85	0.68	1.07	12.10	1.23	4.48	1.64	35.51	0.72	0.52	0.96	78,898	1126.8

layered quills are trimmed to 106.7 cm quills as specified by the world cinnamon market. Quills are packed in 45 kg bales and classified into 10 grades according to diameter and the number of 106.7 cm quills to a pound; permissible amounts of foxing are specified for each grade (see ISO 6539–1983).

Quills harvested for grade 'A' cassia cinnamon must be 1 m long and taken from the main trunk of the tree. The grade 'B' is harvested from the side branches and has an oil content of about 2.25%. Broken pieces are marketed as grade 'C' that contains about 1.5% volatile oil. The US and Mexico are the biggest importers of quills.

The market for quillings is not as extensive in comparison with the other products. Mexico is the main market for this product. Featherings are imported mainly by Argentina, Switzerland and the UK, while countries like Canada, Australia, and the UK are the markets for chips.

Ground or crushed cinnamon is another product which has gained importance since the drive to export value added products. Chips, referred to as 'quillings' and 'featherings', are sold as medium-quality cinnamon for grinding into 'cinnamon powder'. It is sold on its own or as 'pudding spice' in a compound form with nutmeg, clove, cardamom, mace and allspice. The chips are also sold for the distillation of oil.

Table 12.17 gives the spot prices that prevailed during May 1999 for various grades of cinnamon and cassia. Sri Lankan cinnamon (Ceylon H-2 cinnamon) fetched higher price than cassia. Except for ground spices in a few countries, most spices are imported at zero rates of duty under GSP and other preferential trade agreements. A reduction in MFN tariffs as outlined in the Uruguay Round may not make a significant change.

Cinnamon fetched peak prices in 1989 averaging US$7.50/kg for the best quality, but fell to US$5.00/kg in subsequent years. Although the annual average price was US$1.6/kg during 1998, the best quality Sri Lanka 3/6 inch strips fetched a price of US$9.2/kg, while the chips of the same lot fetched only US$0.70/kg. In the beginning of 2001, the average New York price was around US$1.60/kg. In India the average wholesale price in the Mumbai market was Rs 91/kg, while it was Rs 92.08 for cassia in the Chennai market during the same period. The annual average price for Ceylon cinnamon in New York market is given in Table 12.18 for the period 1996–2000.

Table 12.17 Differences in prices of various grades of cinnamon and cassia in the world market (May, 1999)

S. no.	Grade/variety		Price $/lb
1	Vera C/W 'AA' 2-3/4"	spot	1.00
2	Korintji 'A' 3.00 ML/SVO	spot	0.61
3	Korintji 'A' 2.75 ML/SVO	spot	0.57
4	Korintji 'B' 2.50 ML/SVO	spot	0.53
5	Korintji 'B' 2.25 ML/SVO	spot	0.49
6	Korintji 'C' 1.50 ML/SVO	spot	0.45
7	Korintji 'C' 1.00 ML/SVO	spot	0.40
8	Chinese 2.75 ML/SVO	spot	0.68
9	Vietnamese Cassia A 3.50 ML/SVO	spot	1.40
10	Ceylon H-2 Cinnamon	Afloat	2.65

Source: Weekly Market Report, AA Sayia & Company, Inc. May, 1999.

Economics and Marketing of Cinnamon and Cassia   303

Table 12.18 Import price of cinnamon and cassia in to the US and the EU markets by source country (1996–2000 average)

Country of origin	USA market[1]		EU market[2]	
	Unit price	Price ratio	Unit price	Price ratio
Canada	3.52	1.20	–	–
China*	1.63	0.55	1.64	0.77
Germany	4.34	1.47	–	–
Hong Kong	2.03	0.69	–	–
India*	1.51	0.51	–	–
Indonesia*	1.31	0.44	–	–
Mexico	3.82	1.30	–	–
Netherlands	5.03	1.71	–	–
Singapore	2.25	0.77	–	–
Sri Lanka*	4.80	1.63	4.04	1.9
Thailand	2.84	0.96	–	–
Turkey	4.28	1.46	–	–
United Kingdom	2.21	0.75	–	–
Vietnam*	2.35	0.80	1.55	0.73
Seychelles*	–	–	1.42	0.67
Madagascar*	–	–	1.25	0.59
Other	2.20	0.75	2.11	0.99
Average	2.94	–	2.13	–

Notes
Price US$/kg.
* Producing countries.
1 1996–2000 average.
2 1992–96 average.

Cinnamon and Cassia Oils

The most important cinnamon oils in the world trade are:
*   Cinnamon bark and leaf oil from *C. verum*
*   Cassia oil from *C. cassia*

World supply and demand trends

Cinnamon bark oil is a high value oil traded the world over. But the volumes traded are very low. Sri Lanka is the major supplier of cinnamon bark oil. During 1987–1996 the average export was around 2.8 t (FAO, 1995). However, in 1995, the total supply from Sri Lanka was 3.32 t and in subsequent years it went up marginally. Table 12.19 provides the list of destination-wise export of cinnamon bark oil for 1987–96. The major market for cinnamon bark oil is the European Union; France is the biggest importer followed by the US.

Leaf oil

World demand for cinnamon leaf oil has been around 120–150 t pa in recent years, a demand met mostly by Sri Lanka (Table 12.19). Sri Lankan exports have averaged

304 M.S. Madan and S. Kannan

Table 12.19 Exports of cinnamon leaf and bark oil from Sri Lanka (1987–96)

Country	1987	1988	1989	1990	1991	1992	1993	1994	1995	1996
Bark oil										
France	0.6	0.6	0.7	0.2	1.1	1.1	–	0.99	0.99	5.7
US	0.1	0.4	0.4	0.1	0.5	0.4	–	0.63	1.12	4.19
Italy	0.4	0.3	0.4	0.1	0.2	0.3	–	0.53	0.5	0.21
UK	0.5	0.4	0.2	–	0.3	0.2	–	0.36	0.22	1.32
Netherlands, the	0.5	0.2	0.3	0.1	0.3	0.1	–	–	–	–
Germany	0.2	0.4	0.2	0.1	0.2	0.2	–	0.37	0.39	2.7
Switzerland	0.2	0.4	0.2	–	–	0.2	–	–	–	–
Total	2.7	2.7	2.6	0.7	2.8	2.8	–	2.95	3.32	7.55
Leaf oil										
US	38	54	78	13	46	54	–	96.96	125.88	33.61
UK	29	19	18	2	9	11	–	20.8	24.59	17.56
France	19	21	24	8	9	8	–	13.98	13.63	7.6
Hong Kong	11	13	16	14	17	20	–	11.94	10.59	3.5
India	8	5	7	1	2	5	–	–	–	–
Spain	5	4	6	4	6	3	–	6.6	9.3	2.4
Switzerland	5	3	5	1	5	7	–	–	–	–
Germany	5	2	3	–	3	5	–	1.2	1.0	2.0
Total	133	132	162	46	107	119	–	151.4	184.9	66.67

Source: Ministry of Commerce, Government of Sri Lanka.

Note
Qty in tonnes.

141 t during 1987–96, except in the years 1990 and 1996 when the supply was exceptionally low. The US and EU are the largest markets for cinnamon leaf oil. Imports into the UK and France have gone down drastically during the past decade.

World demand

Estimation of world demand for cassia oil is difficult because export data are not available from other producing countries such as China, Madagascar, etc. Added to this problem, the oil is not separately specified in world import statistics except in the US and Japan.

The levels of imports of cassia oil into the US are shown in Table 12.20 for the period 1996–99. Imports into the US have risen in recent years due to the soft drinks market that shows no sign of weakening. Imports from Japan and Hong Kong are almost entirely re-exports of Chinese oil. The Japanese import of cassia oil has averaged 60 t pa during 1988–93, virtually all of it coming directly from the People's Republic of China. A significant proportion of the imports are re-exported (FAO, 1995).

Economics and Marketing of Cinnamon and Cassia 305

Table 12.20 US imports of cassia oil by country-wise sources

Country	1996		1997		1998		1999	
	Qty	Value	Qty	Value	Qty	Value	Qty	Value
Bahamas	1292	41,311	0	0	11,250	471,200	0	0
China	259,644	4,838,195	219,197	3,480,133	364,075	5,609,903	225,930	4,715,315
France	102	7511	401	7848	0	0	95	22,209
Germany	0	0	0	0	55	1656	0	0
Hong Kong	100	10,358	2000	31,650	0	0	8000	118,383
India	1166	17,548	1738	112,605	0	0	1130	46,392
Indonesia	28,740	543,097	0	0	0	0	0	0
Japan	19,494	827,288	98,365	4,327,333	50,753	2,252,654	74,703	2,445,593
South Africa	0	0	0	0	0	0	0	0
Spain	0	0	0	0	0	0	0	0
Sri Lanka	8000	71,259	6000	42,520	6000	47,777	15,400	126,290
United Kingdom	5755	195,846	1047	38,036	59,855	2,931,948	76,829	3,473,247
Others	1996	79,705	1736	8993	0	0	1629	47,852
Total	326,289	6,632,118	330,484	8,049,118	491,988	11,315,138	403,716	10,995,281

Source: USDA, Tropical Products – World Markets and Trade, various issues of Circular Series.

Note
Qty in kgs, value in US$.

Supply sources

Sri Lanka is the only regular supplier of cinnamon bark and leaf oils. With the exception of 1990, when both oils were in short supply, production (as reflected in exports) has remained constant for bark oil, with a slight downward trend for leaf oil. Since internal consumption is small, the production levels are not much greater than exports. Madagascar and the Seychelles have been intermittent suppliers of leaf oil on a very minor scale in the past. India produces a very small quantity of leaf oil for domestic use (Spices Board, 1992).

Most cassia oil in international trade is of Chinese origin. There is believed to be a significant domestic consumption of cassia oil in the country. So, total annual production may be in excess of 500 t. Small quantitites of cassia oil are produced in Indonesia and Vietnam.

Quality and price

There is no international standard for cinnamon bark oil, however, the higher the cinnamaldehyde content the higher the price. (In the US and EOA, standard specifies an aldehyde content of 55–78%.)

International (ISO) standards exist for cinnamon leaf and cassia oils. For cinnamon leaf oil, the minimum eugenol content required for the oil is specified in terms of total phenol content. Oil from the Seychelles is preferred because of its high eugenol content (ca 90%). In practice, Sri Lanka now accounts for almost all of the oil in international trade and the standard specifies a 75–85% phenol content and a maximum level of 5% cinnamaldehyde. Physico-chemical requirements are also given. The US and FMA monograph, which replaces the old EOA standard, specifies the eugenol content of cinnamon leaf oil in terms of its solubility in potassium hydroxide (80–88%). For cassia oil, cinnamaldehyde is the major constituent and a minimum content of 80% is specified in the ISO standard.

Cinnamon bark oil is considerably more expensive than the leaf oil and probably the most highly priced of all essential oils. During 1992 it was being offered at around US$385/kg, largely reflecting the high raw material cost. In 1993 and early 1994 dealers in London were only quoting prices on request. Cinnamon leaf oil, in contrast, has been in the range of US$6.50–7.50/kg during the past three years. The price fell gradually from US$7.50/kg in early 1991 to US$6.50/kg in mid-1993. In late 1993, it had risen again to US$7.30/kg and in early 1994 it was US$8.25/kg. Although it is a comparatively low priced oil, it is still more expensive than clove leaf oil as a source of eugenol (which was approximately US$2.70/kg in early 1994).

Cassia oil, too, has remained fairly level in price over the last few years. During early 1991 to mid-1993, it fetched US$33–35/kg. The price then fell slightly and in early 1994 it was about US$29/kg. These prices are significantly lower than those that prevailed in the early and mid 1980s when there was a shortage of cassia bark in the People's Republic of China. Any appreciable rise in price above the US$30–35/kg level is likely to encourage end-users to blend cheaply available synthetic cinnamaldehyde with natural cassia oil.

Future Prospects

Like all other spices cinnamon and cassia are also subject to vagaries of the market. This could have adverse effects on the cinnamon growers of the developing countries. However, there are also favourable trends, which could be exploited. There is a strong growing market in the Middle East as well as in the Asia Pacific region, where a strong preference for specific flavours exists. The introduction of new snack foods is dependent to a large extent on the difference of flavours. Another factor, which is encouraging for developing countries, is the increase in global travel. This leads people to experiment with different flavours and spicy food. It has also given rise to the increase in ethnic restaurants. The importance of cinnamon is predominantly in the food-processing sector.

The future prospects for cinnamon and cassia can be studied in terms of two major influencing factors: supply and demand. Cinnamon and cassia production is influenced by national as well as international factors. While demand is influenced by many factors including the overall economic development, supply is influenced not only by economic factors but also by agro-climatic, biotic and abiotic stress factors in the growing regions. The product of commerce comes from perennial tree crops, hence responses to price changes get reflected in the form of altened supply after many years. Thus, there are a multitude of factors which are to be considered when forecasting the future of cinnamon and cassia. The kind of data available to us do not permit sophisticated forecasting models which may give correct and reliable predictions. What we have is only the historic data for area, production and export. A suitable model, which can give a reasonable prediction with these data, has been identified and fitted and this must be seen as a step on the road towards a more sophisticated modelling analysis based on superior data once if they become available.

Model identification

A variety of statistical forecasting techniques are available, ranging from very simple to very sophisticated. All of them try to capture the statistical distributions in the data provided and quantitatively present the future uncertainty. Lack of quality data forced us to choose methodologies, which forecast the future by fitting quantitative models to statistical patterns from historic data for several years. Therefore, univariated methodologies based solely on the history of the variable (one at a time) were tried. There are three such models:

Simple moving average models
Exponential smoothing models
Box-Jenkins models.

To identify the right model the data have been explored first.

Exploring the data

The time series data on production and export were plotted/graphed to select an appropriate model. The characteristics observed in the time series data for cinnamon are:

1. There is an overall positive trend (i.e. the trend cycle accounts for over 90%).
2. Non-seasonal in nature.
3. The time series is non-stationary in both mean and variance.

The classical decomposition of the time series data also revealed the fact that the trend cycle accounted for about 95% and above, while the irregularity accounted for the rest. Thus the forecasting model should account for trend, non-seasonality and also the non-stationary factor. Though (Box-Jenkins) models can be used, the models of exponential smoothing were more suitable, as these models were built upon clear-cut features like level, trend and seasonality.

Model selection

In order to identify a suitable model, the data was subjected to autocorrelation and partial autocorrelation analysis. The outcome of the analysis for both production and export separately indicated that AR (auto regression) (1) model was the suitable one. The AR (1) model is identical with exponential smoothing (Box and Jenkins, 1976). Hence exponential smoothing models were selected and tried.

The exponential smoothing, as its name suggests, extracts the level, trend and seasonal index by constructing smoothed estimates of these features, weighing recent data more heavily. It adapts to changing structure, but minimises the effects of outliers and noises. Three major exponential smoothing models are available:

a. Simple exponential smoothing
b. Holt exponential smoothing
c. Winters exponential smoothing

Finally, the Holt exponential smoothing model was selected as the best and the forecasting was done for variables in production and export.

308   *M.S. Madan and S. Kannan*

The model

Holt's (1957) exponential smoothing model uses a smoothed estimate of the trend as well as the level to produce forecasts. The forecasting equation is:

$$Y(m) = S_t + mT_t \qquad (1)$$

The current smoothed level is added to the linearly extended current smoothed trend as the forecast into the indefinite future.

$$S_t = \alpha Y_t + (1 - \alpha)(S_{t-1} + T_{t-1}) \qquad (2)$$

$$T_t = \gamma(S_t - S_{t-1}) + (1 - \gamma)T_{t-1} \qquad (3)$$

Where,

m	forecast lead time
Y_t	observed value at time t
S_t	smoothed level at end of time t
T_t	smoothed trend at end of time t
γ	smoothing parameter for trend
α	smoothing parameter for level of series

Equation (2) shows how the updated value of the smoothed level is computed as the weighted average of new data (first term) and the best estimate of the new level based on old data (second term). In much the same way, equation (3) combines old and new estimates of the one period change of the level, thus defining the current linear (local) trend.

Demand

On the demand side, apart from the increased imports from the traditional importing countries like the US, Mexico, Germany, and the Netherlands, import demand has also increased significantly in the newly emerging markets like the Middle East where the increase in demand for the product is not strong enough to create a demand driven force, but is enough to sustain the present level of production and export mainly for cassia. India, the largest consumer of spices in the world, is also increasing its import of cinnamon and cassia. In light of tariff-free trade among countries of the world, import is bound to increase not only in developed country markets but also in spice producing, developing and under developed countries. But in all these newly emerging markets imports are dominated by cassia. In recognising the increasing demand for 'cinnamon buns' in the US fast food diet, it is considered that this trend is likely to continue. Further, invasion of this western food culture into many emerging economies, along with an increase in income levels, is bound to accelerate the demand for fast food the world over. The growing awareness about the natural flavours and colours among health conscious consumers will also change demand.

In the absence of consolidated data for world imports, the data on exports were taken as the net imported quantity of the world market, as one country's export becomes another country's imports. Future demand for the commodity is forecasted using the model identified. The equation fitted for demand is:

Table 12.21 The forecasted world production and export of cinnamon (canella) up to 2005–06

Year	Production	Export/demand
1999–2000	134,417.36	88,358.05
2000–01	137,312.09	91,209.80
2001–02	140,206.81	94,061.55
2002–03	143,101.53	96,913.30
2003–04	145,996.26	99,765.05
2004–05	148,890.98	10,2616.80

Source: Based on FAO Data.

$$S_t = 0.37Y_t + 0.63S_{t-1}$$

The global demand for cinnamon is expected to be around 102,616 t by 2004–05 (Table 12.21).

Supply

The opposite considerations apply to supply. Here, Indonesia holds the key. Its ability to regulate supply of cassia is likely to dictate the price internationally. Indonesia's management of this issue may be influenced by the proximity of the energetic entrepot Singapore. New exporters of cassia must also be conscious of China and now Vietnam and the speed at which they are expanding production level. Total world production of cinnamon is growing at the rate of 5.6% per annum, which is more than the world population growth rate. Thus, the anticipated supply in the form of increased production is expected to be a positive one. The fitted equation based on the selected model is:

$$S_t = 0.987Y_t + 0.013S_{t-1}$$

Accordingly the projected world supply in the form of production increase up to 2004–05 will be as shown in the Table 12.21. The expected supply (production) by the year 2005–06 will be around 148,890 t. As can be seen from the fitted equation, where α is nearer to one, recent developments have more impact on the future supply.

As concluded by Vinning (1990) the combination of the above two calculations of supply and demand suggest that within the overall pessimistic outlook the prospects for cassia is better than for cinnamon in the years to come.

References

Anonymous (1995) Cassia (Cinnamon) – A Modern Herbal Home Page. Eletronic Newsletter.

Anonymous (accessed in December 2001) Cassia. http://herbal-tonics.com/pc-ingr.html/.

Anonymous (accessed in December 2001) http://www.mekonginfo.org/mrc-on/doclib.net/.

Baruah, A. (2000) *Cinnamomum verum* Presl., the source of inevitable flavouring spicy bark – True Cinnamon. *Spice India* – Tree Spices Special, November, 2000, pp. 5–7.

310   *M.S. Madan and S. Kannan*

Box, G.E.P. and Jenkins, G.M. (1976) *Timeseries Analysis: Forecasting and Control*. San Francisco Bay, USA.

FAO (1995) Cinnamomum Oils (including cinnamon and cassia). In: *Non-wood Forest Products for Rural Income and Sustainable Forestry*. M-37, ISBN 92–5–103648–9. Publications Division, Food and Agriculture Organisation, Rome, Italy.

FAO (2000) Spices. In. Definition and Classification of Commodities (Draft). http://www.fao.org/WAICENT/faoinfo/economic/faodef/fdefioe.htm.

Grieve, M. (1979) *A Modern Herbal*. Dover Publications, Inc., New York.

Holt, C.C. (1957) *Forecasting seasonals and trends by exponential weighted moving averages*. ONR Research Memorandum, Carnige Institute, 52.

International Trade Centre UNCTAD/GATT (1982) *Spices: a survey of the world market*. Geneva, 1982. 2 vols.

International Trade Centre UNCTAD/GATT (1996) *The Global Spice Trade and the Uruguay Round agreements Geneva*: ITC/CS, 1996, xi, 99p.

International Trade Centre UNCTAD/WTO, (1998) *Market Research File on Spices: Overview of the European Union, Poland, Hungary, Czech Republic, Russian Federation*. M.DPMD/98/0103/Rev.1, 83p.

International Trade Centre UNCTAD/WTO, (2000) *Global Spice Markets – Imports 1994–98, 1211*, Geneva, Switzerland.

Leung, A.Y. and Foster, S. (1996). *Encyclopedia of Common Natural Ingredients used in Food, Drugs and Cosmetics*, 2nd ed. New York: John Wiley & Sons, Inc.

Madan, M.S. (2001) The Sensational tree bark: a commercial outlook. *Spice India* (Tamil). Vol. 14, No. 1, January, 2001.

Ratwatte, Fl. (1991) *The Spice of life: cinnamon and Ceylon*. http://www.Infolanka.org.

Rema, J. and Krishnamoorthy, B. (1999). Commercial exploitation of Chinese Cassia (*Cinnamomum cassia*). *Indian Spices*, Vol. 35, No. 4, pp. 2–4.

Spices Board (1988) *Report on the Domestic Survey of Spices (Part – 1)*. Spices Board, Ministry of Commerce, Govt. of India, Cochin-18.

Spices Board (1990) *Production and import of tree spices*. In: Status Paper on Spices. Spices Board, Ministry of Commerce, Govt. of India, Cochin-18.

Spices Board (1992) *Report of the forum for encouraging export of spices*. Spices Board, Cochin-18.

United States Department of Agriculture/Foreign Agricultural Service (1998) *Tropical Products: World Markets and Trade*. Circular Series, FTROP 1–98, March, 1998.

United States Department of Agriculture/Foreign Agricultural Service (1999). *Tropical Products: World Markets and Trade*. Circular Series, FTROP 1–99, March, 1999.

Vinning, G. (1990) *Marketing Perspectives on a Potential Pacific Spice Industry*. ACIAR Technical Reports No. 15, 60pp.

13 End Uses of Cinnamon and Cassia

B. Krishnamoorthy and J. Rema

Introduction

Cinnamon and cassia have been used as spices and medicines since ancient times and have long been held in high esteem as aromatics as well as ingredients of oils and perfumes. Chinese cassia, along with clove, was used by the ancient Egyptians for embalming. Sponges soaked in extracts of cinnamon were used for bathing by the privileged classes of the ancient and medieval world to protect the body from diseases. Cinnamon was also sometimes used as barter money for payment of loans, in place of precious metals. Cinnamon was mentioned in many books by the end of the middle ages as an ingredient of throat and cough medicines. Most western countries do not differentiate much between cinnamon and cassia. In the US, cassia cinnamon is more popular and has almost wholly replaced Ceylon cinnamon in the market. A brief discussion on the end uses of cinnamon and cassia is presented below.

Cinnamon

Cinnamon as a spice

Commercially, cinnamon bark is marketed as quills, which are produced from the dried, inner bark extracted from shoots of *C. verum*. The bark has an aromatic and sweet taste with a spicy fragrance, due to the presence of essential oils. The major use of cinnamon bark is for flavouring processed food, the aromatic ingredient of the bark improving the overall flavour of the food item. A wide range of curry powders are used in Indian and other oriental cusines; the main components in most curry powders are: turmeric for colour; chillies for pungency; and coriander, cumin, fenugreek, ginger, celery and black pepper for flavour. Flavour enhancement is achieved through the addition of cinnamon, nutmeg, cloves, caraway or other spices. Cinnamon forms an ingredient of curry powder used in the preparation of meat, fish and vegetable dishes. Some typical curry powder formulations and the Federal Specifications for curry powder are given in Tables 13.1 and 13.2.

In the food industry, ground spice is essential for flavouring baked products, such as cakes, buns, biscuits, cookies, steamed puddings, pies, candies, chewing gum and desserts. Cinnamon is also added to marinades, beverages and ice creams. In Spain and Mexico, powdered cinnamon is a constituent of chocolate preparations. It is also used, along with other spices, in pickles, sauces, soups, confectionaries and canned fruits, such as Dutch pears or stewed rhubarb.

0-415-31755-X/04/$0.00 + $1.50
© 2004 by CRC Press LLC

312   B. Krishnamoorthy and J. Rema

Table 13.1 Curry powder formulations containing cinnamon

Spice	US standard formula		General purpose curry formula		
	No. 1 (%)	No. 2 (%)	No. 3 (%)	No. 4 (%)	No. 5 (%)
Coriander	32	37	40	35	25
Turmeric	38	10	10	25	25
Fenugreek	10	0	0	7	5
Cinnamon	7	2	10	0	0
Cumin	5	2	0	15	25
Cardamon	2	4	5	0	5
Ginger	3	2	5	5	5
White pepper	3	5	15	5	0
Poppy seed	0	35	0	0	0
Cloves	0	2	3	0	0
Cayenne pepper	0	1	1	5	0
Bay leaf	0	0	5	0	0
Chillies	0	0	0	0	5
Allspice	0	0	3	0	0
Mustard seed	0	0	0	3	5
Lemon peel dried	0	0	3	0	0
	100	100	100	100	100

Source: Farrell, 1985.

Table 13.2 Federal Specifications for curry powder (EE-S-631J)

Ingredient	Limit (%)
Turmeric	37.0–39.0
Coriander	31.0–33.0
Fenugreek	9.0–11.0
Cinnamon	<7.0
Cumin	<5.0
Black pepper	<3.0
Ginger	<3.0
Cardamom	<3.0

Source: Tainter and Grenis, 1993.

The oriental five-spice blend is commonly used in a variety of oriental dishes. This ground spice blend contains cinnamon (25–50%), anise or star anise (10–25%), fennel (10–25%), black pepper (10–25%) and cloves (10–25%). The famous Chinese spice salt is a mixture of salt, ground cinnamon (Chinese cassia) and a pinch of Chinese five-spice powder, the last being an essential ingredient in most spicy Chinese dishes. Chinese-five spice powder is made by mixing equal parts of cassia, Szechuan pepper (*Zanthoxylum piperitum*), cloves, fennel seeds and star anise. Certain pickling and pie formulations are given in Tables 13.3, 13.4 and 13.5.

Masalas are mixtures of spices either in dry or paste form, which are available in ready-to-use formulations. In many masalas, cinnamon and cassia form essential ingredients. Garam masala is used in Indian (both northern and southern) cooking. Unlike other masala mixes, it is sprinkled over a prepared dish before serving, to enhance the flavour.

Table 13.3 Typical western pickling formulation

Ingredient	Range (%)
Coriander	10–40
Mustard	10–40
Powdered bay leaves	10–20
Crushed red pepper	5–10
Whole allspice	5–10
Whole dill seed	5–10
Whole celery seed	5–10
Whole black pepper	0–10
Whole cloves	0–10
Cracked cinnamon	0–10
Cracked ginger	0–5

Note
The components are powdered before use.

Table 13.4 Pumpkin pie spice formulation

Ingredient	Range (%)
Ground cinnamon	40–80
Ground nutmeg	10–20
Ground ginger	10–20
Ground cloves	10–20
Ground black pepper	0–5

Table 13.5 Apple pie spice formulation

Ingredient	Range (%)
Ground cinnamon	60–95
Ground nutmeg/mace	2–15
Ground allspice	2–15
Ground anise/fennel	0–10

Garam masala formulations contain a variety of spices depending upon the particular brand. Premavally *et al*. (2000) found between 11 and 19 spices in the common brands available in the market. The major ingredients in garam masala mixes are coriander, chillies, cumin, pepper, cloves, ginger, cinnamon and turmeric. In addition to cinnamon, cassia is also present in certain brands. Premavally *et al*. (2001) analysed seven common brands of masalas and found that they contained 14–19 spices. In six of the masalas cassia formed a component, while cinnamon was present only in one brand (Badshah Fish Masala). Cinnamon is also a component of the famous *mus-sa-man* curry paste of Thailand; *Lakama*, the popular Moroccan spice mix; and many local spice mixtures in Far East Asia and the Middle East. (e.g. Sri Lankan curry powder, Singapore-style curry powder, Seven Seas curry powder, Kashmiri masala, and Sweet pickling spice mix.)

Cinnamon is also used in the preparation of mulled wines. Mulled wines are prepared by infusing wine with cinnamon sticks, powdered ginger, cardamon fruits and cloves, tied in cheesecloth bags. The bags are suspended in the wine and heated. Drinking small quantities of mulled wine before food aids digestion.

Table 13.6 The replacement ratios for ground cinnamon using oils and oleoresin with other spices

Spice	Replace 1# of ground spice with	
	Oil	Oleoresin
Cinnamon	0.025	0.025
Clove	0.140	0.050
Cardamom	0.030	0.015
Celery	0.010	1.000
Coriander	0.003	0.070
Ginger	0.015	0.035
Mace	0.140	0.070
Nutmeg	0.600	0.080
Pepper	0.015	0.050

Source: Tainter and Grenis, 1993.

Cinnamon bark oil is used in food, pharmaceutical and perfume industries. It is a light yellow aromatic oil with a sweet and spicy odour and has largely replaced ground cinnamon in the processing industry, since it can be measured accurately and imparts a uniform flavour to meat, other processed foods and confectionary. In food industry, it is used to flavour meat and fast food, sauces, pickles, baked foods, confectionery, liqueurs and soft drinks. Cinnamon leaf oil is obtained by steam distillation of cinnamon leaves and the oil yield ranges between 0.5% and 1.8 % (Senanayake and Wijesekera, 1989). More than 47 compounds have been identified from the leaf oil, the most significant being eugenol, which constitutes 65–92% (Senanayake *et al.*, 1978). Cinnamon leaf oil is cheaper than bark oil and is used in flavour industry, to a lesser extent, to flavour confectionery. It is also used as a source of eugenol for the preparation of synthetic vanillin. Cinnamon oleoresin obtained by solvent extraction is a dark brown extremely concentrated and viscous liquid, closely approximating the total spice flavour and containing 50% or more volatile oil. It is used mainly for flavouring food products such as cakes and confectionary. Ground spice has been replaced by oils and oleoresin in food industry. The replacement ratios for ground spice using oils and oleoresins are given in Table 13.6. Spice Extractives-Equivalencies (SEE) of cinnamon in comparison with other common spices are given in Table 13.7, which are useful in the manufacture of foods and seasonings.

Synergistic and suppressive effects

The taste of prepared food can be modified when combined with other food or ingredients. A synergistic effect is produced when the taste of one component in the food is enhanced by association with other food components. Similarly, there is a suppressive effect when a certain taste decreases in strength through combining with other food components. Cinnamon has the effect of enhancing the sweetness of food containing sugar. Due to its synergestic effect, cinnamon has been widely used in cooking baked foods and confectionary. Cinnamon has a sweet aroma and when it is added to sweet food, the sweetness sensation is enhanced because of the synergistic effect between the taste of sugar and the sweet aroma of cinnamon.

Table 13.7 Spice Extractives-Equivalencies of cinnamon in comparison with other common spices

Extractive	Type	G/type equivalent to 1_{Oz} ground spice	Extractives on soluble dry edible carrier (%)	Minimum volume oil in extractive vol/wt. (%)	Federal specification	
					Extraction dry carrier (%)	Volume of oil oil extract vol/wt. (%)
Cinnamon	SR	0.567	2.000	65	6.0	50
Cassia	EO	0.430	1.517	100	–	–
Cardamom	EO	0.851	3.000	100	–	–
Clove	EO	4.260	15.026	100	–	–
	SR	1.701	6.000	70	6.0	7.0
Coriander	EO	0.142	0.500	100	–	–
	SR	0.851	3.000	40	–	–
Cumin	SR	1.418	5.000	60	–	–
Ginger	OR	1.134	4.000	28	4.0	25
Pepper	OR	1.488	5.250	23	4.5	150

Source: Farrel, 1985.

Notes
SR = Superesin; OR = Oleoresin; EO = Essential oil.

Equally important is the suppression of undesirable flavours, for which spices are used widely, in a variety of foods. Ito *et al.* (1962) added a phased concentration of diluted solution of an individual spice to an emulsion of 1% mutton oil and determined organoleptically the amount of dilute solution of spice or mixtures of spices necessary to deodourise the smell of mutton. The deodourising effect of cinnamon in comparison with other major spices is given in Table 13.8. According to the Weber-Fecher law, the strength of an odour perceived by the sense of smell is proportional to the logaritham of the concentration of the compounds in question. In other words, the sensational strength perceived with the five senses is proportional to the logaritham of the actual strength of these stimuli. Thus, even if 99% of the total smelled components are

Table 13.8 Deodourising effects of cinnamon in comparison with other spices

Order of deodourising effect	Spice	DP
1	Sage	0.7
2	Thyme	2.5
3	Cloves	3.0
4	Caraway	4.0
5	Coriander	5.0
6	Garlic	23.0
7	Celery	25.0
8	Cardamom	30.0
9	Allspice	30.0
10	Nutmeg	45.0
11	Cinnamon	50.0
12	Ginger	90.0
13	Onion	190.0
14	Pepper	600.0

Source: Ito *et al.*, 1962.

Note
DP: deodourising point × 100.

eliminated chemically, the sensational strength perceived is reduced to only 66% (Hirasa and Takemasa, 1998). It is easier and more effective to use an aromatic spice to deodourise the remaining 1% via the masking function. Kikuchi *et al.* (1968) evaluated the masking effect of each spice sensorially by adding a spiced solution with phased concentration to trimethylamine solution. This report indicated that cassia oil is effective in masking the odour of trimethylamine, but to a lesser degree compared to oils of onion, bay leaf, sage, thyme, caraway, ginger and clove.

Suitability pattern

Spices are used to fulfil certain functions based on their suitability to flavour, deodourise (mask), colour or add pungency. Each spice has a basic function and often a sub-function. In the case of cinnamon, the basic function is flavouring and its sub-function is deodourising or masking. Cinnamon is mostly used in simmered, baked, fried, deep-fried and pickled food items, being less suitable for steamed food. It is more suitable for meat and mild dishes, grains, vegetables, fruits and beverages and less suitable for beans and seed dishes. Cinnamon is used as spice in South-East Asia, US, UK, Germany and France, but it does not play much of a role in Italian, Chinese and Japanese foods (Hirasa and Takemasa, 1998).

Cinnamon – in cooking

Cinnamon is widely used in sweet dishes, but also makes an interesting addition to savoury dishes, such as stews and curries. The blending of spices and herbs in traditional cooking is very imprecise. In domestic cooking, the experienced hand of the cook or the housewife does the mixing and blending. But, in factory production, processing and in the development of seasoning formulations, a great amount of precision is needed in the approach. A seasoning can inadverterly become the major flavour of the product, masking the primary flavours of meat, fish or vegetables present. A good seasoning should be subtle and should be used specifically to enhance the flavour characteristics of the food (Heath, 1978). Herbs and spices need careful blending, taking into account the flavour profile of all the ingredients in the product, the aim being to fortify those characteristics, which improve the flavour profile of the end product, or suppress those which detract from a pleasant flavour. Such skillful blending needs a good understanding of the flavouring power of spices. The relative flavour strength of some of the spices are listed in Table 13.9. Some condiments play dual roles, both as condiments and seasonings. Cinnamon and cassia do not play any significant role in the preparation of condiments, sauces or seasonings, unlike spices such as chillies, ginger or pepper. Cinnamon and cassia give a flavour strength of 406–425 points on a scale where the highest point is occupied by fresh red chillies (1000 points). In Table 13.10, herbs and spices are listed in the order of their flavour strength to indicate the position occupied by cinnamon and cassia in relation to other spices for flavouring food items. In this flavour triangle table the weakest spice is at the top and the strongest at the bottom. In the right column, the main protein sources are listed in the order of descending flavour strength (Heath, 1978). In this triangle, cinnamon and cassia occupy a position almost one-third from the top, indicating that the flavour tends more towards weak, which indicates the suitability of this spice for flavouring lamb and chicken dishes. Stronger spices are needed to flavour mutton and Game Ham (both have a stronger intrinsic flavour).

Table 13.9 Relative flavour intensity of cinnamon in comparison with other common spices

Spice	Flavour intensity
Fresh red chillies	1000
Cayenne pepper, dried	900
Mustard powder	800
Cloves, dried	600
Garlic, fresh	500
Ginger, dried	475
Black pepper, dried	450
Cassia, dried	425
Cinnamon, dried	400
Nutmeg, dried	360
Mace, dried	340
Coriander seed	230
Cardamom, dried	125

Source: Farrell, 1985.

Table 13.10 The place of cinnamon and cassia in the flavouring/seasoning triangle

Spice		Food
	Parsley	Fish
	Dill	
	Fennel	
	Marjoram	Veal
	Rosemary	Lamb
	Spanish sage	
	Tarragon	
I	English sage	
N	**Cinnamon**	Chicken
C	**Cassia**	
R	Oregano	
E	Dalmatian sage	
A	Origanum	Pork
S	Savoury	
I	Thyme	
N	Caraway	
G	Coriander	
	Fenugreek	
F	Basil	
L	Cardamom	Beef
A	Celery	
V	Cumin	
O	Allspice	
U	Clove	
R	Nutmeg mace	
	Ginger	Mutton
	Pepper	
	Chillies	Game Ham
	(Cayenne)	
	Mustard	

Source: Heath, 1978.

318   B. Krishnamoorthy and J. Rema

In cinnamon, the leaf and bark oils have entirely different flavour profiles and qualities, and one cannot be substituted for the other. There are many blends available which are imitation flavourings and could be distinguished through sensory and instrumental analysis.

Cinnamon – in medicine

Cinnamon bark is known as *Twak* (Sanskrit) in *Ayurvedic* literature (synonyms are: *surabhi, utkala, gudatwak, lataaparna, daarusheeta, tanutwak*). In *Ayurvedic* texts the properties of cinnamon are given as:

Rasa – *Katu, tikta, madhura*
Guna – *Laghu, rooksha, teekshna, pichchila*
Veerya – *Ushna*
Vipaka – *Katu*

Its actions are indicated as "carminative, antispasmodic, aromatic, stimulant, haemostatic, astringent, stomachic and germicide". The oil has no astringency but is a vascular and nervine stimulant; in large doses it is an irritant and narcotic poison (Nadkarni, 1954; Satyavati *et al.*, 1976).

The Greeks and Romans historically never used cinnamon or Chinese cassia in their food, but their use of the substance in perfumes and balms is well documented. In pharmaceutical industries, cinnamon forms an ingredient in medicines administered for colds, asthma and coughs as it has febrifuge and expectorant properties. Cinnamon gives relief from flatulence and, due to its carminative property, it is considered as a good medicine against gastric troubles and other minor intestinal upsets. Cinnamon bark is also a blood purifier, antispasmodic, stimulant, diuretic, haemostatic, soporific and deodorant. It is taken internally as a remedy for cardiac diseases. In modern medicine, cinnamon is combined with other ingredients to treat diarrhoea, internal haemorrhage, impotency, typhoid, halitosis, checking nausea and vomiting and for restoring normal skin colour on the face (Warrier *et al.*, 1994). It acts as a stimulant of the uterine muscular fibre and is hence employed in menorrhagia and in labour depending upon the insufficiency of uterine contractions. As a powerful stimulant it is used to alleviate cramps of the stomach, toothache and paralysis of the tongue (Watt, 1872). Cinnamon is externally used as a rubefacient and is employed to counteract the stings of poisonous insects. In India, some of the *Ayurvedic* medicines prepared from cinnamon bark include *astanga lavana churna, caturjata churna, sitopaladi churna, sudarsana churna, talisadya churna, chandraprabha vati, khadirarista, pippalyadyasava, lavanbhaskara churna* (Dey, 1980), and *vyaghri haritaki* (Asolkar *et al.*, 1994).

Bark oil forms a component of both *Ayurvedic* and *Unani* medicines of India. Its aroma is characteristically warm and spicy. It is employed as an adjuvant in stomachic and carminative medicines and is also administered in cases of anorexia, vitiated conditions of *vata*, inflammation, vomiting and tubercular ulcers (Warrier *et al.*, 1994). When applied in very small quantities to the forehead it gives relief in neuralgic headaches. It is also very effective for rheumatism and inflammation (Wagner *et al.*, 1986). In pharmaceutical preparations, bark oil is also used to mask the unpleasant taste of medicines. Bark oleoresin forms a constituent of sugar-based syrups in medicines. The leaf oil gives relief from rheumatism and inflammation. Cinnamon oil has been reported to cause miscarriages and hence, has to be used with caution during pregnancy.

Cinnamon – in perfumes and beauty care

Cinnamon bark oil is used to impart a woody and musky undertone to perfumes. However, the use of bark oil in the perfume industry is limited since it has a skin sensitising property. Cinnamon bark oil is highly irritant and is therefore not used widely in this way. Occupational allergic contact dermatitis is occasionally observed among those who work with this spice (Kanerva *et al.*, 1996). It is also used to impart fragrance to soaps, flavour dentifrices and mouthwashes. Leaf oil is also used in the perfume industry, as it imparts fragrance to soaps and other such toiletries.

Cinnamon is very useful in beauty care. It is a useful cure for acne, headaches, lumps and pustules when applied as a facial pack mixed with clove and pepper (Bhandari, 1987). A medicinal formulation for treating acne (Sepi control A5) contains cinnamon bark powder mixed with lipoglycine. This product is reported to have high acne curing activity, resulting in clear, less oily skins. The product has high hypo-allergeric action and excellent skin tolerance (Stoltz, 1998).

Cinnamon – an antimicrobial agent

Salt and spice have played very significant roles in preventing food spoilage caused by microorganisms. Cinnamon, like many other spices, possesses antiseptic and anti-microbial properties. *In vitro* studies showed that the bark extracts were active against 27 strains of *Vibrio cholerae* and were also a good antibacterial agent for *Shigella* and *V. cholerae* infections (Islam *et al.*, 1990). The ground spice of cinnamon also exhibited bacteriostatic effects against *Yersinia enterocolitica* (Bara and Vanetti, 1995). Bullerman (1974) noticed that cinnamon bread did not get mould growth, unlike other breads. The effect of ground cinnamon on growth and aflatoxin production by *Aspergillus parasiticus* is given in Table 13.11. The addition of 0.02% cinnamon powder suppressed aflatoxin production to 21–25%. Cinnamon bread usually contains 0.5–1.0% cinnamon, indicating that mould growth and aflatoxin production should be almost completely inhibited. The result was further confirmed in experiments involving alcohol extracts of cinnamon where 0.2% extract was shown to cause 98–99% suppression of toxin production.

Table 13.11 Effect of ground cinnamon on growth and aflatoxin production by *Aspergillus parasiticus*

Level of cinnamon	Strain NRRL 2999				Strain NRRL 3000			
	Mycelia mg	Inhibition %	Aflatoxin μg/ml	Inhibition %	Mycelia mg	Inhibition %	Aflatoxin μg/ml	Inhibition %
Control	2301	–	356	–	1896	–	292.0	–
0.02	1943	16	267	25	1959	(+3)	232.0	21.0
0.2	1768	23	148	58	1557	18	49.0	83.0
2.0	1589	31	11	97	1658	13	2.0	99.0
20.0	ND	100	ND	100	100	95	0.3	99.9

Source: Hirasa and Takemasa, 1998.

Note
ND: Not detected.

320　B. Krishnamoorthy and J. Rema

The aqueous extract of powdered bark exhibits appreciable activity against fluconazole-resistant and susceptible *Candida*, which causes mucosal candidosis. Chloroform extracts of cinnamon exhibited inhibitory activity on the growth and aflatoxin production of *Aspergillus parasiticus* (Sharma *et al.*, 1984). Ueda *et al.* (1982) evaluated the antifungal and antibacterial properties of alcohol extracts of spices against several moulds and bacteria. The Minimum Inhibitory Concentration (MIC) of cinnamon on pathogenic bacteria is given in Tables 13.12, 13.13 and 13.14.

Cinnamon bark oil is a powerful fungicide, germicide, insecticide and also inhibits aflatoxin production. Cinnamaldehyde, the major constituent of cinnamon and cassia barks, possesses antibacterial and antifungal activities. At 0.33 mm concentration it causes complete inhibition of *Candida albicans, A. fumigatus, A. niger, Pencillium frequentans, P. decumbens* and *Cladosporium bantiannum*.

Cinnamon leaf oil is antimicrobial and is sensitive to yeast and filamentous fungi, including dermatophytes and moulds (Mangiarotti *et al.*, 1990). Maruzzella and Lichtenstein (1958) tested a variety of volatile oils and found that volatile oils of

Table 13.12 Minimum Inhibitory Concentration (%) of cinnamon in comparison with other major spices

Spice	pH	B.s.	S.a.	E.c.	S.t.	S.m.	P.a.	P.v.	P.m.
Cinnamon	7.0	4.0	2.0	4.0	4.0<	4.0<	4.0	2.0	4.0
	5.0	0.5	2.0	2.0	4.0	4.0	4.0	1.0	2.0
Cardamom	7.0	2.0	2.0	4.0<	4.0<	4.0<	4.0<	4.0<	4.0<
	5.0	0.1	0.5	4.0<	4.0<	4.0<	4.0<	2.0<	4.0
Celery	7.0	4.0	1.0	4.0<	4.0<	4.0<	4.0<	4.0<	4.0<
	5.0	0.5	1.0	4.0<	4.0<	4.0<	4.0<	2.0<	4.0<
Cloves	7.0	1.0	1.0	1.0	1.0	1.0	2.0	1.0	1.0
	5.0	0.5	2.0	1.0	1.0	1.0	1.0	0.5	0.5
Mace	7.0	0.2	0.05	4.0<	4.0<	4.0<	4.0<	4.0<	4.0<
	5.0	0.1	0.5	4.0<	4.0<	4.0<	4.0<	4.0<	4.0<

Source: Hirasa and Takemasa, 1998.

Notes

B.s. – *Bacillus subtilis*; S.a. – *Staphylococcus aureus*; E.c. – *Escherichia coli*; S.t. – *Salmonella typhimurium*; S.m. – *Salmonella marcescens*; P.a. – *Pseudomonas aerugenosa*; P.v. – *Proteus vulgaris*; P.m. – *Proteus morgani*.

Table 13.13 Minimum Inhibitory Concentration (%) of cinnamon for fungi in comparison with other major spices

Spice	S.c.	C.p.	C.k.	P.sp.	A.o.
Cinnamon	1.0	1.0	1.0	1.0	1.0
Cardamom	4.0	4.0	4.0<	4.0<	4.0<
Caraway	4.0<	4.0<	4.0<	4.0<	4.0<
Celery	4.0	4.0	4.0	10.0	10.0
Cloves	0.5	0.5	0.5	0.5	0.2
Mace	4.0<	4.0<	4.0<	4.0<	4.0<

Source: Hirasa and Takemasa, 1998.

Notes

S.c. – *Saccharomyces cerevisiae*; C.p. – *Candida parakrusei*; C.k. – *Candida krusei*; P.sp. – *Pencillium* sp.; A.o. – *Aspergillus oryzea*.

Table 13.14 Minimum Inhibitory Concentration (%) of hexane extract of cinnamon on pathogenic bacteria

Spice	E.c.	S.sp.	S.a.	B.s.	C.
Cinnamon	5	10	2.5	2.5	1.3
Clove	10	10	5.0	5.0	2.5
Allspice	10	>10	10.0	10.0	10
Marjoram	>10	>10	10.0	10.0	>10
Organo	2.5	5	1.3	2.5	1.3
Rosemary	>10	>10	0.31	0.1	>10
Sage	>10	>10	0.83	0.31	>10

Source: Hirasa and Takemasa, 1998.

Notes
E.c. – *Escherichia coli*; S.sp. – *Salmonella* sp.; S.a. – *Staphylococcus aureus*; B.s. – *Bacillus subtilis*; C. – *Campylobactor*.

cinnamon, cumin, dill and thyme exhibited relatively strong antibacterial activities. Of the bacteriae tested, *Bacillus subtilis* was the most susceptible and *Escherichia coli* was relatively resistant. Growth and production of *Alternaria alternata* can be inhibited by cinnamon oil and it is suggested that the oil can be used as a preservative in processed tomato products (Hasan, 1995). Eugenol, the major component of cinnamon leaf oil, was mainly responsible for the above effect. Hitokoto *et al.* (1980) have shown that 125 µg/ml of eugenol can lead to a high inhibition of mycotoxin production by *Aspergillus flavus*, *A. ochraceus* and *A. versicolor* (100, 76 and 95%, respectively), while the corresponding values for 250 µg/ml were 100, 83 and 100%. Cinnamon is also useful as an insecticide to a limited extent.

The essential oil obtained from cinnamon seed exhibited antimicrobial activity (Chaurasia and Jain, 1978). The possibility of industrial application of seed oil as an antimicrobial agent needs to be explored. Root bark of cinnamon yields about 0.9–2.8% oil (Senanayake and Wijesekera, 1989), which is different from bark and leaf oils. The oil has a camphoraceous odour, and camphor separates out on standing. The major component (60%) of root bark oil is camphor (Senanayake *et al.*, 1978).

Cinnamon – an antioxidant

Rancidity of food is caused mainly by oxidation of the lipid fraction, which in turn is related to the formation of the peroxide radicals. Spice oils are effective in preventing peroxidation leading to rancidity. Though cinnamon has antioxidative properties, among the spices and herbs, rosemary and sage are the most effective antioxidants having the capacity to prevent food spoilage (Table 13.15). Cinnamon and its petroleum ether and alcohol soluble fractions, are found to delay the oxidation of lard (Chipault *et al.*, 1952).

Cinnamon – in home remedies

Cinnamon forms a part of many home remedies as it has many medicinal properties. It is included as a part of most Indian dishes, since it is a carminative. The Chinese take cinnamon as a remedy for excess gas in the stomach and to reduce fever. Some of the home remedies where cinnamon is used are listed below.

322   B. Krishnamoorthy and J. Rema

Table 13.15 Antioxidative activity of cinnamon in comparison with other spices (con.added 0.02%) against lard

Spice	Ground spice POV (mEq/kg)	Pet.ether soln. fraction POV (mEq/kg)	Pet.ether insoluble fraction POV (mEq/kg)
Cinnamon	324.0	36.4	448.9
Cardamom	423.8	711.8	458.6
Black pepper	364.5	31.3	486.5
Chillies	108.3	369.1	46.2
Clove	22.6	33.8	12.8
Ginger	40.9	24.5	35.5
Turmeric	399.3	430.6	293.7
Mace	13.7	29.0	11.3
Nutmeg	356.6	31.1	66.7
Rosemary	3.4	6.2	6.2
Sage	2.9	5.0	5.0

Source: Saito *et al.*, 1976.

Note
POV: Peroxidase value.

1.   Cinnamon taken in hot rum gives relief from the common cold.
2.   Ground cinnamon added to hot milk or stirred into a mixture of lemon juice, honey and hot water and taken at bedtime lessens the miserable effect of the common cold.
3.   A tablespoon of cinnamon water taken half an hour after meals is effective against flatulence and indigestion.
4.   A paste of cinnamon powder in water, applied to the forehead, is effective against headache.
5.   Coarsely powdered cinnamon boiled in a glass of water with a pinch of pepper and honey is an effective medicine for sore throats and colds.
6.   Gargling with cardamon seeds and cinnamon boiled in water, gives instant relief from sore throats and prevents further infection.
7.   Cinnamon has the ability to control blood sugar levels. Though it has not been proved clinically to control diabetes, it has been reported that adding a quater of a teaspoon cinnamon per day to the daily diet has shown favourable results in many diabetics.
8.   Cinnamon can be used as an ingredient in spice mixtures used to ward off insects and as room freshners. Sweet bags are used to perfume linens when keeping them in cupboards. Sweet bags are prepared by mixing half a cup dried rosebuds, a third of a cup of ground orris root, one cup of coriander seeds bruised in mortar, 1 teaspoon ground cinnamon, ten slightly bruised whole cloves, half a cup of dried orange flowers, half a teaspoon of common salt all filled in small cotton bags.
9.   Moths are a bane in many households and a variety of scented mixtures have been developed to deter insects from laying their eggs amongst fabrics and clothes. Moth bags can be made by mixing equal amounts of ground caraway, cloves, nutmeg, mace, cinnamon and tonka beans and by adding the same quantity of ground orris root.

Chinese Cassia

Chinese cassia – as spice

Bark of Chinese cassia resembles that of cinnamon in appearance but it has a more powerful aroma. The bark powder is reddish brown in colour, unlike cinnamon powder, which is tan. The general composition of Chinese cassia is similar to that of cinnamon, but the mucilage content of various plant parts is higher in cassia. Cinnamaldehyde is the major component (70–95%) of cassia bark.

Since ancient times, Chinese cassia has been used as a spice, due to its pleasant aroma and taste. The major uses of cassia bark, both in whole and ground forms, are for culinary purposes and for processed foods. It is an essential spice in Chinese and other cuisines in South-East Asian countries. Cassia bark is used to flavour all kinds of meat dishes, in which it is ground and usually mixed with other spices. The ground spice is used in the flavouring of bakery products, sauces, pickles, puddings, curry powders, beverages and confectionery. It is ideal for spiced cakes, pies, sticky buns, pumpkin bread, cheese cake and apple strudel. It can even be sprinkled on top of French toast, oatmeal or hot cocoa. Commercial cassia oil is obtained by steam distillation of leaves, leaf stalks and twigs. It is widely used to flavour baked foods, confectionery, meat, sauces, pickles, soft drinks and liqueurs.

Dried unripe fruits of *C. cassia* are known as cassia buds, which resemble small cloves in appearance. They have an aroma and flavour similar to that of cinnamon bark and contain about 2% essential oil which in turn contains about 80% cinnamaldehyde. The buds possess properties similar to those of the bark and were formerly used widely in Europe as a spice for preparing a spiced wine called *Hippocras*. Cassia buds are usually used in savoury dishes rather than in sweets. The dried cassia buds are used in the East as a spice to flavour pickles, curries, spicy meat dishes, confectionary, chocolates and in making pot pourri.

Cassia oleoresin has similar applications as those of the ground spice in the flavouring of processed foods.

*Cassia – in medicine**

Dried bark and twigs of cassia are not only important as spices but are also used as an important crude drug in oriental medicine. Medicinal use of cassia was first mentioned by Tao Hunkin (AD 451–536). The therapeutic effect is regarded as being due to tannins present in its bark (Yazaki and Okuda, 1990). Tannins of cassia were recorded to have both antiviral and cytotoxic activity. Dried stem bark of cassia is used to treat inflammation, headaches and pyrexia (Kanari *et al.*, 1989), diarrhoea, nausea and flatulence and as a tonic. In Yemenite folk medicine, cassia is an ingredient of the compounds used against headache and melancholy (Asolkar *et al.*, 1994). It is also used in the traditional medicines of Tibet and folk medicine in Mongolia and the Tran-Baikal region. Cassia also forms one of the components of Japanese herbal medicines such as "TJ 960", which is recommended for hippocampal neuron damage (Sugaya *et al.*, 1991). Chinese herbal medicine formulations contain the bark of cassia and are used to treat blood hyperviscosity, hyperlipemia, hypercoagulability (Toda *et al.*, 1989), gynaecological

---

* See also Chapter 6 on Chinese cassia for more information.

disorders like hypermenorrhea, dysmenorrhea and infertility (Sakamoto *et al.*, 1988). It is also administered for gonorrhea and the leaves are used in rheumatism as a stimulant. The bark was prescribed by the Hakims for the enlargement of spleen, disorders of nerves and for the retention of urine. Cassia was held in considerable repute by the ancients for its stomachic and soporific properties. It is also given as a decoction or powder for the suppression of lochia after childbirth. The bark is also a minor constituent of *Unani* medicine *jawarish jalinoos*, a drug prescribed for gastro-enterological complaints (Asolkar *et al.*, 1994).

Trans-cinnamaldehyde isolated from the cortex showed antimutagenic activity in *Escherichia coli* (Kakinuma *et al.*, 1984). A number of diterpenes were isolated from cassia bark and a few of them were reported to exibit antiallergic activity. 3-(2-hydroxy phenyl)-propanoic acid and its o-glucoside isolated from the stem bark is reported as having potent antiulcerogenic constituents (Tanaka *et al.*, 1989). Cassia oleoresin is stomachic, carminative, mildly astringent, emmenagogue and capable of decreasing the secretion of milk. It is used mainly to assist and flavour other drugs used against diarrhoea, nausea, vomiting, uterine haemorrhage and menorrhagia, and to relieve flalutence. It is a strong local stimulant and a powerful germicide. Accidental intake of considerable proportions or overdoses of cassia may result in acute poisoning and inflammation of the gastro-intestinal mucous membrane.

Cassia – in industry

Cassia oil obtained by the distillation of leaves, stalks and twigs is used for purposes similar to that of cinnamon bark oil, in perfumery and flavouring. The oil finds extensive use in flavouring soft drinks and other beverages. As in the case of cinnamon bark oil, the use of cassia oil in the perfume industry is limited due to its skin sensitising property. Cassia buds once formed a part of the "spicy wedding gift box". In China, including cassia bud twins in the gift box symbolises two joining as one spiritually.

(Refer to Chapter 6 for more details on the uses of Chinese cassia.)

References

Asolkar, L.V., Kakkar, K.K. and Chakre, O.J. (1994) *Glossary of Indian Medicinal Plants with Active Principles* Part – I (A–K), Publication and Information Directorate (CSIR), New Delhi.

Bara, M.T.F. and Vanetti, M.C.D. (1995) Antimicrobial effect of spices on the growth of *Yersinia enterocolitica. J. Herbs Spices Medicinal Plants*, 3(4), 51–58.

Bullerman, L.B. (1974) *J. Food Sci.*, 39, 1163 (cited from Hirasa and Takemasa 1998).

Chaurasia, S.C. and Jain, P.C. (1978) Antibacterial activity of essential oils of four medicinal plants. *Indian J. Hosp. Pharm.*, 15(6), 166–168.

Chipault, J.R., Mizuno, G.R., Hawkins, J.W. and Lundberg, W.O. (1952) *Food Res.*, 17, 46 (cited from Hirasa and Takemasa 1948).

Dey, A.C. (1980). *Indian Medicinal Plants Used in Ayurvedic Preparations*, Bishen Singh Mahendra Pal Singh, Dehra Dun, India.

Farrell, K.T. (1985) *Spices, Condiments and Seasonings*, Avi Pub. Co., USA.

Hasan, H.A.H. (1995) *Alternaria* mycotoxins in black rot lesion of tomato fruit: conditions and regulations of their production. *Mycopathologia*, 130(3), 171–177.

Heath H.B. (1978) *Flavour Technology*, Avi Pub. Co., USA.

Heda, S., Yamashita, H., Nakagima, M. and Kuwahara, S. (1982) *Nippon Shokuhin kogyo Gakkaishi*, 29. III (cited from Hirasa and Takenasa 1998).

Hitokoto, H., Morozumi, S., Wauke, T., Sakai, S. and Kurata, H. (1980) *Appl. Environ. Microb.*, 39, 818. Cited from Hirasa and Takemasa (1998).

Hirasa, K. and Takemasa, M. (1998) *Spices Science and Technology*, Marcel Dekker, New York.

Islam, S.N., Ahsan, M., Ferdous, A.J. and Faroque, A.B.M. (1990) *In vitro* antibacterial activities of commonly used spices. *Bangladesh J. Botany*, 19(1), 99–101.

Ito, Y., Miura, H. and Miyarga, K. (1962) *Gyoniku Sausage Kyoukaisi*, 85. Cited from Hirasa and Takemasa (1998).

Kakinuma, K., Koike, J., Kotani, K., Ikekawa, N., Kada, T. and Nomoto, M. (1984) Cinnamaldehyde: Identification of an antimutagen from a crude drug cinnamomi cortex. *Agricultural and Biological Chemistry*, 48(7), 1905–1906.

Kanari, M., Tomoda, M., Gonda, R., Shimizu, N., Kimura, M., Kawaguchi, M. and Kawabe, C. (1989) A reticuloendothelial system activating arabinoxylan from the bark of *Cinnamomum cassia*. *Chemical and Pharmaceutical Bulletin*, 37(12), 3191–3194.

Kanerva, L., Estlander, T. and Jolanki, R. (1996) Occupational allergic contact dermatitis from spices. *Contact Dermatitis*, 35(3), 157–162.

Kikuchi, T., Hirai, K. and Sudarso, A.S. (1968). *Eiyo to Shokuryo*, 21: 253 (cited from Hirasa and Takemasa (1998).

Mangiarotti, A.M., Del, F.G. and Caretta, G. (1990) Note on the action of some essential oils on fungi. *Boletin Micologico*, 5(2), 1–4.

Maruzzella, J.C. and Lichtenstein, M.B. (1958) *J. Am. Pharm. Assoc. Sci. ed.*, 47, 250. Cited from Hirasa and Takemasa (1998).

Nadkarni, K.M. (1954) *Indian Materia Medica*. Popular Prakashan, Bombay.

Premavally, K.S., Majumdar, T.K. and Malini, S. (2000) Quality evaluation of traditional products. II. Garam masala and puliyodara mixmasala. *Indian Spices*, 37(2), 10–13.

Premavally, K.S., Majumdar, T.K. and Leela, R.K. (2001) Quality evaluation of spice mixtures for non-vegetarian dishes. *Beverage & Food World*, 28(2), 33–35.

Saito, Y., Kimura, Y. and Sakamoto, T. (1976). *Eiyo to Syokurya*, 29, 505. Cited from Hirasa and Takemasa (1998).

Sakamoto, S., Kudo, H., Kawasaki, T., Kuwa, K., Kasahara, W., Sassa, S. and Okamoto, R. (1988) Effects of a Chinese herbal medicine, Keishi-bukuryogan, on the gonadal system of rats. *J. Ethnopharm.*, 23(2–3), 151–158.

Satyavati, G.V., Raina, M.K. and Sharma, M. (eds) *Cinnamom* (Lauraceae) (1976) *Medicinal Plants of India*. Vol. 1., ICMR, New Delhi, pp. 232–235.

Senanayake, U.M., Lee, T.H. and Wills, R.B.H. (1978) Volatile constituents of cinnamon (*Cinnamomum zeylanicum*) oils. *J. Agric. Food Chem.*, 26(4), 822–824.

Senanayake, U.M. and Wijesekera, R.O.R. (1989) Volatiles of the *Cinnamomum* species. Proc. 11th International Congress Essential Oils, Fragrances and Flavours, New Delhi, November, 1989.

Sharma, A., Ghanekar, A.S., Desai, P.S.R. and Nadkarni, G.B. (1984) Microbiological status and antifungal properties of irradiated spices. *J. Agric. Food Chem.*, 32(5), 1061–1063.

Stoltz, C. (1998) Sepicontrol A5. *SEPIC Newsletter*, pp. 152–156.

Sugaya, E., Ishiga, A., Sekiguchi, K., Yuzurihara, T., Iizuka, S., Sugimoto, A., Takeda, S., Wakui, Y., Ishihara, K. and Aburada, M. (1991). Protective effects of TJ 960 herbal mixture on hippocampal neuron damage induced by cobalt focus in the cerebral cortex of rats. *J. Ethnopharm.*, 34(1), 13–19.

Tainter, D.R. and Grenis, A.T. (1993) *Spices and Seasonings*, VCH Pub. Inc., USA.

Tanaka, S., Yoon, Y., Fukui, H., Tabeta, M., Akira, T., Okana, K., Iwani, M., Iya, Y. and Yokoyama, K. (1989) Anti ulcerogenic compounds isolated from chinese cinnamon. *Planta Medica*, 55(3), 245–248.

Toda, S., Ohnishi, M. and Kimura, M. (1989) Actions of Chinese herbal medicines Keishibukuryo-gan and Tougakujyouki – to on the hemolysis and lipid peroxidation of mouse erythrocytes induced by hydrogen peroxide. *J. Ethnopharm.*, 27(1–2), 221–225.

Ubl, S.R. (2000) *Handbook of Spices, Seasonings and Flavorings.* Technomic Pub., USA.

Wagner, H., Wierer, M. and Bauer, R. (1986) *In vitro* inhibition of prostaglandin biosynthesis by essential oils and phenolic compounds. *Planta Medica*, 3, 183–187.

Warrier, P.K., Nambiar, V.P.K. and Ramankutty, C. (1994) *Indian Medicinal Plants A Compendium of 500 Species* (Vol. II). Orient Longman Ltd., Madras, India.

Watt, G. (1872). *Dictionary of the Economic Products of India.* (Vol. II), Periodical Experts, New Delhi, India.

Yazaki, K. and Okuda, T. (1990) Condensed tannin production in callus and suspension cultures of *Cinnamomum cassia. Phytochem.*, 29(5), 1559–1562.

14 Cinnamon and Cassia – The Future Vision

U.M. Senanayake and R.O.B. Wijesekera

Cinnamon and cassia are two established natural food flavours. Food technologists and food manufacturers find it very difficult to imitate them completely with synthetic substitutes. The earliest attempts to imitate cinnamon or cassia flavours were by using synthetic cinnamic aldehyde. Although cinnamic aldehyde gives a very crude imitation of either cinnamon or cassia, it does not closely resemble the natural flavour. Using advanced analytical methods such as gas chromatography, Ter Heide (1972) was able to study in detail the chemical composition of *Cinnamomum cassia*. Ter Heide claimed that Germany no longer had to depend on imported cassia as his laboratory could compose it from synthetic chemicals. To prove this Ter Heide used GLC methods that exactly matched natural cassia oil, whose GLC charts were the same. He challenged observers to distinguish between the GLC chart of synthetic cassia oil from that of natural cassia oil. Even though many observers failed to distinguish between the two charts, later organoleptic tests revealed a different story. The synthesised cassia oil had a crude resemblance to natural cassia oil, but it was nowhere near the delicate and subtle flavour of the natural oil. It became evident that the ultimate judgment of the flavourist was more sensitive and discriminating. During the 1970s, many essential oil producers feared that their industry may come to a halt as most of their oils were being studied in detail and could possibly have been totally synthesised in the laboratory. But, soon their fears were dispelled as synthetic essential oils were rejected by leading food manufacturers.

Cinnamon bark oil contains no less than 90 identified compounds and over 50 very minute unidentified compounds (Senanayake and Wijesekera, 1989). The mild and mellow aroma of both cinnamon bark and cassia oils is due to the total synergistic effect of all these compounds. Many of the natural chemicals are laevo-rotatory, while the synthetic chemicals are dextro-rotatory. The olfactory sites in the nose are capable of distinguishing the olfactory manifestation of this difference. Cinnamic aldehyde isolated from natural cinnamon bark oil has a pleasing odour, while synthetic cinnamic aldehyde has only a crude resemblance to this pleasing odour. The synthetic version could replace the natural flavour to a certain extent but not entirely. Because of this reason, the natural essential oils will always have a market.

Even natural cinnamon or cassia oils have different flavours depending upon the origin of the source (see Chapters 3 and 4). In general, spices and spice oils contain complex mixtures of volatiles – aromatic flavouring compounds. Some of the volatile compounds affect the olfactory centres. Since the odour and the effect on the taste buds determines what is called flavour, the flavouring characteristics of a spice oil are, in part, directly related to the nature and the relative amounts of its volatile compounds.

0-415-31755-X/04/$0.00 + $1.50
© 2004 by CRC Press LLC

Cinnamon, cassia and their products have long been recognised for their delicate flavours, and are widely used in food, pharmaceutical, soap and cosmetic industries. As such, the development of methods for the assessment of quality and adulterants are required. The complex interplay of producers, buyers, users and consumers is possible only if all the parties concerned use the same and consistent technical language. The standarisation of spices, such as cinnamon, cassia and their products is a difficult problem. Often the quality depends on environmental influences, such as climate and soil conditions; botanical variety; methods of production; and, harvesting conditions.

Future visions on cinnamon and cassia largely depend on the answers to the above issues. Even though true cinnamon (*C. verum*) has been introduced to countries with suitable climates, such as east African countries, and grows in the Seychelles, the Malagasy Republic, and South Vietnam, the quality of the bark and bark oils produced by them are different from those cultivated in Sri Lanka. Also, it has been noted that cinnamon from the same root stock but grown in different climatic conditions, even within Sri Lanka, have different chemical and organoleptic properties. Thus, both producers and consumers suffer from being unable to demand the quality to be consistent in different consignments.

In Sri Lanka this problem has been somewhat overcome by confining the products from one micro-climatic zone away from those of others. Differences have been amply demonstrated in assessing the cinnamon bark and bark oils from the three main zones of Sri Lanka. The most acceptable cinnamon is generated from the sea belt area, from Colombo to Matara. The cinnamon grown in area north of Colombo has a higher safrole content, while that grown in the hill country has a higher cinnamic aldehyde content and fewer minor terpenoid constituents.

The same situation applies to cassia. Various categories of cassia, such as Chinese cassia, Saigon cassia, Burmese cassia and Taiwan cassia are thus named to distinguish crops originating from the respective geographic regions. Lawrence (1967) proposed to use the country of origin as a prefix to indicate the source. In nomenclature, true cinnamon or *C. zeylanicum* or *C. verum* is known as **cinnamon** and *C. cassia* is known as **cinnamon cassia**, whilst the country of origin denotes the quality.

There will always be a demand for natural cinnamon and cassia. Recently, Sri Lanka and other cinnamon and cassia growing countries, have embarked on programmes to propagate the best varieties using tissue culture techniques. There is a prospect of large scale plantations evolving from selected plants through tissue culture, thereby producing a product of consistent quality.

Both cinnamon and cassia are labour-intensive crops. Harvesting is similar to tea plucking, where each bud has to be hand picked. In the case of cinnamon quills, each quill has to be produced by hand peeling. As each bush and stick is unique, it is not conceivable to devise a machine for this purpose. In countries where cinnamon and cassia are grown, plenty of relatively cheap labour can be available, but the situation seems to be changing, as in Sri Lanka. The making of quills requires skilled labour of a specialised kind. There is a shortage of such skilled labour and situation is worsening each year. Cinnamon is not a profitable crop for countries where labour is costly.

One immediate matter that should be considered in future visions is the depletion of supply due to natural causes, such as rain, drought and disease. In many countries the present supply comes from small holders. As a long-term solution, organised cultivation of cinnamon and cassia on a plantation basis, with government support, should be encouraged.

Another problem is the fluctuation of market prices. Whenever prices fall below a certain limit, the plantations are neglected. A guaranteed floor price scheme would be helpful to maintain a steady supply of cinnamon bark, cassia bark and their oils. Good storage facilities for products is an urgent need that could limit the fluctuation in prices.

The cinnamon and cassia industry produces a range of unique natural products, and an international obligation exists to ensure that prices are maintained at steady levels, so as to maintain their existence in world markets. Continued attention to Research and Development is vital in this obligation. The research component has been neglected, except for some clonal selection work. Little improvement has been made on production technology issues over and above what has been practiced traditionally. Growers have not turned their attention seriously on these crops. There is a definite scope for improvement through cross-breeding, to improve yield as well as quality. Novel flavour profiles can be evolved by the use of such breeding strategies. Interspecific hybridisation between *C. verum* and *C. cassia* can be useful in evolving hybrids with more vigour, and improved or novel flavour characteristics.

Good agricultural practices, better harvesting and processing methods, and improved relationships between buyers and sellers, will help to keep this trade active in the future.

References

Lawrence, B.M. (1967). A review of some commercial aspects of cinnamon. *Perf. Essent. Oil Record*, **59**, 236–241.

Ter Heide, R. (1972) Qualitative analysis of essential oil of cassia. *J. Agric. Food Chem.* **20**, 747–751.

Senanayake, U.M. and Wijesekera, R.O.B. (1989) Volatiles of *Cinnamomum* species. In: *Proceedings 11th International Congress of Essential Oils, Fragrances and Flavours.*

15 Other Useful Species of *Cinnamomum*

M. Shylaja, P.N. Ravindran and K. Nirmal Babu

Introduction

Though only cinnamon (*C. verum*) and cassias (Chinese, Indonesian and Indian) are of commercial importance, there are other species that yield aromatic oils and are of local importance, both as spice and as an ingredient of local medicine. The types of spices and herbs used by the forest tribes and local people, are often known only by them. *Cinnamomum* is a large genus and consists of around 450 species, and many of them may have some localised use. Little is known about these species, although some have been studied in more detail. The following is a brief consideration of those species with economic value, other than the cinnamon and cassias.

C. angustifolium *Rafin*

C. angustifolium Rafinesque, Sylv. Tellur.135, 1838; Merrill, Index Rafin. 127, 1949: It is a species occurring in Madagascar. Chalchat and Valade (2000) analysed the leaf oil of this species and obtained 39 compounds. The percentage composition of the various compounds is given in Table 15.1. This species has a high proportion of hydrocarbons in its leaf oil, with α-phellandrene and p-cymene the most prominent.

C. bejolghota *(Buch-Ham) Sweet (syn.* C. obtusifolium *Nees)*

C. bejolghota (Buch-Ham) Sweet, Hort. Brit. Ed 1:334, 1827; Balak, Fl. Jowai 2:407, 1983; Haridasan and Rao, Forest Fl. Meghalaya 2:720, 1987; *Laurus bejolghota* Buch-Ham. in Trans. Linn-Soc. 13:559, 1822; *L.obtusifolia* (Roxb.) Nees in Wall. As. Rar 2:73, 1831; HKf. Fl.Brit. India. 5: 128, 1886; Gamble, Man. Indian Timbers, 561, 1902; Brandis, Ind. Trees 533, 716. 1906; Kanjilal *et al*. Fl.Assam, 4:56, 1940.

Baruah *et al*. (2000) found that the leaves of this species are also traded and used as *tejpat*. This is a moderate sized evergreen tree, attaining a height of 6–7 m, leaf buds silky tomentose, leaves aromatic, glabrous above, sparsely hairy beneath, narrowly elliptic-obovate-lanceolate to oblong-lanceolate, triplinerved, lateral veins ascending to the tip. Panicle pseudoterminal, axillary to solitary, lax-flowered, purple brown to pale brown, minutely pubescent, shorter or equal to the leaves, up to 10 cm in length, longer at fruiting stage, up to 25 cm. Flowers 6–7 mm long, perianth 3 + 3, subequal, oblong-lanceolate, silky tomentose on both surfaces. Floral structure is similar to other species. Occurrence is not as common as other species in the eastern Himalayas and Myanmar.

0-415-31755-X/04/$0.00 + $1.50
© 2004 by CRC Press LLC

Other Useful Species of Cinnamomum   331

Table 15.1 Chemical composition of leaf oil of *C. angustifolium*

Compound	%
α-Pinene	4.8
α-Thujene	1.9
α-Fenchene	0.3
Camphene	0.1
β-Pinene	1.2
Sabinene	1.0
δ-3-Carene	1.8
α-Phellandrene	23.4
Myrcene	2.6
α-Terpinene	0.3
Limonene	1.8
1,8-Cineole	9.7
(Z)-β-ocimene	0.2
γ-Terpinene	0.5
(E)-β-ocimene	1.1
p-Cymene	17.6
Terpinolene	0.3
Camphor	0.1
Linalool	3.3
β-Caryophyllene	6.2
Terpinen-4-ol	0.7
α-Humulene	0.2
Salicylaldehyde	0.2
Viridiflorene	0.4
α-Terpineol	2.1
Germacrene D	1.8
Bicyclogermacrene	3.0
(Z)-cinnamaldehyde	0.9
δ-Cadinene	0.3
Geranyl acetate	0.3
p-Menth-1(7),5-dien-2-ol	0.2
p-Cymene-8-ol	0.3
Caryophyllene oxide	1.0
(E)-cinnamaldehyde	0.2
(E)-nerolidol	0.4
Guaiol	0.4
Spathulenol	1.4
(E)-cinnamyl acetate	0.4
Carvacrol	0.1

Source: Chalchat and Valade, 2000.

It is used as a spice by local people and is known by names such as *Pati-Hunda, Naga-dalchini, Seerang-esing, Sami-Jong* and *Tejpat-manbi* (Baruah *et al.*, 1997). The bark is sold in local markets. The bark (as well as its infusion) is used as medicine for the treatment of coughs, colds, toothaches, liver complaints, gall stones and as a mouth freshner (Rao, 1981). The leaves can also be used for the preparation of a kind of rice-beer known as *Apong* (Hajra and Baishya, 1981).

Chowdhury *et al.* (1998) reported the composition of bark and flower oils from two locations of Assam (India) using GC and GC-MS. The bark from both areas yielded

0.08% essential oil. The essential oil yield from flowers was 0.13% from Jorhat and 0.6% from Sibsagar. These authors have identified 32–75 compounds. The major constituents are the following:

Bark oil:	Jorhat	Sibsagar
1,8-Cineole	31.3%	7.2%
α-Terpineole	21.3%	12.7%
Linalool	20.0%	19.9%

The predominant components in the flower essential oils from Jorhat and Sibsagar were α-pinene (42.9 and 17%) and β-pinene (24.9 and 17.2%).

Baruah and Nath (2000a) reported the occurance of three chemically distinct types of *C. bejolghota*. They were (i) linalool – α-terpineol type; (ii) linalool – α-phellandrene type; and (iii) linalool type. Morphologically the linalool – α-phellandrene type exhibited distinct variations, the other two types were more or less similar. Baruah and Nath (1998, 2000a) provide micromorphological descriptions of these three variants. Baruah *et al.* (1997) analysed the composition of the leaf oil of the linalool type by GC, and recorded the following compounds (Table 15.2).

Nath *et al.* (1999) reported the chemical composition of another chemotype having a distinctly different chemical make up. The leaf oil of this sample had 32.82% alpha-phellandrene, 24.45% linalool, 11.75% α-farnesene, 5.55% 1,8-cineole and 5.30% α-pinene (Table 15.3). Bark oil contains α-terpineol, linalool, nerolidol and methyl cinnamate as the major constituents.

C. capparu-coronde *Blume*

C. capparu-coronde Blume, Rumphia, 1 :34; 1836; Nees, Syst. Laur. 665–66, 1836; Thw. Enu. Pl-zey. 252, 1861; Miq. Ann. Mus. Bot. Lugduno Batavum, 1: 256, 1864; Meissn, in DC Prodr. 15(1): 20, 1864., Kostermans, Bib. Laur. 276, 1964.

This species is endemic to Sri Lanka. The bark has a distinct smell of nutmeg or cloves (this species is known locally as *kapparu kurundu* – camphor cinnamon). The bark is used in tribal and local medicines and is also sold in the market.

A medium sized tree, with smooth to rough bark, light brown to dark brown having a strong smell of nutmeg or clove. Leaves 7–14 × 2–5.5 cm, opposite, ovate – oblong

Table 15.2 Major components (%) in the essential oil of *C. bejolghota*

Methyl cinnamate	0.17	α-Humulene	0.15
Linalool	57.41	β-Phellandrene	1.20
1,8-Cineole	10.20	Borneol	0.70
α-Farnesene	9.30	(z)Methyl isoeugenol	0.56
β-Caryophyllene	2.26	α-Terpeneol	0.43
Linalyl acetate	1.75	Camphene	0.15
Terpinen-4-ol	1.75	Myrcene	0.50
β-Pinene	1.30	p-Cymene	0.23
α-Pinene	1.30	γ-Terpinene	0.15
α-Phellandrene	1.24	Terpinolene	0.05
Camphor	0.09	Methyl eugenol	0.24
(E)-cinnamaldehyde	0.04	Ethyl cinnamate	0.13
Eugenol	0.23	Methyl isoeugenol	0.56

Source: Baruah *et al.*, 1997.

Other Useful Species of Cinnamomum 333

Table 15.3 Constituents of leaf oil of a chemotype of *C. bejolghota*

Component	%
Benzaldehyde	1.18
α-Pinene	5.30
Camphene	0.36
β-Pinene	2.06
Myrcene	1.10
α-Phellandrene	32.82
p-Cymene	0.80
1,8-Cineole	5.55
β-Phellandrene	1.60
γ-Terpinene	0.60
Terpinolene	2.38
Linalool	24.45
Borneol	t
Terpinen-4-ol	0.30
Benzyl acetate	t
Guaiacol	t
α-Terpineol	0.26
Linalyl acetate	t
(E)-cinnamaldehyde	t
Eugenol	0.60
(E)-methyl cinnamate	0.36
Methyl eugenol	0.26
β-caryophyllene	2.00
(E)-ethyl cinnamate	t
(2) Methyl isoeugenol	0.70
α-Humulene	0.60
α-Furnasene	11.75
Caryophyllene oxide	0.40

Source: Nath *et al.*, 1999.

Note
t: trace.

to ovate, initially densely microscopically sericeous, glabrescent, three-veined, lateral veins terminating 1–2 cm below leaf-tip. Panicles axillary or pseudoterminal, densely sericeous, flowers 3 mm long, cup 2 mm high, tepals 1.5–2 mm long, stiff, upright, oblong, subacute. Fruit 11 × 8 mm, ellipsoid, apiculate with resinous odour, cotyledons red (Dassanayake *et al.*, 1995). Wijesekera and Jayewardene (1994) analysed this species and found the main chemical constituents to be 1,8-cineole (15%), linalool (29%) and eugenol (23%).

C. cecidodaphne *Meissner*

C. cecidodaphne Meissner in DC Prodr. 15(1), 25,1864; Drury. Ind. Fl. 3, 55, 1869; Gamble, Man. Ind. Timb. 305, 1881; List of trees, shrubs Darjeeling Distr. Ed. 2, 65, 1896; Hooker f. Fl.Brit. India 5, 135, 1886; Staub, Geschichte Genus *Cinnamomum* t 2,3, 1905; Brandis, Ind. Trees 528, 534 et 716, 1906; Troup, Eco. Prod. Ser., Ind. For Mem.,1, 114, 1909; Fischer in Rec. Bot. Sur. Ind. 12(2), 128, 1938; Kanjilal, Fl. Assam 4, 58, 1940; Howard, Man. Timb. World, ed 3, 147, 1951; Bor, Man. Ind. Forest Bot. 53, 1953; Rao, Trees Duars & Terai 85, 1957.

334   M. Shylaja, P.N. Ravindran and K. Nirmal Babu

An evergreen tree reaching 8–10 m in height, distributed in the eastern sub-Himalayan tract, mainly in Assam and Manipur, up to an elevation of 1300 m. It is cultivated in the Dehra Dun area (Anon, 1950). The wood of *C. cecidodaphne* is known as Nepal camphor or Nepal Sassafras. Its wood is yellowish or light grey in the outer layers, grading to light brown towards the centre. The wood has a strong camphoraceous odour. Its timber is valued for cabinet making. The wood on distillation gives about 2.5–4% of a clear, pale yellow essential oil. Dixit *et al.* (1989) carried out GC analysis of this oil and found that the main constituents were: cineole (23.0%), methyl cinnamate (22.0%), caryophyllene (9.5%), p-cymene (6.5%), linalool (5.6%), borneol (4.4%) and lesser quantities of α-pinene (1.5%), β-pinene (2.1%), Δ^3-carene (1.6%), α-terpenene (2.7%), borneol (4.36%), citral (0.7%) and eudesmol (1.1%). Safrole, myristicin and elemicin have been identified in the wood oil (Anon, 1950). The seed contains a semisolid fat that contains lauric acid (83%), oleic acid (12%), linolenic acid (3%) and linoleic acid (2.5%).

Khanna *et al.* (1988) studied the essential oil from the fruit rind. The fruits of this species (known by the name of "*sugandh kokila*", though the true *sugandha kokila* is *C. glaucescens* from Nepal) are used in a traditional medicine perscribed as a demulcent and stimulant. Tobacco manufacturers use this fruit in the manufacture of *kimam* – scented tobacco (Khanna *et al.*, 1988). The fruit rind contains about 4.5% essential oil, having a yellow colour and sweet camphoraceous odour. Its physicochemical properties are given in Table 15.4. GLC analysis of the oil led to the identification of 17 compounds, the major ones were γ-terpinene (26.74%), safrole (17.42%), nerol (8.94%), elemicin (6.9%) and methyl cinnamate (6.7%). The oil when tested against two pathogenic fungi (*Sclerotium rolfsii* and *Colletotrichum capsici*) totally inhibited the growth of the former at 1000 ppm, and resulted in 73% inhibition of the latter at 2000 ppm (Khanna *et al.*, 1988).

C. citriodorum *Thw.*

C. citriodorum Thw. Enum. Pl. Zey. 253, 1861; Meissn. in. DC Prodr. 15(1): 22, 1864; Miq. Ann. Mus. Bot., Logduno Batavum, 1: 258, 1864; Hook. f. Fl. Br. Ind., 5: 134, 1886; Trimen, Hert. Zey. 69, 1888; Trimen, Hanb. Fl. Ceylon 6: 247, 1931; Kostermans, Bib. Laur. 284, 1964.

This species occurs in Sri Lanka and south western Ghats. The bark and leaves have the characteristic smell of lemon grass (citronella), and are used by tribes both as a spice and as a medicinal plant. A decoction of bark and leaves is used in various stomach ailments.

A medium sized tree with smooth, pale grey/brown bark having a strong odour of lemon grass. Branchlets opposite, stiff, slender, smooth, almost glabrous; terminal bud silvery-sericeous. The wood is heavy, close-grained, yellowish, having the smell of

Table 15.4 Physico-chemical characteristics of the essential oil from the fruit rinds of *C. cecidodaphne*

Specific gravity (nd^{25})	0.9399
Refractive index d^{25}	1.495
Specific rotation (in alcohol) [α]^{25}	$+5°68'$
Acid number	8.32
Ester value	72.3
Ester value after acetylation	91.1
Carbonyl percentage (by oxidation as $C_{10}H_{20}O$)	1.89
Phenol percentage (by absorption)	7.5

cloves. Leaves measure 3–12 × 1.5–4 cm, lanceolate or oval to subovate to lanceolate, gradually tapered, obtuse or shortly accuminate, base cuneate, rigidly coriaceous, glabrous, smooth and glossy. Leaves are pinnately veined, midrib slender, prominent on both surfaces, lateral veins four to eight pairs, scarcely visible above, obscurely visible beneath, two subbasal veins at times reaching one-third of the length of lamina. Panicles 8–10 cm long, axillary on long peduncle. Flowers greenish-white, sericeous, tube short, tepals 3 mm long, narrowly ovate, obtuse, sericeous within; stamens 2 mm long, anthers of the outer two whorls narrowly ovate, sub-acute, upper cells very small; inner anthers two-celled, extrorse, basal glands large, sessile. Staminodes smaller than stamens, sagitate, acute; ovary sub-globose, with very short style and small subpeltate stigma. Infructescence up to 10 cm long, fruit 12 × 7 mm; ellipsoid or ovate-ellipsoid; cup hemispherical, up to 7 mm diameter, rim entire (Dassanayake *et al.*, 1995).

The major constituent of bark oil is cinnamaldehyde, while that of leaf oil is citronellal (Sritharan *et al.*, 1994). This citronellal imparts a distinctive lemongrass odour to the leaves of this species.

C. cordatum *Kosterm.*

It is a shrub (or a small tree) of about 4–6 m in height. This species is common to Malaysia and is abundantly found in the forests of Perakand and Pahang states. The leaves are opposite, triplinerved. This species does not have any commercial uses, but is used in traditional medicine mainly to reduce pain. Jantan *et al.* (2002) analysed the oil of this species. It contains about 0.8% leaf oil and 0.7% bark oil. The oil analysis by GC-MS revealed 55 components, out of which 43 representing 96.5% of the oil were identified by the authors (Table 15.5). Monoterpinoids and phenylpropanoids constituted 53.8% and 21.6% of the oil, respectively. The major constituents of leaf oil were linalool (17.3%) and (E)-methyl cinnamate (17.1%). The bark oil was made up almost entirely of methyl euginol (92.1%). It is an uncommon constituent in *Cinnamomum*, and was earlier reported only from *C. partheroxylon* and *C. rigidissimum*.

C. culitlawan *(L.) J S Presl*

C. culitlawan Presl, Priroz. Rostlin 2, 36, 1925., Blume, Bijdr. Fl. Ned. Ind., II stuk: 571, 1826; Rumphia 1, 26, t.9, fig.1, 1836; Th. Nees and Ebermayer, Handb. Med. Pharm. Bot. z: 429, 1831; C.G. Nees in flora 15(2): 587, 601 et 602, 1831; Syst. Laur. 71 et 668, 1836 (see Kostermans 1964). (Syn. *C. culilaban* (L.) J.S. Presl; *C. culitlawan* (Roxb.) J.S. Presl.; *C. culitlawan* Bl., *Laurus culitlawan* L.)

This species is found in the Indonesian Islands (Moluccas, Amboina and adjacent Islands). It is a small to medium sized tree with slender branchlets. Leaves opposite, petiole up to 1 cm long; blade lanceolate – oblong to ovate-elliptical, coriaceous, glabrous; size 9–12 cm × 2.5–4.5 cm. Inflorescence is an axillary panicle, 6 cm long, few flowered. Flowers are densely pilose.

The bark has the smell of clove (and is hence known as *kulitlawang* in the Indonesian language). It is used as a spice locally. The bark as well as its oil (lawang oil) are used medicinally as a constipating agent and in the treatment of cholera. The root bark has the flavour of fennel and has been used as a substitute for sassafras bark, *Sassafras albidum* (Nutt.) Nees (Dao *et al.*, 1999).

Table 15.5 Composition of essential oils of *C. cordatum*

Compound	Leaf oil (%)	Bark oil (%)
α-Thujene	0.2	—
α-Pinene	1.7	—
Sabinine	1.3	—
β-Pinene	1.1	—
Myrcene	1.2	—
α-Phellandrene	2.6	—
α-Terpinine	0.6	—
P-Cymene	1.5	—
β-Phellandrene	9.0	—
Limonene	2.4	—
γ-Terpinene	1.5	—
Terpinolene	0.5	—
Linallol	17.3	1.8
Camphor	0.4	—
Citronellal	0.3	
Terpinen 4.ol	7.0	
α-Terpineol	3.6	0.2
Phellandrene epoxide	0.4	—
cis-Piperitol	0.2	—
Nerol	0.2	—
Geraniol	0.8	—
Safrole	0.1	1.1
Ethyl-2$-$6 Xylenol	0.1	—
δ-Elemene	0.2	—
(*E*)-methyl cinnamate	17.1	—
Methyl eugenol	4.4	92.1
β-Caryophyllene	0.3	—
Aromadendrene	0.3	—
α-Humulene	0.2	—
Allo Aromadendrene	0.1	—
(*E*)-Methyl isoeugenol	—	0.1
γ-Muurolene	0.2	—
α-Elemene	0.2	—
γ-Cadinene	0.1	—
Calamenene	0.1	—
δ-Cadinene	0.9	—
Elemicin	—	2.5
(*Z*)-nerilidol	0.7	—
Caryophyllene alcohol	—	0.3
Spathulenol	1.6	—
Caryophyllene oxide	—	0.2
Globulol	1.6	—
Viridiflorol	1.4	—
β-Eudismol	0.5	—
α-Eudismol	0.3	—
Benzyl benzoate	7.6	—
Benzyl salicylate	4.7	—
Hexadecanoic acid	—	0.1
Total	96.5	98.4

Source: Jantan *et al.*, 2002.

C. deschampsii *Gamble*

C. deschampsii Gamble in Kew Bull. 219, 1910; J. Asiat. soc. Bengal, 80, 1912; Ridley in Agri. Bull. Straits & FMS,10, 235, 1911; Fl. Malay Pen., 3, 93, 1924; Calder *et al.* in Rec. Bot. sur. India 11(1), 29, 1926; Burkill, Dict. Eco. prod. Malay Pen. 1, 550, 1935; Corner in Gard. Bull. Straits Settl.10, 276, 1939.

This species occurs in the Malaysian region. All parts of the tree are aromatic, having a pleasant flavour similar to the Chinese cassia. It is used as a substitute for cassia and cinnamon by natives. It is a bushy tree, leaves coriaceous, blade oblong to elliptical ovate 7.5–15 cm × 5–7 cm. Inflorescence is a lax spreading panicle with silky flowers; fruit is about 1 cm long. The chemical composition has not been worked out.

C. glanduliferum *Nees (Nepal Camphor wood or Nepal Sassafras)*

C. glanduliferum Nees in Wallich, Pl. Ag. var. 2, 72, 1831; Meissner in DC., Prodr. 15(1) 25, 1864; Baillon, Hist. Pl. 2, 461, 1872; Steward and Brandis, For. Fl. N.W. India 376, 1874; Gamble, List of Trees and Shrubs in Darjeeling Dist, 64,1878; Man. Ind. Timb; 306, 1881; Horkerf. Fl. Brit. India 1886; Staub, Geschichte Genus *Cinnamomum* 1905; Strachey, Cat. Plants Kumaon 154, 1906; Brandis, Ind. Trees 534, 1906; Bassu, Ind. Med. Plants, 830, 1918; Burkill in Rec. Bot. Surv. India 10(2), 351, 1925; Dict. Eco. Pro. Malay Pen. 1, 550, 1935 (See Kostermans (1964) for full citation).

It is a small tree distributed in the central Himalayas and Khasi hills of north India. Its bark is rough and pale brown, the wood when cut has a strong smell of camphor. The wood contains d-camphor and is a substitute for sassafras. The timber is used for cabinet making and is also useful for carving. Leaves and twigs on distillation gives 0.6–1% of an essential oil having a strong smell of camphor. The oil from various sources has the following characteristics (Anon, 1950): Sp. gravity. 0.9024–0.9058; $[\alpha]^{20}{}_D - 23°0'$ to 24°57'; $n^{20} - 1.4685$; saponification value – 11.1; sap. value after acetylation–39.7; solubility: in 80% alcohol.

The chief constituents are cineole (34%), α-terpineol (10%) and camphor. The seed kernels contain a fat to the extent of 27–30%, having d = 0.9156; n^{40}-21.5–22.0°; solidifying temp. – 14.5–19.0°; m.p – 21.5–22.0°; acid value – 1.94; saponification value – 287.02; iodine value – 3.76 (Anon, 1950).

The oil can be used as a substitute for sassafras oil (from the roots of *Sassafras officinale*) in the soap and perfume industries.

C. glaucescens *(Wall.ex Nees) Drury.*

C. glaucescens Drury, Handb. Ind. Fl.3:55, 1869; St. Lager in Ann. Soc. Bot. Lyon 7:122; Yamada in Trans. Nat. Hist. Soc. Formosa 17, 440 1927.

This species occurs in Nepal and the adjoining areas and is the well-known Nepal *Sugandha kokila* used extensively by local people as a spice, in medicine, in flavouring tobacco, etc. The commercial *Sugandha kokila* oil is steam distilled from the fruits. Adhikary *et al.* (1992) studied the oil hydrodistilled from pericarp as well as from the whole fruit using chromatographic and spectroscopic methods.

Major components of the whole fruit oil are 1,8-cineole (13%), methyl cinnamate (14%), alphaterpineol (7%), and many mono and sesqinterpene hydrocarbons. The pericarp oil consists mainly of 1,8-cineole (50%) and alphaterpineol (10%), along with minor amounts of monoterpene hydrocarbons, other monoterpene alcohols and methyl cinnamate (Sthapit and Tuladhar, 1993).

C. impressinervium *Meissn.*

C. impressinervium Meissner in DC. Prodr. 15 (1) 25, 1864; Hkf. Fl.Brit. India, 5: 129, 1885; Kanjilal *et al*. Fl. Assam, 4:58, 1940.

C. impressinervium grows wildly in the north-east of India, where it is also cultivated occasionally. It is sold as the best quality *tejpat* (Nath and Baruah, 1994). It is a medium sized evergreen tree, 6–8 m tall, with rough bark, aromatic leaves, smells like *tejpat* leaves, glabrous, shining above, triplinerved, lateral nerves reaching the base of the acumen, leaf buds silky, panicles subterminal to axillary, shorter than leaves, up to 6.5 cm long, glaborate, perianth 3 + 3, subequal, minutely puberulous, truncate cup-shaped fruiting tepals, pedicel obconic, pedicel with fruiting tepal up to 8 mm long.

This species occurs mainly in the North Cachar hills and Cachar districts of Assam at an elevation of 800–1050 m. Leaves are sold and used as the best quality *tejpat* by the people of the North Cachar Hill district of Assam. Nath and Baruah (1994) and Nath *et al*. (1999) carried out chemical analysis of wild and cultivated plants (Table 15.6) and reported 23 components in the oil of a sample, the important ones being eugenol (88.3%) and limonene (4.1%) (Table 15.7). Fresh leaves yielded 1.8–2% of oil, having $n_d^{25} = 1.5236$, $d^{25} = 1.0332$ and $[\alpha]_d^{25} = +15°$. The oil content is more than that in *C. tamala* (Indian cassia) and the oil has a higher eugenol content, which may be the reason for considering this as the best quality *tejpat*.

C. japonicum *Sieb.*

C. japonicum Siebold in Verhandel. Batav. Genootselv.kunst & Wentenselv. 12, 23,1830; Nees, Syst.Laur.79, 1836; Steudel, Nom. ed.2, 1, 366, 1840; Siebold & Zuccarini, Fl. Jap. Fam. Nat.2,78, 1846; Zollinger, Syst. Verg.2, 113, 1854; Meissner in DC, Prodr. 15(1), 16, 1864; Miquel, Ann. Miss. Bot. Lugd. bat.1, 268, 1864; 2, 195, 1867; Stuart, Chinese Mat. Med., 109, 1911; Nakai in Bot. Mag. Tokyo, 517, 1927; Fl. Syl. Koreana, 22, 25, 1939; Sasaki, List pl. Formosa, 192, 1928; Makinol Hemoto, Fl. Jap. ed.2; 365,

Table 15.6 Major chemical constituents of the leaf oils of
C. impressinervium

Component	Wild %	Cultivated %
Eugenol	83.2	88.3
α-Pinene	1.2	0.5
β-Pinene	0.2	0.1
δ-3-Carene	7.2	1.6
Limonene	2.3	4.1
p-Cymene	0.7	0.6
Guaicol	0.4	0.3
α-Terpineol	0.4	0.3
β-Caryophyllene	0.2	0.1
Eugenol acetate	1.0	1.1

Source: Nath and Baruah, 1994.

Other Useful Species of Cinnamomum   339

Table 15.7 Composition of essential oil from *C. impressinervium*

Component	%
Benzaldehyde	t
α-Pinene	0.50
Camphene	t
β-Pinene	0.10
Myrcene	t
α-Phellandrene	1.3
p-Cymene	0.60
1,8-Cineole	0.20
β-Phellandrene	t
γ-Terpene	t
δ-3-Carene	1.60
Limonene	4.10
Guaiacol	0.10
α-Terpineol	0.30
Eugenol	88.30
(E)-methyl cinnamate	t
β-Caryophyllene	0.10
Eugenyl acetate	1.10
(E)-ethylcinnamate	0.50
(Z)-methylisoeugenol	0.70
α-Furnescene	0.20
Caryophyllene oxide	0.10

Source: Nath *et al.*, 1999.

Note
t: trace.

1931; Sonohara *et al.* Fl. Okinowa 57, 1957; Fl. Japan 555 1953 (see Kostermans (1964) for complete citation).

This is a Japanese species studied in detail chemically by Masada (1976). He reported around 49 compounds, though the prominent ones were p-cymene and 1,8-cineole in the twigs and leaves, while l-linalool formed an important component in branchlets (Table 15.8).

Okada (1975) investigated the cytology of this species. The somatic chromosome number is 2n:24, and the chromosomes are distinguishable by their morphological characteristics. The cytological characteristics exhibited bear resemblance to *C. camphora*, except for the variation in heterochromatin distribution of the seventh and tenth pairs of chromosomes.

C. kanahirai *Hay.*

C. kanahirai Hayata, Icon. Pl. Formos, 3: 159,1913; Matsumura, Shokubutsu-Mei.,12:1915; Kanehirai, Formosan Trees, 424,1917; 203, 1936; Atlas of Formosan Timbers, 34, 1940.

This species occurs in China and Taiwan (known as *sho-Gyu* in Chinese), is important as it is one of two host trees, (the other being *C. micranthum* Hay.) of the famous medicinal fungus, *Antrodia camphorata* (NIU-Chang-Chih, syn. *Antrodia cinnamomea* – Polyporaceae). The fungus fructification is highly valued in the traditional medicine of China, Taiwan and Thailand (Sheng Hua *et al.*, 1997). *C. kanahirai* yields an essential oil, the main constituent of which is terpinen-4-ol.

Table 15.8 The composition of the essential oils of *C. japonicum*

Peak no.	Compound	Twigs (%)	Leaves (%)	Branchlets (%)
1	α-Pinene	2.7	3.7	0.3
2	Camphene	0.1	0.5	0.1
3	β-Pinene	2.4	3.0	0.2
4	β-Myrcene	4.7	4.7	0.9
5	*l*-α-Phellandrene	2.8	7.7	0.2
6	Limonene	2.0	3.0	0.2
7	1,8-Cineole	23.7	16.5	12.0
8	*p*-Cymene	45.2	40.0	24.1
9	3-Hexen-1-ol	0.4	0.3	0.2
10	Unidentified ketone	trace	trace	trace
11	*trans*-Linalool oxide	0.1	0.1	1.3
12	*cis*-Linalool oxide	0.1	0.2	1.2
13	*l*-Linalool	3.3	2.8	23.5
14	*l*-Copaene	0.3	0.9	1.4
15	Camphor	0.2	0.4	0.6
16	Unidentified alcohol	0.2	0.3	0.5
17	SHC[1]	trace	0.2	0.2
18	Terpinen-4-ol	1.6	2.4	4.1
19	β-Elemene	0.1	0.2	0.4
20	*l*-Caryophyllene	1.2	0.6	0.1
21	Unidentified alcohol	–	–	0.6
22	SHC[1]	0.3	0.1	–
23	*l*-α-Terpineol	3.7	1.6	3.2
24	α-Terpinyl acetate	0.4	2.0	7.0
25	*l*-Carvotanacetone	0.1	0.5	0.2
26	α-Humulene	0.2	0.5	0.3
27	SHC[1] (s-Cadinene?)	trace	trace	0.1
28	Citronellol	trace	trace	1.5
29	SHC[1]	0.1	0.5	0.5
30	*d*-δ-Cadinene	0.4	1.0	1.1
31	Nerol	0.2	trace	trace
32	SHC[1]	trace	0.2	trace
33	*l*-*trans*-Yabunikkeol	0.5	0.5	2.5
34	Geraniol	trace	trace	1.3
35	Unidentified alcohol	trace	trace	trace
36	Calamenene	0.3	0.1	0.4
37	Safrole	trace	trace	trace
38	*d*-*cis*-Yabunikkeol	0.4	0.6	1.3
39	α-Calacorene	trace	0.1	0.1
40	β-Calacorene	trace	trace	0.1
41	Unidentified alcohol	trace	trace	trace
42	Methyleugenol	0.2	0.1	0.5
43	Elemol	0.7	1.3	2.6
44	Unidentified ketone	trace	trace	0.1
45	Unidentified alcohol	trace	trace	trace
46	Eugenol	0.7	1.5	1.4
47	α-Cadinol	0.5	1.2	2.2
48	Unidentified	trace	trace	1.0
49	*l*-Kaurene	trace	trace	0.5

Source: Masada, 1976.

Note
1 SHC: Unidentified sesquiterpene hydrocarbon.

The centre of origin of this tree is regarded as Taiwan, where it is now an endangered species (Lin *et al.*, 1997). Lin *et al.* (1997) reported the allozyme variations in this species in a study comprising 164 clones from four geographical areas. Seven out of the 11 loci examined by them were polymorphic.

C. malabatrum *(Burman f.) Bercht & Presl.*

C. malabatrum Berchthold & Presl. in Priroz. Rostlin, 2. p 46, 1825; Batka in Nov. Act. Nat. Cur. 17(2):618 t. 45, 1835; Nees, Syst. Laur, 38 et 663, 1836; Blume, Rumphia, 1.38, 1835; Heynold, nom.bot. nort., 197, 1840; Dietrich, Syn 2: 1334, 1840; Steudel, Nom. Ed. 2, 1: 366, 1840 et 2: 16, 1841; O'shaughnessy, Bengal Pharmacop. 81, 1844; Miqual, Fl. Ind. Bat. 1(1), 97, 1858; Ann. Miss. Bot. Lugd. Bat. 1, 258, 1864; Thwaites, Enum. Pl. Zeyl. 253, 1861; Meissner, in DC., Prodr. 15(1) 20, 1864; Hooker, Fl. Brit. Inida, 5, 130, 1886; Kostermans, Bibl. Lau. 318, 1964; Rheede, Hort. Ind, Malabaricus, 5, 5, 53, 1685 ('kattukaruva').

Basionym *Laurus malabatrum* Burm. F. Burman, Index alter Rheede, Hort. Malab. (t.53) 5, 1769; Bergius, Mat. Med. 1:318, 1778; Linnaeus. Mat. Med.ed. 5:125, 1787; Lamarck, Encycl. Bot. 3: 445, 1793; de Candolle, Essaie Propr-med. Pl. 66, 1804; Stokes, Bot. Mater. Med. 2:411, 1812; Dennstedt, Schleussel, Hort. Ind. Malab., 12, 22, 31, 1818; Th. & C.G. Nees, de Cinnamomo Disput. 55, 1823; Wallich, Cat. No. 2583-A, 1830; Lindley, Introd. Nat. Syst. 29, 1831; Fl.med. 331, 1838; Nees in Flora, 15(2): 597, 1831; Syst.Laur 35 et 38, 1836; Meissner, in DC., Prodr. 15(1): 14, 18, 20, 1864 (see Kostermans (1964) for detailed citations).

This species is a close relative of *C. verum* and is a highly variable species occurring in south India, and can be found in the Western Ghats and adjoining areas. Kostermans (1980) was of the opinion that the *Kattu karuva* described by Van Rheede in his *Hortus Malabaricus* (1685) was nothing but *C. malabatrum*. According to him this was the species described under *C. iners* by Hooker (1886) as well as Gamble (1925). This is a moderately sized tree, sometimes attaining larger proportions. The leaves are very variable, opposite or subopposite, coriaceous, elliptic to oblong or elliptic lanceolate, size highly variable, 5–10 cm wide, 15–30 cm long, base acute, tip acuminate, petiole 5–20 mm, leaves glabrous above, minutely hairy below, triplinerved, the outer pair arising from the base or slightly above the base, sometimes obliquely placed, the lateral nerves reaching to the tip of the leaves. Panicles pseudoterminal, lax, many flowered, up to 25 cm long, branched, minutely hairy, pedicels slender, flower tube shallow, tepals fleshy, ovate, acute, 3–3.5 mm long, stamens 2–2.5 mm, anthers four-celled, filaments pilose, whorls one and two introrse, three extrorse and glandular, fourth whorl staminodes. Ovary ellipsoid, style cylindrical, small, peltate. Fruit ellipsoid, up to 8–10 mm, cup deep, rather fleshy, the base merging to the pedicel, the rim, with persistent thickened tepals of which the apical part drops off, the basal part rounded and pilose (Kostermans, 1983).

In the past, immature fruits of this species were traded in place of cassia buds, both as a substitute and as an adulterant. It was used in *pan* as a substitute for clove buds. *Pan* chewing is a habit among many people in South Asia – the basic ingredient of *pan quid* consists of betel leaves (*Piper betel* L.), areca fruit (*Areca catechu* L.) lime and clove bud for flavour, with or without tobacco. Local practioners of *Ayurveda* use the dry, young fruit as '*Nagakesar*' (in place of the dried buds of *Mesua ferrea*, used in other parts of the country). The bark is extracted, dried, powdered and used as a base material in the manufacturing of incense sticks (*agarbathi*). The leaves are used in households as

342   M. Shylaja, P.N. Ravindran and K. Nirmal Babu

a holding substratum for steaming during the preparation of certain dishes (called 'thirali' or elayappam in the local language). The wood is a good quality firewood. Indiscriminate felling led to drastic reduction in the population of this species.

Shylaja (1984) and Ravindran *et al.* (1991) carried out detailed study of this species, and recorded much variability in vegetative morphology. Leaf morphology is the most variable, especially the leaf size; the L/B values vary from 2.19 to 4.25. The leaves are minutely hairy, the hair frequency can be sparse or moderately dense. Leaf anatomical characteristics such as leaf thickness, relative thickness of the palisade and spongy tissues, stomatal frequency and size also showed variations. Inflorescences of this species were found to fall under three categories: (i) equal to or slightly longer or shorter than the leaves, few or many flowered; (ii) distinctly longer than the leaves, large, showy, many flowered; and (iii) distinctly shorter than the leaves, and few flowered.

Shylaja (1984) and Ravindran *et al.* (1991, 1996) carried out studies using cluster analysis and principle component analysis and found that *C. malabatrum* collections got into different clusters. A chemotaxonomic study (Shylaja, 1984; Ravindran *et al.*, 1992) also indicated terpenoid and flavonoid variability at the infraspecific level. Shylaja (1984) distinguished three distinct varieties of *C. malabatrum*, based on inflorescence characteristics (*C. malabatrum*, var. *malabatrum*; var. *giganteum*, var. *pauciflorum*). Unfortunately information on the essential oil composition of this species is not available.

C. osmeophleum *Kan.*

C. osmeophleum Kanehira, Formos. Trees 428, 1917; id 206, 1936; Hayata Icon. Pl. Formos. 10, 29, 1921; Sasaki, List of pl. Formosa 193, 1928; Catal. Gvt. Herb. Taihoku 221, 1930; Makino & Nemoto Fl. Jap. ed.2. 366,1931; Nakai in J. Jap. Bot. 16(3)130, 1940; Chen Yung, Chinese trees, 24, 1957.

This species occurs mainly in Taiwan. The leaf of this species is much sweeter and aromatic than other species occurring in the region. Hussain *et al.* (1986) carried out a chemical analysis of the leaves. They found that an 80% methanolic extract of the leaf is highly sweet; the compound responsible for this characteristic was identified as *trans*-cinnamaldehyde, present to the extent of 1.03% w/w. This compound was estimated to be around 50 times sweeter than a 0.5% w/v solution of sucrose. GC-MS analysis of the leaf extract led to the identification of the major compounds listed in Table 15. 9.

In addition to being sweet, *trans*-cinnamaldehyde has a pungent and spicy taste. The leaf of this species is used in place of Chinese cassia leaves both for spicing dishes and medicinal uses.

C. porrectum *(Roxb.) Kosterm.*

C. porrectum (Roxb.) Kostermans in J. Sci. Res. Indon. 1, 126 (27), 1952; Commun. For Res. Inst; Bogor 57, 24, 1957; Stern in Trop Woods, 100, 24, 1954.

This Chinese species, known as Jiang-Zhang, is used in Chinese medicine. Liang-fang *et al.* (1984) analysed the chemical composition. The leaves of this species contains 0.5–0.8% oil. The main constituents of the oil is citral (α-β-citral – 64.11%). Forty-seven compounds were identified by the above workers (Table 15.10).

Xiang Dong *et al.* (1996) identified a new type of ribosome inactivating protein (RIP) from the seeds of *C. porrectum*. This RIP (porrectin) is a glycoprotein having a mw of

Other Useful Species of Cinnamomum   343

Table 15.9 Compounds identified in the essential oil of *C. osmeophleum*

Compound	% Occurrence
α-Pinene	0.53
β-Pinene	0.14
Benzaldehyde	10.32
Camphene	0.27
p-Cymene	0.24
Limonene	0.21
3-Phenyl propanal	5.38
Estragole (methyl chavicol)	1.52
Eugenol	1.18
Terpinen-4-ol	0.33
trans-Cinnamaldehyde	76.96
Unidentified compounds	2.92

Source: Hussain *et al.*, 1986.

Table 15.10 Composition of the essential oil of *C. porrectum*

Compound	Content (%)	Compound	Content (%)
α-Thujene	0.06	Geranyl formate	0.09
α-Pinene	2.42	α-Copaene	0.12
Camphene	1.26	*trans*-Methylcinnamate	0.02
Sabinene	0.21	n-Dodecane	0.20
β-Pinene	1.38	β-Elemene	0.12
Myrcene	0.38	Caryophyllene	4.67
α-Phellandrene	0.32	α-Guaiene	0.04
Δ³-Carene	0.01	β-Guaiene	0.06
p-Cymene	0.21	β-Selinene	0.97
α-Limonene	1.57	Azulene	0.38
1,8-Cineole	0.82	β-Cubebene	0.30
β-Phellandrene	0.10	n-Pentadecane	0.05
cis-Linalool oxide	0.07	β-Gurjunene	0.04
Linalool	8.43	*epi*-β-Santalene	0.46
Epicamphor	0.26	Aremophilene	0.05
Camphor	1.10	Alloaromadendrene	0.03
Borneol	1.07	α-Elemene	0.26
β-Citral (neral)	28.28	*trans* β-Farnasene	0.04
Geraniol	0.25	(z)-β-Farnasene	0.30
Nerol	0.47	γ-Elemene	0.03
α-Citral/geranial	35.83	β-Bisaboiol	0.41
Methyl citronellate	0.12	Cedrol	0.16
n-Undeceane	0.18	Methyl geranate	0.2
Safrole	0.02		

Source: Liang-Fang *et al.*, 1984.

64,500 daltons, and a sugar content of 2.5%. It consists of an A-chain (MW 30,500) and a B-chain (MW 33,500) linked by the disulphide bond. The terminal sugar of glycan in the B-chain is a mannose residue. Porrectin is a potent inhibitor of eukaryotic protein synthesis in the rabbit reticulocyte lysate system. The molecular mechanism of action of porrectin on rat liver ribosomes was shown to be specific for RNA-N-glycosidase.

344 M. Shylaja, P.N. Ravindran and K. Nirmal Babu

The clevage site in the adenosine at position 4324 (rat liver 28 SrRNA) is embedded in the highly conserved ricin/alpha-sarcin ('R/S') domain.

C. pauciflorum *Nees*

C. pauciflorum Nees in Wallich, Pl. Asiat. rar.2, 75, 1831; in Flora 15(1), 587, 596, 601, 1831; Syst. Laur. 68, 1836; Meissner in DC Prodr. 15(1)17, 1864; Miquel, Ann. Mus. bot. Lugd. bat. 1, 268, 1864; Gamble, Man. Ind. Timbers, 305, 1881; Hooker f. F.Br.Ind.5, 129, 1886, Brandis, Ind. Trees 533,1906, Kanjilal *et al.* Fl. Assam 4, 47, 1940.

This is a small tree occurring in north-eastern India, in Assam, and Khassia hills. This is a species with small leaves, glaucous and firmly reticulated beneath, and having a strong aroma and flavour of cumin. The panicles are small, very few flowered, sometimes reduced to three flowers. The fruiting calyx is very small. Nath *et al.* (1996) examined the essential oil derived from leaf and stem bark and found that the main component was cuminaldehyde, 94%, 92.4% and 85%, respectively, in leaf, root bark and stem bark oils. This is the only *Cinnamomum* species having cuminaldehyde as the predominant component.

C. parthenoxylon (*Jack.*) *Nees*

C. parthenoxylon (Jack.) Nees in Wallich, Pl. Asia. rar.2, 72, 1831; Meissner in DC Prodr.15(1), 26 *et* 504, 1864; Bentham, Fl. Hongkong, 290,1861; Baillon, Hist. Pl. 2, 461, 1872; For. Fl. Br. Burma, 2, 289,1877; Gamble, Man, Ind. Timbers, 305, 1881, Hookerf. Fl. Brit. India, 5, 135, 1886; Watt. Dict. Eco. Prod. India, 2, 318,1889; Brandis, Ind. Trees, 534, 1906; Matsumura, Index. Plants of Jap. 2(2), 135, 1912; Burkill, Dict. eco. prod. Mal. pen.1, 554, 1935 (See Kostermans (1964) for full citation).

This species occurs in Vietnam, China, Indo-Malayan and the adjoining regions such as the Java-Sumatera Islands of Indonesia. The wood of this species is an orange-brown. It is scented and moderately hard and used for construction and as a cheap cabinet wood. Hooker included this species under section camphora. The leaves are alternate, elliptic ovate or oblong, subcaudate-accuminate, penninerved, often glaucous beneath; panicles are short, nearly glabrous, shorter than the leaves, perianth nearly glabrous, not pubescent within, fruit 8–9 mm in diameter; globose. Dung *et al.* (1997) have analysed the composition of trees growing in Vietnam and identified more than 30 compounds in the root bark oil and 20 compounds in the wood oil. The main constituent in the root bark oil is benzyl benzoate (52.0%), whereas the wood oil consists mainly of Safrole (90.3%). They also found that the oil yield and safrole content of trees from different geographic regions of Vietnam did not vary much.

Baruah and Nath (2000b) investigated two chemically distinct plant populations from the north-east of India. They were morphologically alike, but different in their odour characteristics. They were medium large evergreen trees having aromatic plant parts. The compounds present in the leaf and bark oils of two variants are given in Table 15.11. The first variant yielded 3% leaf oil and 0.2% bark oil. The second variant yielded 5% leaf oil and 0.5% bark oil. In the leaf and bark oils of the first variant, the predominant component was 1,8-cineole (64.7 and 53%, respectively) while in the second variant the predominant compound was linalool (86.9 and 87.12%, respectively, in leaf and bark).

Table 15.11 Percentage composition of the essential oils of *C. parthenoxylon* variants

Components	Variant I		Variant II	
	Leaf	*Stem bark*	*Leaf*	*Stem bark*
Benzaldehyde	0.45	0.50		
α-Pinene	3.26	1.35	0.24	0.56
Camphene	–	t	0.14	0.35
β-Pinene	14.83	1.15	0.16	0.40
Myrcene	0.80	1.00	0.06	0.10
α-Phellandrene	–	–	–	0.06
p-Cymene	–	0.90	0.54	–
1,8-Cineole	64.70	53.00	0.20	1.35
α-Phellandrene	0.34	1.60	–	0.44
β-Terpinene	0.90	–	0.50	–
Terpinolene	–	0.70	1.81	–
Linalool	10.70	t	86.90	87.12
Camphor	–	3.08	–	–
Borneol	10.90	1.44	–	0.50
Terpinen-4-ol	0.65	6.06	–	0.72
β-Terpinolene	11.68	16.40	–	–
α-Elemene	–	3.03	–	–
Linalylacetate	–	–	–	0.44
Cinnamaldehyde	–	–	–	0.21
Eugenol	–	–	–	0.57
Methyl cinnamate	–	–	–	0.70
α-Caryophyllene	–	0.60	0.30	1.40
Ethyl cinnamate	–	–	–	0.09
Methyl isoeugenol	–	–	0.20	0.50
α-Humulene	0.80	1.45	2.85	–
Bycyclogermacrene	–	1.60	–	–
a'-Farnesene	–	1.90	–	0.17
Caryophyllene oxide	–	0.65	0.93	1.15
Total	99.90	96.77	94.63	96.83

Source: Baruah and Nath, 2000b.

Note

t: trace.

C. sulphuratum *Nees*

C. sulphuratum Nees in Wall. Pl. As. Rar. 2, 74: 1832; Syst. Laur. 55: 1836; Hooker f. Fl. Brit. Ind. 5:132, 1885; Kostermans, Bull. Bot. Sur. India, 25:114, 1983; Nath and Baruah, J. Eco. Tax. Bot. 18, 211–212, 1994.

This is a medium sized tree distributed in the Western Ghats as well as in north-eastern India. The leaves and bark are aromatic. Baruah *et al.* (1999a,b) carried out an ethano-botanical study and chemical examination of many plants belonging to this species and identified four chemically distinct types (chemovars). They are linalool, citral, cinnamaldehyde and methyl cinnamate types. Morphoplogically, linalool type is closely related to citral type, while cinnamaldehyde and methyl cinnamate types are similar. Linalool and citral types always have comparatively smaller leaves and their panicles are comparatively shorter than or rarely equal to the leaves in comparison to the leaves and panicle characteristics of the other two types. Baruah *et al.* (1999a) also found

346 M. Shylaja, P.N. Ravindran and K. Nirmal Babu

micromorphological differences among the chemotypes. Linalool and citral types have comparatively larger epidermal cells and aereoles/mm^2. Cinnamaldehyde and methyl cinnamate types have smaller epidermal cells and aereoles and more number/mm^2. In the leaf essential oil of linalool type, 92.66% is linalool (Nath *et al.*, 1994). In the oil of citral type, 45.40% is citral. In the cinnamaldehyde type, the oil contains 50.0% cinnamaldehyde. In the methylcinnamate type 72.42% is methylcinnamate (Table 15.12). Baruah *et al.* (1999a) reported the result of a GC/MS analysis of leaf and bark oils of a particular chemotype, the bark oil of which contained 65.6% cinnamaldehyde and the leaf oil contained geranial (27.8%) geraniol (23.2%), and neral (17.6%) (Table 15.13).

Nath *et al.* (1994) reported the composition of a chemotype that contained 60.73% linalool, 10.5% α-pinene, 10.42% β-pinene and smaller amounts of benzaldehyde, camphene, limonene (Table 15. 14).

C. tenuipilis *Kosterm*

This species is from the Yunnan province of China. Xue-Jian and Bi-quang (1987) identified the components listed in Table 15.15. It is a unique species in that it has a very high content of linalool (97.51%) in the leaf oil. This species is a good source of linalool and is now being commercially planted in China for linalool extraction. Linalool is the starting material for the synthesis of a variety of perfumery compounds.

C. wightii *Meissn*

C. wightii Meissner in DC Prodr. 15(1), 11, 1864; Drury, Handb. Indian Fl. 3, 52, 1869; Beddome, Fl. Sylv. T. 262, 1872; For. Man. 154, 1872; Stewart & Brandis, For.Fl. N.W. India 375, 1874; Hooker f. Fl. Brit. Ind. 5, 132, 1886; Trimen, Handb. Fl. Ceylon, 3,440, 1895;Bourne, List Pl. S. Ind., 27, 1898; Brandis, Ind. Trees 533, 1906; Rao, Fl. Pl. Travancore, 342, 1914, Fyson, Pl. The Nilgiri and Pulney Hill Tops, 1, 347, 1915; Fisher in Rec. Bot.Suv. Ind. 9, 153, 1921; Gamble, Fl. Presidency, Madras, 1225 (rep. 1957).

Table 15.12 Chemotypes of *C. sulphuratum*

Chemotypes	Salient characteristics	Major component in leaf oil (%)
Linalool	Leaves comparatively smaller in size (1.7–4 × 6.5–14.5 cm) and glabrous	Linalool, 92.66%
Citral	Same as above	Citral, 45.40%
Cinnamaldehyde	Leaves comparatively larger in size (2–5.4 × 5.5–16 cm); sparsely distributed, microscopic, unicellular, simple and filiform hairs present on lower surface	Cinnamaldehyde, 50.00%
Methyl cinnamate	Same as above	Methyl cinnamate, 72.42%

Source: Baruah *et al.*, 1999a,b.

Table 15.13 Percentage composition of the leaf and stem bark oils of *C. sulphuratum*

Compound	Leaf	Stembark
Hexanal	t	—
(E)-2-hexenal	t	—
(Z)-3-hexenol	0.2	—
(E)-2-hexenol	0.1	—
3-Methyl-1-pentanol	0.2	—
Xylene*	0.1	—
Styrene	0.2	—
Benzaldehyde	—	0.1
α-Pinene	1.2	0.3
Camphene	0.4	0.1
6-Methyl-5-hepten-2-one	1.4	—
β-Pinene	2.4	0.2
Myrcene	1.1	—
p-Cymene	0.2	0.2
1,8-Cineole	0.2	1.6
β-Phellandrene	0.5	—
Limonene	1.7	0.3
cis-Sabinene hydrate	t	—
trans-Linalool oxide (furanoid)	0.1	—
cis-Linalool oxide (furanoid)	0.1	—
Terpinolene	t	—
Linalool	3.0	0.2
Perillene	0.2	—
α-Fenchol	t	0.3
cis-p-Menth-2-en-1-ol	0.1	—
3-Phenylpropanal	—	0.3
Citronellal	0.2	—
Borneol	0.3	1.9
Terpinen-4-ol	0.7	1.3
Myrtenal	t	—
α-Terpineol	0.6	1.7
(Z)-cinnamaldehyde	—	0.8
Nerol	0.1	—
Neral	17.6	—
Piperitone	0.2	—
(E)-cinnamaldehyde	—	65.6
Geraniol	23.2	—
Geranial	27.8	0.4
Bornyl acetate	2.0	0.7
Geranyl formate	0.4	—
Eugenol	0.1	—
δ-Elemene	—	0.4
Neryl acetate	0.1	—
Geranyl acetate	6.3	—
α-Copaene	—	1.5
(E)-cinnamyl acetate	—	5.4
β-Caryophyllene	0.3	1.3
trans-α-Bergamotene	—	0.7
α-Humulene	t	t
ar-Curcumene	—	1.3
α-Muurolene	—	0.5
β-Bisabolene	—	1.5
δ-Cadinene	—	0.3

Table 15.13 (Continued)

Compound	Leaf	Stembark
(E)-nerolidol	0.7	—
Spathulenol	0.7	—
Caryophyllene oxide	0.6	1.0
Tetradecanal	—	4.6
Benzyl benzoate	t	—
trans-Phytol	0.2	—
Other compounds	3.8	5.6

Source: Baruah *et al.*, 1999b.

Notes
* correct isomeric form not identified; t = trace (<0.1%).

Table 15.14 Chemical composition (%) of a linalool-chemotype of C. sulphuratum

Component	%
Benzaldehyde	1.40
α-Pinene	10.50
Camphene	0.36
β-Pinene	10.42
Myrcene	0.08
α-Phellandrene	0.40
Limonene	3.21
Linalool	60.73
Benzyl acetate	0.11
α-Terpineol	0.24
Geraniol	0.24
Linalylacetate	0.30
(E)-cinnamaldehyde	0.24
Eugenol	0.85
(E)-methyl cinnamate	t

Source: Nath *et al.*, 1999.

Note
t: trace.

Table 15.15 Composition of *C. tenuipilis* leaf essential oil

Compound	%
Linalool	97.51
3-Hexen-1-ol	0.02
Ocimene	0.02
Geraniol	0.09
α-Copacene	0.04
$C_{13}H_{24}$ (ui)	0.05
δ-Humulene	0.02
$C_{15}H_{24}$ (ui)	0.05
δ-Cadinene	0.03
(z)β-Farnesene	0.21
Diphenylamine	0.49
Farnesol	1.39

Source: Xue-Jian and Bi-qiang, 1987.

Note
ui: unidentified.

Other Useful Species of Cinnamomum   349

C. wightii is a high elevation species endemic to the Western Ghats, especially in the Nilgiris above 2000 m elevation.

The dried infructescenes, immature fruits and even the leaf galls are reportedly used as *Madras nagakesara* (Anandkumar *et al.*, 1986). *Nagakesara* is a crude drug in *Ayurveda*, used as deodorant, diaphoretic and stimulant, as an appetiser, brain tonic, anti-medic, anthelmintic, aphrodisiac, diuretic and antidote. *Mesua ferrea* Linn. (Clusiaceae) is regarded as the real '*Nagakesara*' (north Indian *nagakesara*) by most practitioners, while the dried inflorescence and immature fruits of *C. wightii* is traded as '*Madras nagakesara*'. The dried fruits of *Dillenia pentagyna* Roxb. are also sold as '*Malabar nagakesara*'. Anandkumar *et al.* (1986) have made a comparative pharmacognostic study of the three sources of *nagakesara*.

C. wightii is a small to medium sized tree. The leaf and young shoots have a large number of leaf galls. The flower pedicel has a single layered epidermis containing tannin deposits. The cortex is parenchymatous, and stone and oil cells are found scattered in the cortical tissue. Pericyclic fibres form a cap over the vascular bundles. There are 12–17 vascular bundles arranged in a circle. The pith is parenchymatous and contains oil cells. Perianth lobes in surface view are sclerenchymatous with thick-walled unicellular trichomes. In CS the epidermis of perianth is sclerenchymatous with thick-walled unicellular trichomes. The ground tissue is parenchymatous having many oil cells. The vascular strands are concentric, and the tracheids are spirally thickened with simple perforation. The fruit wall in surface view shows small epidermal cells. The ground tissue is parenchymatous with scattered stone cells. Oil cells are abundant, and the endosperm has spirally thickened parenchyma cells. The peduncle of the fruiting panicle has unilayered cutinised epidermis with rarely unicellular trichomes. The cortex is parenchymatous with numerous cells containing tannin intermingled with oil cells. Pericyclic fibre is discontinuous. The stele is made of collateral bundles in a continuous ring. Phloem has many oil cells. Vessels are spirally thickened with scalariform or simple perforation. The pith is parenchymatous with oil cells.

The stem galls have cutinised single-layered epidermis, with a parenchymatous cortex having many oil cells. Pericycle fibre is discontinuous. The vascular strand is made of very few vessels and mostly of fibres that contain starch granules. Rays converge to protoxylem ends where fibres and stone cells are present. The pith is sclerosed. Parenchyma cells are few and pitted. The stone cells are scattered. The central portion of the pith is almost dissolved. The gall tissue contains starch grains in the cells.

No information is available on the chemical composition of *C. wightii*.

Less Known Taxa

High safrole containing taxa

Li-Xi Wen and Bi-Qiang (1997) described a new taxa of *Cinnamomum*, known in Chinese as '*xia –ye-gui*' which is regarded by Chen *et al.* (1997) as *C. burmanni f. heyneanum*. Li-Xi Wen and Bi-Qiang (1997) named it as a new species, *C. heyneanum* (but this name is not sustainable as this species epithet has already been used for another species).

This is a wild plant that begins to flower four to six years after being planted. Three- to four-year old branches and leaves can be collected for oil distillation. Steam distillation of fresh leaves yields 0.54–0.85% oil. The oil contains 96.38 to 99.7% safrole. As, an important source of safrole, the exploitation and utilisation of this species are promising.

350 M. Shylaja, P.N. Ravindran and K. Nirmal Babu

Recently its cultivation has been encouraged in China. Safrole is a very valuable chemical and is used in the perfume industry.

2-Methylene-3-buten-1-yl benzoate containing Cinnamomum sp. from Vietnam

Dung *et al.* (1997) reported an unusual chemical composition of a *Cinnamomum* species (known as '*Re Gung*' in Vietnamese) occurring in the Ha Bac province of Vietnam. This species contained 2-methylene-3-buten-1-yl benzoate as the major component in the leaf and stem essential oils. The content of this unusual chemical was 85.9% in leaf oil, 92.4% in stem oil and 44.2% in wood oil. Camphor was the second most abundant constituent of wood essential oil (13%). Root essential oil contained mainly safrole (63.8%) and camphor (17.9%). Dung *et al.* (1997) have elucidated the structure of 2-methylene-3-buten-1-yl benzoate from spectral data.

Cinnamomum spp. from Hubei Province, China

Guang-Fu and Yang (1988) investigated the *Cinnamomum* spp. occurring in the Hubei province of China (Table 15.16). These workers have done a cluster analysis based on the occurrence of chemical constituents, and observed good correlation among morphological and chemical characteristics. They found that the species were clustered in three groups: Group (1) – *C. appelianum*, *C. pauciflorum*, *C. wilsonii* (2), *C. wilsonii* (1); Group (2) – *C. bodinieri* (1) and (2), *C. septentrionale* and *C. platyphyllum*; Group (3) – *C. camphora*, *C. camphora* var. *linoolifera*, *C. parthenoxylon*, and *C. longepaniculatum*. Based on Q clustering analysis the above workers also proposed the probable relationships among the species (mainly in the light of chemical relationships).

C. macrocarpum Hooker f.

C. macrocarpum Hooker f. Fl. Brit. Ind. 5, 132 et 133, 1886; Gamble, Fl. Madras Pt 7, 1225, 1925 (reprint 2, 857, 1957).

This species is a medium-large tree occurring in higher elevations of Western Ghats (>1500 m) of south India. The fruit of this species is the largest in the genus.

Table 15.16 *Cinnamomum* sp. from China and their major constituents

Sp.	Major components
C. bodinieri var. *hupehanum* (1)	Citral
C. bodinieri var. *hupehanum* (2)	Camphor
C. camphora	Cineole
C. camphora var. *linaloolifera*	Linalool
C. pauciflorum	α-Pinene
C. appelianum	Cineole
C. longepaniculatum	Bulnesol
C. platyphyllum	*trans*-Methyl-isoeugenol
C. parthenoxylon	Linalool
C. septentirionale	*trans*-Methyl-isoeugenol
C. wilsonii (1)	Citral
C. wilsonii (2)	Cinnamicacetate

Source: Guang-Fu and Yang, 1998.

The leaves, when crushed, smell of aniseed (safrole) and cloves (eugenol). The leaves and bark are used as a spice for flavouring certain dishes. Powdered bark is used as a base material in the manufacturing of incense sticks

C. nicolsonianum *Manilal and Shylaja*

C. nicolsonianum Manilal and Shylaja, Bull. Bot. Sur. India, 28, 111–113, 1986.

This is a very rare and endangered species having large leaves and very small axillary panicles that are few flowered. The species originates from a low elevation forest of Western Ghats. The tree is extremely rare and is possibly on the verge of extinction (Manilal and Shylaja, 1986). Biosystematics of this species was studied by Shylaja (1984).

C. oliverii *FM Bailey*

C. oliverii FM Bailey is Bull –18, Dept. Agri. & Stock, Brisbane (Dept. Agri. Bot. Bull.10) 24, 1895; Catal. Queensland woods No.315, 1899; in Proc. Roy. Soc. Queensland 11(1), 24, 1895; Compreh. Cat. Queensland Pl. 431, f-418, 1913. This species is indigenous to Australia (Queensland). The bark is aromatic and is used as a spice in Australia, where it is used as a substitute for cassia bark. It is commonly known as Olivers bark.

The other *Cinnamomum* species having local importance in South Asia are listed in Table 15.17.

In South-East Asia and the Far East regions many species of *Cinnamomum* are used locally as spice or for timber purposes. The important ones (other than those described earlier) are given in Table 15.18.

White cinnamon (Canella alba, *Canellaceae*)

White cinnamon is not a member of the genus *Cinnamomum*, but is *Canella alba*, a member of Canellaceae. It is a medium sized tree, occurring in the Caribbian Islands (West Indies). The bark of this tree is whitish in colour. The leaves are alternate, oblong, thick, shining and laurel green. The flowers are small, and cluster on the shoot apex. The fruit is an oblong berry containing four kidney shaped seeds.

Table 15.17 Useful *Cinnamomum* species from South Asia

Species	Remarks
C. assamicum Baruah	Newly reported from N.E. region of India. Main component of essential oil is benzyl benzoate.
C. perrottettii Meissn.	A high elevation species in Western Ghats, leaves and young twigs are tomentose.
C. riparium Gamble	Found at low elevation forests in south India, in valleys and river banks; its timber is useful.
C. travancoricum Gamble	A mountain species, on Western Ghats; leaves are pilose; its timber is useful.
C. dubium Nees (=*C. multiflorum* (Roxb.) Wight *C. villosum* Wight, *C. thwaitesii* de Lukmanoff)	Wild cinnamon of Sri Lanka used as a timber tree.
C. litsaefolium Thw.	Timber tree, tallest of the Sri Lankan *Cinnamomum* spp.

Table 15.18 Useful species of *Cinnamomum* in South-East Asia and other regions

C. cambodianum H. Lec	Spice: major constituent of wood oil are alpha-terpineol, linalool and terpinen-4-ol.
C. deschampsii Gamble	Spice
C. eugenenifolium Kosterm. (*C. gigaphyllum* Kosterm.; *C. hentyi* Kosterm.)	Timber
C. albiflorum Nees	Spice aromatic: major constituents eugenol, 1-8, cineole, geraniol, alpha-terpineol and terpinen-4-ol.
C. glanduliferum Kosterm. (=*C. massria* Schewe)	Timber
C. iners Reinw. ex Blume (=*C. eucalyptoides* T. Nees;	Timber
C. nitidum Blume (=*C. paraneuron* Miq.)	Timber
C. javanicum Blume (=*C. neglectum* Blume)	Timber
C. massoie (Beca.) Schewe (Massoi bark)	Spice, substitute for cassia, indigenous to Papua, New Guinea
C. mercadoi S. vidal	Timber
C. micranthum Hay.	Important host tree for *Antrodia cinnamomea*, the famous medicinal fungus. Spice: oil is an insect repellant whose main ingredient is safrole.
C. mollissimum Hooker f.	Timber
C. pendulum Cammerl. (=*C. endlicheriaceearpum* Kosterm.)	Timber
C. puberulum Ridley	Spice
C. rhynchophyllum Miq. (=*C. lampongum* Miq.)	Spice
C. scortechinii Gamble (=*C. velutinum* Ridley)	Timber
C. sintoc Blume (=*C. calophyllum* Reinw ex. Nees; *C. camphoratum* Blume, *C. cinereum* Gamble)	Timber
C. subavenium Miq. (=*C. borneense* Miq.; *C. crytopodum* Miq.; *C. ridleyi* Gamble)	Timber
C. tetragonum A. Cher.	Spice: leaf used for preparing a type of tea, beverage, locally consumed, and is medicinal.

Source: Compiled from various sources.

The tree is aromatic. Dried flowers when put in warm water produce a fragrance similar to musk. The Spaniards located the tree during their exploration in the sixteenth century and mistook it for cinnamon, naming it white cinnamon.

The inner bark of this tree is aromatic and is used in flavouring. The outer corky layer is removed by gentle beating and the inner bark is extracted. The commercial product looks like quills or twisted pieces, having a pale orange-brown colour. The bark has an agreeable odour resembling cinnamon and clove, and the taste is pungent, bitter and

Other Useful Species of Cinnamomum   353

acrid. The local people use it as a spice, and also for flavouring tobacco. The bark contains a volatile oil having a pungent aromatic taste, and contains eugenol, cineol and terpenes. It also contains an alkaloid Canellin. The bark is a traditional medicine, used in curing stomach pain and indigestion. Along with aloes it is used as a purgative (Grieve and Leyel, 1931).

Conclusion

Many species of *Cinnamomum*, in addition to those that are important spices (*C. verum, C. cassia, C. burmannii, C. tamala*), are sources of essential oil and are economically useful. Information on the chemical composition of only a few species is available.

Many species occurring in the South-East Asian regions have been investigated to some extent, while little information is available on species occurring in other regions. The existence of many species is under severe threat, and some may even be extinct. A typical case is that of *C. heynianum* Nees, reported to be from the Western Ghat region, which is known only from the single original type collection (Kostermans, 1983). *C. nicolsonianum*, a new species described by Manilal and Shylaja (1986), could not be located subsequently, and may be on the verge of extinction. Just like any other forest taxa, the severe threat on forestland – encroachment, urbanisation, indiscriminate exploitation – are leading to the decline of the species population. In the Western Ghat regions of Kerala for example, the bark of wild cinnamon trees was extracted indiscriminately for use as a base material in agarbatti (incense stick) manufacturing which has led to a sharp decline of cinnamon trees. This practice was going on for many decades though in the recent past this has been checked to some extent. Species such as *C. heynianum* and *C. nicolsonianum* are so rare that special efforts are needed for locating them and for their conservation.

Genus *Cinnamomum* also plays an important role in the search for new sources of aroma chemicals. This is a very important area in which the world perfume industry is interested. Recently *C. burmannii f. heyneanum* (*C. heyneanum*, Li-Xi Wen and Bi-Qiang, 1997) has gained importance as a valuable source of safrole and this species is now being considered for commercial exploitation. *C. tenuipilis* is another source of the valuable chemical linalool. Further investigations on other species may lead to the discovery of more species having unique flavour characteristics and compounds that can be commercially exploited.

References

Adhikary, S.R., Tuladhar, B.S., Sheak, A., Van Beek, T.A., Posthumus, M.A. and Lelyveld, G.P. (1992) Investigations of Nepalese essential oils. 1. The oil of *Cinnamomum glaucescens* (Sugandha Kokila). *J. Essential Oil Res.*, 4, 151–159.
Anandakumar, A., Balasubramanian, M. and Muralidharan, R. (1986) Nagakesara – a comparative pharmacognosy. *Ancient Science of Life*, 5, 263–268.
Anonymous (1950) *The Wealth of India, Raw materials*, Vol. II, CSIR, New Delhi, pp. 173–183.
Baruah, A., Nath, S.C., Hazarika, A.K. and Sarma, T.C. (1997) Essential oils of leaf, stem bark and panicle of *Cinnamomum bejolghota* (Buch-Ham) Sweet. *J. Essent. Oil Res.*, 9, 243–245.
Baruah, A. and Nath, S.C. (1998) Diversity of *Cinnamomum* species in north-east India: a micromorphological study with emphasis to venation pattern. In A.K. Goel, V.K. Jain and A.K. Nayak (eds) *Modern Trends in Biodiversity*, Jaishree Prakashan, Muzzafarnagar, pp. 147–167.

354   M. Shylaja, P.N. Ravindran and K. Nirmal Babu

Baruah, A., Nath, S.C. and Boissya, C.L. (1999a) Taxonomic discrimination amongst certain chemotypes of *Cinnamomum sulphuratum* Nees, with emphasis to foliar micromorphology. *J. Swamy Bot. Cl.*, 16, 3–7.

Baruah, A., Nath, S.C. and Leclercq, P.A. (1999b) Leaf and stem bark oils of *Cinnamomum sulphuratum* Nees from North-East India. *J. Essent. Oil Res.*, 11, 194–196.

Baruah, A. and Nath, S.C. (2000a) Certain chemo-morphological variants of *Cinnamomum bejolghota* (Buch-Ham) Sweet and their relevance to foliar micromorphology. *J. Swamy Bot. Cl.*, 17, 19–23.

Baruah, A. and Nath, S.C. (2000b) Morphology and essential oils of two *Cinnamomum parthenoxylon* variants growing in North-East India. *J. Medicinal and Aromatic Plant. Sci.*, 22, 370–376.

Baruah, A., Nath, S.C. and Boissya, C.L. (2000) Systematics and diversities of *Cinnamomum* species used as "Tejpat" spice in North-East India. *J. Eco. Taxon. Bot.*, 24, 361–374.

Brophy, J.J., Goldsack, P.J. and Foster, P.I. (2001) The leaf oils of the Australian species of *Cinnamomum* (Lauraceae). *J. Essent. Oil Res.*, 13, 332–335.

Chalchat, J-C., and Valade, I. (2000) Chemical composition of leaf oils of *Cinnamomum* from Madagascar: *C. zeylanicum* Blume, *C. camphora* L. *C. fragrans* and *C. angustifolium*. *J. Essent. Oil Res.*, 12, 537–540.

Chen-B.Q., Xu-Y., Zeng-F.X., Yu-X.J., Ding-J.K.D. and Wu-Y. (1997). Studies on the introduction and chemical constituents of essential oil of *Cinnamomum burmanii f. heyneanum*. *Acta Bot. Acta Bot. Yunnanica*, 14, 105–110.

Chowdhury, S., Ahmed, R., Barthel, A. and Leclercq, P.A. (1998) Composition of the bark and flower oils of *Cinnamomum bejolghota* (Buch-Ham) Sweet from two locations of Assam, India. *J. Essent. Oil Res.*, 10, 245–250.

Dassanayake, M.D., Fosberg, F.R. and Clayton, W.D. (eds) (1995) *A Revised Handbook to the Flora of Ceylon*. Oxford & IBH Pub. Co., New Delhi, pp. 112–129.

Dao, N.K., Hop, T. and Siemonsma, J.S. (1999). *Cinnamomum* Schaeffer. In C.C. De Guzman and J.S. Siemonsma (eds) Plant Resources of South-East Asia, Vol. 13, Spices. Backhuys Pub., Leiden, pp. 95–99.

Dixit, B.S., Srivastava, S.N., Sirkar, K.P. and Bhatt, G.R. (1989) Chemical examination of the fruits of *Cinnamomum cecidodaphne* Meissn. by GLC method. *J. Eco.Tax. Bot.*, 13, 221–222.

Dung, N.X., Sothy, N., Lo, V.N. and Leclercq, P.A. (1994). Chemical composition of the wood oil of *Cinnamomum albiflorum* Nees from Kampuchia. *J. Essential oil Res.*, 6, 201–202.

Dung, N.X., Khien, P.V., Vinh, N.T.Q., Le, H.T., Ven, L.J.M. Vande., Leclercq, P.A. (1997) 2-Methylene-3-buten-1-yl benzoate: the major constituent of the leaf, stem and wood oils from "Re Gung" a Vietnamese *Cinnamomum* sp. tree. *J. Essential Oil Res.*, 9, 57–61.

Gamble, J.S. (1925) *Flora of the Presidency of Madras*. Vol.II, Bot. Sur. India (Reprint).

Gildemeister, E. and Hoffmann, F. (1959) cited from Tetenyi (1970).

Grieve, M. and Leyel, C.F. (1931) *A Modern Herbal*. Jonathan Cape, London.

Guang-Fu, T. and Yang, Z. (1988) Study on numerical chemo taxonomy of *Cinnamomum* in Hubei provinces. *Acta Phytotaxonomica Sinica*, 26, 409–417.

Hajra, P.K. and Baishya, A.K. 1981. Ethnobotanical notes on the *Miris* (Mishings) of Assam. In S.K. Jain (ed.) *Glimpses of Indian Ethnobotany*, Oxford IBH Publ., New Delhi, pp. 161–169.

Hooker, J.D. (1886) *Flora of British India*, Vol. 5. Bishan Singh, Mahendrapal singh, Dehra Dun (Reprint).

Hussain, R.A., Kim, J., Hu, T.W., Soejarto, D.D. and Kinghorn, A.D. (1986) Isolation of a highly sweet constituent from *Cinnamomum osmophloem* leaves. *Planta Medica*, 5, 403–404.

Jantan, I.B., Ayop, N., Hiong, A.B. and Ahmad, A.S. (2002). Chemical composition of the essential oils of *Cinnamomum cordatum* Kosterm. *Flavour Fragr. J.*, 17, 212–214.

Khanna, R.K., Jouhari, J.K., Sharma, D.S. and Singh, A. (1988) Essential oil from the fruit rind of *Cinnamomum cecidodaphne* Meis. *Indian Perfumer*, 32, 295–300.

Kostermans, A.J.G.H. (1980) A note on two species of *Cinnamomum* (Lauraceae) described in Hortus Indicus Malabaricus. In K.S. Manilal (ed.). *Botany and Histrory of Hortus Malabaricus*, Oxford IBH, New Delhi, pp. 163–167.

Kostermans, A.J.G.H. (1983) The South Indian Species of *Cinnamommum* Schaeffer (Lauraceae). *Bull. Bot. Sur. India*, 25, 90–133.

Liang-Fang, Z., Bi-yao, L. and Yu-ying, L. (1984) Studies on chemical constituents of essential oil from leaves of Jiang-Zhag. *Acta Bot. Sinica*, 26, 639–643.

Lin, T.P., Cheng, Y.P. and Huang, S.G. (1997) Allozyme variation in four geographic races of *Cinnamomum kanihirai*. J. Hered. (USA), 88, 433–438.

Li-Xi Wen and Bi-Qiang, C. (1997) A discussion about the botanical name of xia-ye-gui. *Acta Botanica Yunnanica*, 19, 27–28.

Manilal, K.S. and Shylaja, M. (1986). A new species of *Cinnamomum* Schaeffer (Lauraceae) from Malabar. *Bull. Bot. Sur. India*, 28, 11–113.

Masada, Y. (1976) *Analysis of Essential Oils by Gas Chromatography and Mass Spectrometry*. John Wiley & Sons, New York, pp. 193–199.

Nath, S.C. and Baruah, K.S. (1994) Eugenol as the major component of the leaf oils of *Cinnamomum impressinervium* Meis. *J. Essent. Oil Res.*, 6, 211–212.

Nath, S.C., Hazarika, A.K. and Baruah, K.S. (1994) Major component of the leaf oils of *Cinnamomum sulphuratum* Nees. *J. Essent. Oil Res.*, 6, 77–78.

Nath, S.C., Hazarika, A.K. and Baruah, A. (1996) Cinnamaldehyde, the major component of leaf, stem bark and root bark oil of *Cinnamomum pauciflorum* Nees. *J. Essent. Oil Res.*, 8, 421–422.

Nath, S.C., Baruah, A. and Hazarika, A.K. (1999) Essential oils of leaves of *Cinnamomum* Schaeffer members. *Indian Perfumer*, 43, 182–190.

Okada, H. (1975) Karyomorphological studies of woody polycarpicae. *J. Sci Hiroshima Uni.*, *Ser. B*(2), 15(2), 115–200.

Rao, R.R. (1981). Ethnobotanical studies on the flora of Meghalaya – some interesting reports of the herbal medicines. In S.K. Jain (ed.) *Glimpses of Indian Ethnobotany*, Oxford IBH Publ., new Delhi, pp. 137–148.

Ravindran, S., Balakrishnan, R., Manilal, K.S. and Ravindran, P.N. (1991) A cluster analysis study on *Cinnamomum* from Kerala, India. *Feddes Reper.*, 102, 167–175.

Ravindran, S., Krishnaswamy, N.R., Manilal, K.S. and Ravindran, P.N. (1992) Chemotaxonomy of *Cinnamomum* Schaeffer occurring in Western ghats. *J. Ind. Bot. Soc.*, 71, 37–41.

Ravindran, S., Manilal, K.S. and Ravindran, P.N. (1996) Numerical Taxonomy of *Cinnamomum*. II Principal component analysis of major spices from Kerala. In K.S. Manilal and A.K. Pandey (eds) *Taxonomy and Plant Conservation*, CBS, Pub., New Delhi, pp. 153–164.

Sheng Hua, W., and Ryvarden, L. and TunTschu, C. (1997) *Antrodia camphorata* (NIU-chang-chih) new combination of a medicinal fungus in Taiwan. *Bot. Bull. Acad. Sinica*, 38, 273–275.

Shylaja, M. (1984) *Studies on Indian* Cinnamomum. Ph.D. Thesis, Univ. Calicut, Calicut.

Sritharan, R., Jacob, V.J. and Balasubramanian, S. (1994) Thin layer chromatographic analysis of essential oils from *Cinnamomum* species. *J. Herbs, Spices and Medicinal Plants*, 2(2), 49–63.

Sthapit, V.M. and Tuladhar, P.M. (1993) Sugandha Kokila, *C. cecidodaphne* oil. *J. Herbs, Spices and Med. Plants*, 1(4), 31–35.

Wijesekera. R.O.B. and Jayewardene, A.L. (1994) Essential Oils. III. Chemical constituents of the volatile oil from the bark of a rare variety of cinnamon. *J. Natn. Sci. Coun. Sri Lanka*, 2, 141–146.

Xiang Dong, L., Wang Yi, L. and Niu Ching, I. (1996) Purification of a new ribosome-inactivating protein from the seeds of *Cinnamomum porrectum* and characterization of the RNA N-Glycocidase activity of the toxic protein. *Biol. Chem. Hoppe-Seyler*, 377, 825–831.

Xue-Jian, Y. and Bi-qiang, C. (1987) Study on the chemical constituents from the essential oil of *Cinnamomum tenuipilis* Kosterm. *Acta Bot. Sinica*, 29, 537–540.

Index

air layering 62, 125
analytical methodology 81
anatomy 22
  bark 34
  floral 46
  leaf 23
  petiole 32
  wood 43

bark anatomy 34
bark oils 137
  composition *see under* various species
  extraction 136–9
bark regeneration 41
biosynthesis
  cinnamic aldehyde 93
  eugenol 94
  phenylpropanoids 91
  sesquiterpenoids 97
  terpenoids 95
biosystematics 54
botany 14
  anatomy *see* anatomy
  biosystematics and inter-relationship 54
  breeding behaviour 44
  chemotaxonomy 55
  cytology 59
  embryology 48
    carpel 49
    embryo 50
    embryo sac 50
    endosperm 51
    ontogeny 49
    ovule 49
  floral anatomy 46
  floral biology 44
  floral morphology 44
  foliar epidermal characters 27
    cell inclusions 32
    epidermis 27
    stomata 30
    stomatogenesis 31
    venation pattern 27

fruit and seed 52
  cassia buds 53
leaf anatomy 23–4
leaf morphology 22
oil and mucilage cells 25–8
petiole anatomy 32
physiology 65
propagation 59
  micropropagation 63
  tissue culture 65
  vegetative 61
taxonomy
  C. burmannii *see* Indonesian cassia
  C. camphora *see* camphor
  C. cassia *see* Chinese cassia
  C. tamala *see* Indian cassia
  C. verum *see* cinnamon
wood anatomy 43

camphor 21, 228
  preparation 235
  synthetic 233
camphor tree 228
  Chinese camphor industry 221
  composition 224
    acids 224
    alcohols 225
    aldehydes 224
    diterpenes 225
    ketones and oxides 225
    phenols and phenol ethers 225
    sesquiterpenes 225
    terpenes 225
  cultivation in other countries 228
  distillation 217
  ecology 214
  harvesting 216
  husbandry 214
  Japanese camphor industry 216
  oil
    composition 224
    distillation 217
    fractionated camphor oil 223

Index 357

location 90
physico-chemical properties 222
properties 233
uses 233
yield 218
processing 216
propagation 214
properties 233
sub-specific division 213
synthetic camphor 233
Taiwan camphor industry 219
Taiwan camphor trees 220
taxonomy 211
uses 233
Canella alba 351
cassia 9, 10, 14, 54, 55, 59, 61, 62, 63, 65,
71, 80, 81, 107
biosystematics 54
cassia buds 53
chemistry 80
analytical methodology 81
secondary metabolites 107
chemotaxonomy 55
interrelationship 54
oil 11, 168
composition 171, 172, 175, 176
safety data 151
propagation 59
micropropagation 63
vegetative 61
cell inclusions 32
chemical composition 115, 204
chemistry 80, 193
analytical methods 81
biosynthesis
cinnamic aldehyde 93
eugenol 94
phenylpropanoids 91
sesquiterpenoids 97
terpenoids 95
component structures 112–15
GLC 81
HPLC 88
IR 89
location of volatiles 90
mass spectrometric studies 97
oil
camphor tree 228
Chinese cassia 168–74
cinnamon 80–120
composition 82, 115
Indian cassia 204–6
Indonesian cassia 193–6
other species 331, 332, 333, 336, 338,
339, 340, 343, 345, 347, 348
secondary metabolites 107
chemotaxonomy 55

Chinese cassia 16, 19, 71, 156
bark 174
biological effects 174
properties 174
diseases and pests 167
ecology 158
economic value 166
end use 323
industry 324
medicine 323
spice 323
harvesting 162
handling after harvest 163
improvement 71
international trade 164
adulteration and substitute 165
yield 165
oil 168
composition 168
physiochemical properties 168
production 166
planting and husbandry 159
cassia forest 161
protection and management 162
production 164
Ramulus cinnamomi 179
uses 179
chips 135, 142
cinnamic aldehyde 93
Cinnamomum 15, 16, 18, 19, 20, 21, 31,
56, 57, 58, 60
chemical variability 56
chemovar 56–9
cytology 59
nomenclature 15
taxonomy 16
C. angustifolium 330
C. appelianum 58
C. aromaticum 17
C. bejolghota 56, 330
C. bodinieri 58
C. burmannii 19, 185
C. camphora see camphor
C. capparu-coronde 332
C. cassia see cassia
C. cecidodaphne 56, 333
C. citriodorum 334
C. cordatum 335
C. culitlawan 56, 335
C. daphnoides 6
C. deschampsii 337
C. glanduliferum 56, 337
C. glaucescens 337
C. impressinervium 203, 338
C. japonicum 338
C. kanahirai 339
C. kiamis 57

358 *Index*

C. longepaniculatum 58
C. loureirii 57
C. macrocarpum 31
C. malabatrum 31, 341
C. nicolsonianum 351
C. oliverii 351
C. osmophloem 342
C. parthenoxylon 57, 58, 344
C. pauciflorum 58, 344
C. pedunculatum 57
C. perpetuoflorens 15
C. perrottettii 31
C. platiphyllum 58
C. porrectum 342
C. riparium 31
C. septentrionale 58
C. sieboldii 60
C. sintok 57
C. sulphuratum 57, 203, 345
C. tamala 20, 57, 199
C. tenuipilis 346
C. verum *see* cinnamon
C. wightii 346
C. wilsonii 58
C. zeylanicum *see* cinnamon
cinnamon
  anatomy 22
  bark anatomy 34
    bark regeneration 41
  biosynthesis
    cinnamic aldehyde 93
    eugenol 94
    phenylpropanoids 91
    sesquiterpenoids 97
    terpenoids 95
  biosystematics 54
  botany 14
  breeding behaviour 44
  cell inclusions 32
  chemistry 80
    analytical methodology 81
    volatile oil 82–103
    see also oil, essential
  chemotaxonomy 55
  composition 115
  crop improvement 14, 66
    genetic resources 66
    variability and association 68
    varieties 66
  cultivation 121
    field planting 123, 125
    propagation 123
  cytology 59
  development of mucilage cells 25
  development of oil cells 25
  economics and marketing 285
    cinnamon oil 303
    demand 308

  export 296
  market 290
  price analysis 300
  price formation 300
  supply 309
  trade and commerce 289
embryology 48
  embryo 50
  embryo sac 50
  endosperm 51
  ontogeny of carpel 49
  ovule 49
epidermis 27
floral anatomy 46
floral biology 44
floral morphology 44
foliar epidermal character 27
fruit 52
future vision 327
history, early 1
  modern 7
interrelationship 54
leaf anatomy 22
leaf morphology 23
less known taxa 349
  C. macrocarpum 350
  C. nicolsonianum 351
  C. oliverii 351
  Cinnamomum sp. from China 350
  Cinnamomum sp. from Vietnam 350
  high saffrole containing taxa 349
  white cinnamon 351
management 121
  bark oils 137
  chips 135
  cinnamon oils 136
  cleanliness specifications 153
  diseases 127, 128
  extraction and processing of bark 131
  featherings 134
  fertilizer application 126
  harvesting 128
  leaf oils 139
  maintenance 126
  oleoresin 140
  pests 127
  pharmaceutical preparations 141
  processing 128, 131
  products 132
  quality assessment 128
  quillings 132
  replanting 127
  standards and specifications 141
  storage and transport conditions 151
  supercritical fluid extraction of
      cinnamon 139
  training of plants and pruning 127
  types 146

Index   359

mass spectrometric studies   97
  extraction   138, 139
  oil   *see* oil, essential
oil, cinnamon   *see* oil, essential
other useful species   *see Cinnamomum*
pests and diseases   239
  major diseases   250
  major insect pests   239
  management   247
  minor diseases   252
  minor insect pests   247
petiole anatomy   32
pharmacology   260
  action on cardiovascular system   261
  analgesic actions   260
  antiallergenic activity   266
  antibacterial activity   268
  anticancer activity   267
  anticonvulsant effect   264
  antifungal activity   272
  anti-inflammatory actions   261
  antimicrobial activity   268
  antioxidant action   263
  antipyretic actions   260
  antiulcerogenic effects   264
  diaphoretic actions   260
  hypocholestrolemic action   262
  hypoglycemic activity   261
  immunological effects   265
  insecticidal activity   275
  metabolic studies   267
  safety data for cassia leaf oil   151
  sedative effect   264
  therapeutic uses   278
  toxicology   277
physiology   65
present scenario   10
propagation   59, 123
  micropropagation   63
  vegetative   61
properties   *see* uses
research and development   9
secondary metabolites   107
seed   52
soil and climate   121
Sri Lankan cinnamon   14
standards and specifications   141
stomata   30
  stomatogenesis   31
terminology   2
true cinnamon   14
uses   311
  antimicrobial agent   319
  antioxidant   321
  cooking   316
  home remedies   321
  medicine   318
  perfumes and beauty care   319

  spice   311
  suitability pattern   316
  synergistic and suppressive effects   314
  varieties   122
  venation   27
  volatiles   90
  wood anatomy   43
cinnamon powder   147, 149, 150
Cinnamosma
  C. fragrans   3, 5
    Cinnamon fern   3
    Cinnamon rose   3
    Cinnamon vine   3
    Cinnamon wattle   3
classical methods   81
cleanliness specifications   153
climate   121
coefficient of variation   71
crop improvement   14, 66
  Chinese cassia   71
  genetic resources   66
  path analysis   70
  variability and association   68
  varieties   66
cultivation and management   121
cultivation and processing   204
cytology   59

development of mucilage cells   25
development of oil   25
direct sowing   124
diseases   127, 128, 192
distillation of oil   137, 139
  see also oil, essential

ecology   158, 214
economics and marketing   286
  area and production   285
  future vision   327
  trade and commerce   289
    cassia oils   303
    demand   308
    export   296
    market   290
    price analysis   300
    price formation   300
    supply   309
embryo   50
embryo sac   50
embryology   48
end uses   *see* uses
endosperm   51
epidermis   27
extraction of bark   131

featherings   134
field planting   123
floral anatomy   46

360 *Index*

floral biology 44
floral morphology 44
foliar epidermal characters 27
fruit 52

gall forming mites 128
genetic resources 66
GLC 81
grades 143
growth and yield parameters 68

harvesting 128, 162, 189, 216
history 1, 7
HPLC 88
husbandry 159, 214

Indian cassia 20, 199
  C. tamala type 1 200
      type 2 201
      type 3 201
      type 4 202
  chemical composition 204
  cultivation and processing 204
  end uses 208
  other species 203
Indian standard 142
Indonesian cassia 19, 185
  chemistry 193
  diseases 192
  habit 185
  harvesting 189
  pests 192
  post-harvest handling 189
  production 186
    nursery and planting 187
  soil conservation 195
Indonesian cinnamon 185
instrumental techniques 81
international standard 145
interrelationships 54
intra-specific chemical variability 56
introduction 1
IR 89

Java cassia 19
jumping plant louse 128

karyotype 60

Laurus cinnamomum 15
Laurus malabatrum 15
layering 63
leaf anatomy 22
leaf morphology 23
leaf oils 139
  composition *see under* different species
  extraction 119

specifications 141
standards 141
leaf spot 128

marking 144, 150
mass spectrometric studies 97
micropropagation 63

nomenclature 15
nursery and planting 187
nursery method 124

oil, essential
  bark oil 87, 98, 116, 119
  biosynthesis 54, 91–7
  camphor 217
  cassia, Chinese 168
  cassia, Indian 204
  cassia, Indonesian 195
  cinnamon 82–91
  composition *see under* cinnamon, cassia and
      camphor tree
  distillation 136–9
  flower oil 99, 102
  fruit oil 98, 102, 119
  leaf oil 87, 98, 116, 119
  root oil 87, 98
  safety data 151
  structures, components 121–2
oleoresin 140
ontogeny of carpel 49
ovule 49

packing 144, 149
peeling 130
pest 127, 192
petiole anatomy 32
pharmaceutical preparations 141
pharmacology 260
  action on cardiovascular
      system 261
  analgesic actions 260
  antiallergenic activity 266
  antibacterial activity 268
  anticancer activity 267
  anticonvulsant effect 264
  antifungal activity 272
  anti-inflammatory actions 261
  antimicrobial activity 268
  antioxidant action 263
  antipyretic actions 260
  antiulcerogenic effects 264
  diaphoretic actions 260
  hypocholestrolemic action 262
  hypoglycemic activity 261
  immunological effects 265
  insecticidal activity 275

metabolic studies 267
sedative effect 264
therapeutic uses 278
toxicology 277
phenylpropanoids 91
physiology 65
planting 159
post-harvest handling 189
present scenario 10
processing 128, 131
production 186
products 132
propagation 59
pruning 127

quality assessment 141–155
quality characteristics 71
quillings 132, 142
quills 142

Ramulus Cinnamomi 179
replanting 127
research and development 9
root ball method 124

safety data for cassia leaf oil 151
Saigon cassia 19
sampling 144
secondary metabolites 107
cinnamolaurine 110
cinncassiol 108–9
cinnzeylanine 108
corydine 110
norcinnamolaurine 110
proanthocyanidin 110–11
reticuline 110
seed 52
shoot cuttings 125
soil conservation 195
soil management 126
sowing and nursing 160

Sri Lankan cinnamon *see* cinnamon
standards and specifications 142
stomata 30
stomatogenesis 31
storage and transport conditions 151
sub-specific division 213
supercritical fluid extraction (SCFE) 139

taxonomy 16, 18, 211
tejpat 21, 203
terminology 2
tests 144
tissue culture studies 65
toxicology 277
training of plants 127
true cinnamon *see* cinnamon

uses
camphor 234
cassia 179, 323
Cinnamomum
useful species 352
cinnamon 311
antimicrobial 319
antioxidant 321
cooking 316
home remedies 321
medicine 318
perfumes 319
spice 311

variability and association 68
varieties 66, 122
vegetative propagation 61, 125
venation 27
Vietnam cassia *see* Indonesian cassia
volatiles 90

white cinnamon 351
whole cinnamon 149, 150
wood anatomy 43

eBooks – at www.eBookstore.tandf.co.uk

A library at your fingertips!

eBooks are electronic versions of printed books. You can store them on your PC/laptop or browse them online.

They have advantages for anyone needing rapid access to a wide variety of published, copyright information.

eBooks can help your research by enabling you to bookmark chapters, annotate text and use instant searches to find specific words or phrases. Several eBook files would fit on even a small laptop or PDA.

NEW: Save money by eSubscribing: cheap, online access to any eBook for as long as you need it.

Annual subscription packages

We now offer special low-cost bulk subscriptions to packages of eBooks in certain subject areas. These are available to libraries or to individuals.

For more information please contact webmaster.ebooks@tandf.co.uk

We're continually developing the eBook concept, so keep up to date by visiting the website.

www.eBookstore.tandf.co.uk